"Sopwith—the Man and his Aircraft"
is published with the co-operation and
approval of Sir Thomas Sopwith, C.B.E.

Camel Combat

A Sopwith Camel pilot of No. 10 (Naval) Squadron prepares to use the turning capability of his aircraft to counter an attack by Albatros Scouts. This authentic scene is part of the action which occurred over the Western Front on January 23rd, 1918, when a No. 10 Squadron patrol engaged three enemy two-seaters and six Albatros Scouts. In the ensuing fight, three enemy aircraft were destroyed for the loss of one Camel

SOPWITH–THE MAN AND HIS AIRCRAFT

Text compiled and written by
BRUCE ROBERTSON

Line Tracings by **W. F. HEPWORTH**, *M.S.I.A.*

Based on Original Drawings by **PETER G. COOKSLEY**

Colour painting on dust jacket and insert by
J. D. CARRICK

Produced by
D. A. RUSSELL, *C.Eng., F.I.Mech.E.,* and **G. R. DUVAL**, *A.F.M.*

First published Autumn 1970 by
AIR REVIEW LTD,
LETCHWORTH, HERTFORDSHIRE, ENGLAND

Text copyright © 1970 by Bruce Robertson

Colour painting and line drawings copyright © 1970 by **AIR REVIEW LTD**

Made and printed in England by
THE SIDNEY PRESS LTD, BEDFORD

ISBN 0 900 435 15 1

United States of America Library of Congress Catalog Card No. 70-84076

Sopwith—The Man . . .

The pilot, designer and constructor, is seen seated at his desk (about 1916) in the Canbury Park Road Works of the Sopwith Aviation Company. The signature of Sir Thomas was written on 8th October 1970, when he wrote the publishers after seeing proofs of the book.

Foreword

This book has been designed to give the story of Sir Thomas Sopwith and provide a reference work on Sopwith aircraft. To do this without clogging the text with too many aeronautical facts and figures, the book has been divided into parts to which the Contents List, pages 8 and 9, provide the guide.

The main narrative, in chapters, traces Sir Thomas's life from the time he first became interested in aviation, to the end of the Sopwith Aviation Company and then briefly to today. In this story, the progress of each and every prototype from this most enterprising company could well fog the main issues so that a review, type by type, of every Sopwith aircraft is given separately and this review also acts as an index to where, in other parts of the book, information may be found on these aircraft.

Since only a minority of Sopwith aircraft were built by the Sopwith Company, another separate part, titled 'Fortyfold', provides a brief, in alphabetical sequence of contractors' names, of every constructor of Sopwith aircraft. A production summary table at the end (page 164) provides a quick reference to 'who made what'.

To avoid repetitive qualifying terms and values for

An aspect of the Cobham, the largest Sopwith aircraft built.

T. O. M. Sopwith lands his Howard Wright in the grounds of Windsor Castle and is met by King George V.

rate of climb and maximum speed etc., constructional and performance data is tabled, and in this form provides easier comparisons between types. Most of the important types are further documented by general arrangement drawings, all to a common scale of 1/72nd and the Contents List provides an index to these. The object of linking the reference material in this way has been to avoid repetitions and utilise the space available to the full.

In preparing this book I am greatly indebted to Sir Thomas Sopwith, Wing Commander John Crampton, D.F.C., A.F.C. & Bar, A.F.R.Ae.S., of Hawker Siddeley Aviation Limited, W. O. Bentley Esq, M.B.E., and former Sopwith employees, particularly the late Jack Pollard and Jim Whitehorn. Others to whom I am grateful for contributing information are (listed alphabetically): L. A. Anderton Esq., M.A., Michael Goodhall Esq., J. Helme Esq., W. M. Lamberton Esq., R. R. Soar Esq., D.S.C., Frank White Esq., and Frank Yeoman Esq.

London 1970   BRUCE ROBERTSON

ACKNOWLEDGMENT FOR PHOTOGRAPHS

The publishers and author gratefully acknowledge the following sources of photographs (listed alphabetically): Miss Jean Alexander, R. C. B. Ashworth Esq., R. C. Bowyer Esq., Public Archives of Canada, Roy Cross Esq., G. A. Cull Esq., Ministry of Defence, André ver Elst, Flight International, Messrs Hawker Siddeley Aviation Limited, T. S. Hollis Esq., Imperial War Museum, Denny R. May Esq., the late Jack Pollard, Alex Revell Esq., Messrs Real Photographs Limited, the late H. H. Russell, Sir Thomas Sopwith, C.B.E., Stuart Leslie/J. M. Bruce Collection, Wing Commander L. M. Stewart, O.B.E., R.A.F. (Rtd), Z. Titz, Jim Whitehorn Esq., and Frank Yeoman Esq.

About This Book

By **D. A. RUSSELL**, *C.Eng., F.I.Mech.E.*

Sir Thomas and Lady Sopwith in the grounds of Compton Manor, their home near Winchester, after discussing this book with D. A. Russell (left) and Bruce Robertson (right).

As a pioneer of British Aviation, producer of the most famous of the fighting aircraft of the 1914-1918 War, and currently Founder and President of one of the world's largest aircraft organisations, Sir Thomas Sopwith, C.B.E., is an outstanding figure in British Industry. His position in British aviation is unique. For precisely fifty years he has been continuously at the head of one of the most well-known aircraft manufacturing groups in the Commonwealth.

It was in 1963, at the age of 75, that he expressed a desire to take a less active part in aviation and relinquished his chairmanship of the Hawker Siddeley Group, which then had an annual turnover of £300 million and embraced aviation, industrial, electrical and general engineering. Thus, for well over sixty years, Sir Thomas had been associated with British aviation. It may, therefore, seem strange that no biography of him has before appeared. In fact, several London publishing houses had approached him, knowing that only with his co-operation could his story be adequately presented. He refused them all.

Ten years ago, whilst engaged in producing *Fokker— The Man and the Aircraft*—it occurred to me that to publish a similar work on a British pioneer aircraft constructor would be something well worth aiming for. There was an intriguing parallel with Sir Thomas Sopwith whose name came immediately to mind. Both men had been interested in sailing from an early age. Both had commenced a practical association with aeroplanes at the same time, October 1910, and both were soon thereafter designing, constructing and organising their own aircraft factories to cope with large scale expansion due to the countries war.

While in the 1914-1918 War, Fokker produced aircraft for the Germans, Sopwith produced them for Britain and the Allies. Thus their respective products were pitted against each other. Both companies excelled in the design and construction of fighter aircraft. The only British operational triplane was a Sopwith; while the most successful triplane of the Germans was a Fokker. Both companies produced an interrupter gear of their own design to enable a machine-gun to fire through a propeller arc without hitting the blades. At the end of the 1914-1918 War, both Companies were producing in vast quantities 'the best fighter aircraft of the war' for their respective Air Forces.

Post-War, both Sopwith and Fokker retained an interest in sailing and at one time, in the early thirties, they met when their respective yachts were berthed side-by-side in an American harbour. It was but a casual acquaintance and it was not repeated. For all that these two men had in common, their lives were lived in very different ways. Fokker was an extrovert and flamboyant in the extreme. Sopwith, typically British, avoided the lime-light and lived up to the motto on his Family Coat of Arms—'Work without Talk' (or, as he translated the Latin to me when we first met—'get on with the job, don't talk about it').

After the book on Fokker was published, I sent a copy to Sir Thomas and he read it on his way by liner (not airliner!) to South Africa. On his return he wrote me and said that he had 'read it with great interest' . . . I had sown a seed . . . so followed up with a request for an interview for myself and Mr. Bruce Robertson, whom I had invited to compile and write this book provided that Sir Thomas would give us his co-operation, without which we could not have produced this book.

At one of several conferences at Compton Manor, Sir Thomas discusses his records and photographs with publisher and author, D. A. Russell and Bruce Robertson.

Under date of May 14th 1968, Sir Thomas wrote:

'... There have been several offers to do a S. history *all* of which I have turned down, but if you and Mr. Robertson think it might be worth spending a day on coming here to discuss it, my wife and I would be delighted to give you lunch and you could look over what material is available; but I must warn you that you would be dealing with a not very willing customer who has been extremely remiss in keeping evidence of a doubtful past...'

If, in view of the warning, you still would like to come, we shall be delighted to see you...'

Sir Thomas and Lady Sopwith gave us a most cordial welcome at their home—Compton Manor in Hampshire. After an excellent lunch and a lively discussion on a variety of subjects, we were invited to return to the lounge, when Sir Thomas rather casually remarked—'Well what would you like to see?' To my reply—'Everything, please, Sir Thomas', he produced a mountainous pile of albums, photographs and records which he offered on loan for our perusal. The book was on its way!

We made two further visits to Sir Thomas's home, listening to his personal reminiscences, his views on the accuracy of the draft chapters, and his ready answers to a 'barrage' of questions from Bruce Robertson. He also identified many personalities in photographs.

There was another pioneer in aviation engineering from whom we sought co-operation, Mr. W. O. Bentley, M.B.E., who first introduced the aluminium piston into motor car engines and who designed the B.R.1 and B.R.2 rotary aero engines used in large numbers on Sopwith aircraft. 'W. O.', better known for his famous Bentley cars, lives in retirement in a small village near Guildford. I contacted him and outlined the Sopwith book project. He willingly agreed to help and again we were invited to make a personal visit.

We drove to the village concerned. We did not know the precise address, but anticipated no difficulty in finding so-well-known a man and in this surmise, we were correct.

Sir Thomas peruses an early draft chapter to check the facts. On the table, the Gold Trophy awarded to Sir Thomas for winning the First Aerial Derby, 1912.

Joining in conference with Sir Thomas about the book is (centre) Squadron Leader John Crampton, D.F.C., A.F.C. & Bar, A.F.R.AeS., R.A.F. (Ret), Technical Sales Manager (Harrier), of Hawker Siddeley Aviation.

Soon his house was traced and a warm welcome was given us by Mr. and Mrs. Bentley. After an excellent lunch we successfully drew 'W. O.' back to the days when, as a naval Lieutenant he designed aero engines. Moreover, we were permitted to borrow records and photographs... the book was beginning to 'come alive'...

An essential link, for our records, was the goodwill of Messrs Hawker Siddeley Aviation Limited, whose origins can be traced to Sopwith Aviation and who now build Harriers in shops where once Snipes were built. Sir Thomas Sopwith himself suggested that Squadron Leader John Crampton, D.F.C., A.F.C., & Bar, A.F.R.AeS, R.A.F. (Ret.), should be our guide and mentor; and we are indebted to him for his many helpful suggestions and above all for arranging contacts with former Sopwith employees. A meeting was arranged in his office at which Mr. Robertson, my co-Director Mr. G. R. Duval, A.F.M., R.A.F. (Ret.) and myself met Messrs Jack Pollard and Jack Whitehorn, two *very* 'old-timers' from the Sopwith days.

These two veterans had had unique experiences. Jack Pollard had assisted in the rigging and assembly of T. O. M. Sopwith's first Howard Wright monoplane and later joined him for his 1911 American tour. After becoming a works foreman in the Sopwith factory at Canbury Park Road, he was placed in charge of the experimental shop. In later years he became Chief Inspector of Hawker Engineering Ltd. Jack Whitehorn, the original 'Buttons' of Kingston's Skating Rink was engaged (as tea-boy at two shillings-and-sixpence a week) by T. O. M. Sopwith when he took over the rink for his works. He is still (in 1970) 'working for Sopwith', in the Transport Department at the Kingston upon Thames factory of Hawker Siddeley Aviation Limited.

With John Crampton in the chair at the Hawker Siddeley Aviation, Richmond Road offices (which stand immediately in front of the old Sopwith Ham Works) a lively exchange of information was engendered by circulating copies of Sir Thomas Sopwith's photographs and the Sop-

Three ex-Sopwith employees at the Richmond Road, Kingston upon Thames Offices of Hawker Siddeley Aviation Limited in April 1970. They are, left to right, Messrs. V. Derrington, J. Whitehorn and L. A. Pollard. Vic Derrington joined Sopwith's as an apprentice at the age of fifteen on August 1st 1914. An athlete and 'leading light' in the Sopwith Sports Club, he served in the Drawing, General and Buying Offices in turn, but spent most of his time in the Experimental Department. Jack Whitehorn was born at Kingston on August 22nd 1895, on a site subsequently bought by T. O. M. Sopwith for an extension to his works. Born on June 9th 1890, Jack Pollard first started work in the Battersea Park Station railway arches for Howard Wright Limited, after an apprenticeship with James Simpson Limited. He joined the Sopwith team in 1912.

with elements of the Hawker Siddeley and 'Harleyford' photo collections. Personalities were queried and identified, whilst aircraft photographs evoked stories of their production and testing.

We were taken to the old skating rink, the original Sopwith factory building which became the experimental shop when the firm expanded. Here the veterans pointed out where the brickwork had been removed to get the Bat Boat out of the works... Where the various shops and offices were... and how T. O. M. Sopwith himself would come in and fire questions such as—How?—When?—Why? to sum up a situation!

After an excellent luncheon at a local hotel during which many 'reminiscences' were retold, we all returned to John Crampton's office, for a further review of photographs, records and documents. Additionally, the veterans lent us their own photographs and records for copying. All that was said and transpired was taken down by Mr. Robertson, checked and cross-checked before inclusion in the appropriate parts of the manuscript of this book.

I think it is worth recording that this meeting commenced at 10 a.m. and went on, virtually non-stop, to 5.30 p.m. For myself, it was one of the most interesting days in my life, the 'highlight' being provided by Jack Pollard's vivid description of a flight he made with Orville Wright when the Sopwith Team was competing in America in 1911.

There have, of course, been many other sources from which we obtained information, not only from men who built Sopwith aircraft, but from those who flew and maintained them. These have included men from the rank of Air Marshal to Sergeant Pilot and Air Mechanics.

In regard to photographs, those of over 150 Sopwith aircraft have already been published in earlier 'Harleyford' books, and every effort has been made to avoid duplication in this book. In fact, the well-known words—'hitherto unpublished' were never more appropriate than to the many which illustrate it. Acknowledgement to the British Broadcasting Corporation for the words 'HIGH ADVENTURE' would surely make a fine sub-title for its earlier chapters, in which are described the many exciting, at times daunting and at times amusing, experiences of one of Britain's leading air pioneers, long before the expression 'barn-storming' was ever thought of!

The Camel is, of course, the most well-known of the Sopwith designs and Peter G. Cooksley has, therefore, prepared (from original works drawings) the most detailed drawings of this famous aircraft. Even wheel spoke nipples, and fabric-laced eyelets are drawn to scale! These drawings occupy no less than six pages, and will enable the finest, and most detailed (whether 'solid', flying or non-flying) scale models to be built. With all these, and the other twenty-

Where it all started... the Skating Rink at Canbury Park Road, Kingston upon Thames. Opening day was 23rd September 1909. "Buttons", tea-boy at half-a-crown a week, was J. Whitehorn.

The only way to really find out—ask the chaps who were there at the time! Left to right: Bruce Robertson and G. R. Duval of Harleyford Publications Ltd, talk with Jack Pollard who was in charge of the Sopwith experimental shop in the ex-skating rink, and Jack Whitehorn, who was "taken over" and helped build Sopwith aircraft in the Rink. He is still working with Hawker Siddeley Aviation Ltd. In the chair is John Crampton who kindly arranged this meeting of veteran Sopwith employees in his office.

nine 1/72 scale 3-view drawings, traced by Mr. W. F. Hepworth, M.S.I.A., it will be seen that everything possible has been done to support fully the very substantial efforts made by Bruce Robertson in paying full tribute to Britain's most well-known aircraft of the 1914-1918 War.

When I first met Sir Thomas Sopwith and sought and

W. O. Bentley is perhaps better known as the designer of the famous Bentley cars. He is at the wheel of a 1928 4½-litre Bentley, the property of Michael A. Benkert, Esq.

obtained his agreement to our project—I promised that every word written by Bruce Robertson would be submitted for his approval. This promise has been faithfully kept—with the close co-operation of John Crampton—who, despite heavy responsibilities as Technical Sales Manager (Harrier) of Hawker Siddeley Aviation Ltd.—has *also* read and checked every word of the text—including making a number of helpful suggestions. In this connection I am happy to say that Sir Thomas was *not* the 'very unwilling customer' he said we should find him. On the contrary he was, throughout the writing of this book, a 'very willing' co-operator and it is with sincere thanks to him I record that in his documents we found no sign of 'the doubtful past' to which he referred in his letter! We found instead much evidence of a very thorough hard-working young man who made the most of every opportunity (some of which he made himself) that came his way.

That Sir Thomas having turned down *all* of several offers earlier received, should give 'Harleyford' his full support is a privilege that is much appreciated; in particular his permission to reproduce at front of this book his Family Coat of Arms and to be allowed to print thereunder—'This book has been produced with the co-operation and approval of Sir Thomas Sopwith, C.B.E.'

Born in 1902, and spending my boyhood living on the North-Eastern banks of the river Thames, I saw many Sopwith Pups and 1½-Strutters flying from the aerodrome at Rochford (which was a Home Defence Station) in the 1914-1918 War period. Little did I know that some fifty years later, I should lead the 'Harleyford' team to produce this book. It has been a most rewarding project, and one on which, with due modesty, I trust that I may be allowed, on behalf of our united efforts, to quote the Royal Air Force motto—*Per Ardua ad Astra.*

CONTENTS

		Pages
Painting	by J. D. Carrick	Dust Cover and Plate
Frontispiece	Onwards and Upwards	2
Foreword	Bruce Robertson	3
About this book	by D. A. Russell, C.Eng., F.I.Mech.E.	4–7
Plate	Sopwith the man . . .	10
Index		244

TEXT

Chapter		Pages	Chapter		Pages
One	*Sopwith takes to the Air*	11–13	Twenty-one	*Set for Success*	84–85
Two	*A Record by a Novice*	15–17	Twenty-two	*Sopwiths in the Aegean, 1917*	86–89
Three	*Command Performance*	18–20	Twenty-three	*Camel Crisis*	90–93
Four	*American Adventures*	21–27	Twenty-four	*Bentley and his rotaries*	94–95
Five	*Derby Winner and Trainer*	28–30	Twenty-five	*Sopwith Defenders*	96–99
Six	*A Round of Records*	31–34	Twenty-six	*Shipboard Sopwiths*	100–101
Seven	*Prizes and Enterprises*	35–39	Twenty-seven	*Discourse on the Dolphin*	102–106
Eight	*Sopwith Schneider Success*	40–41	Twenty-eight	*The Camel helps to avert a Crisis*	107–111
Nine	*Harry Hawker at Home*	42–43	Twenty-nine	*Trench Fighters*	112–114
Ten	*The Weeks before the War*	44–46	Thirty	*Quest for Superiority*	115–117
Eleven	*Sopwith at War 1914*	47–52	Thirty-one	*Superiority with the Snipe*	118–119
Twelve	*Sopwiths go East—1915*	53–55	Thirty-two	*Sopwith Night Fighters*	120–121
Thirteen	*Sopwiths at Sea—1915*	56–58	Thirty-three	*Anti-Submarine Sopwiths*	122–124
Fourteen	*Manifold Expansion—1916*	59–61	Thirty-four	*Torpedoplane Transact*	125–127
Fifteen	*Sopwiths at Sea—1916*	62–63	Thirty-five	*Sopwith the Fleet*	128–129
Sixteen	*Schneider Triumph—1916*	64–66	Thirty-six	*Sopwith Universal*	130–133
Seventeen	*Naval Sopwiths in France, 1916-1917*	67–71	Thirty-seven	*Transatlantic Venture*	134–136
Eighteen	*Army & Navy Sopwiths on the Western Front*	72–76	Thirty-eight	*Postwar Military Sopwiths*	137–139
Nineteen	*Sopwith Sailors (No. 3 Wing R.N.A.S.)*	77–80	Thirty-nine	*Civil Sopwiths*	141–144
Twenty	*French Strutters*	81–83	Forty	*Forward as Hawker*	145–147

1/72 SCALE THREE-VIEW TONE PAINTINGS

	Pages		Pages
Sopwith Wright	171	Camel 2F.1	186
Bat Boat 1A	172	Camel Seaplane	187
Gun Bus	173	Cuckoo	188
Churchill Sociable	174	B.1	189
Tabloid	175	Dolphin	190
Schneider	176	Snipe	191
Three-Seat Biplane	177	Buffalo	192
Foeder Seaplane 807	178	Antelope	193
Baby (110 Clerget)	179	Swallow	194
1½ Strutter	180	Gnu	195
Pup (Ships Pup)	181	Salamander	196
Triplane	182	Grasshopper	197
Triplane (150 Hispano)	183	Dove	198
Camel F.1	184	Wallaby	199
	Atlantic	200–201	

PAGES OF PHOTOGRAPHS

	Pages		Pages
First Record	14	From Planks to Planes	170
Civil Sopwiths	140	Camel Carriers	185
Family Album	148–9	Camel Crashes	202
From Beginnings at Brooklands	165	Camel Cameos	203
Sopwith Employees 1914	166–7	A Sopwith Pup Contractor—Whitehead	232–3
Aerial View of Canbury Road Works	168–9	Bred in Tradition	243

FEATURES

	Pages		Pages
Forty-Fold	150–164	Type-by-Type Review	210–231
Camel Drawing Details	204–231	Data Tables	234–241
	List of Surviving Aircraft	242	

Sopwith's Successes—Forwards and Upwards . . .

T. O. M. Sopwith is seen above making his first solo flight in a Howard Wright Monoplane in 1910. Today, sixty years later, Sir Thomas Sopwith, C.B.E. is President of the Hawker Siddeley Group whose aeronautical products currently include the Harrier G.R.1 strike aircraft, shown below demonstrating its vertical take-off capability.

CHAPTER ONE

Sopwith takes to the Air

Thomas Octave Murdoch Sopwith had four great assets: a private income from his father, a wealthy civil engineer; a bevy of devoted sisters conveniently placed, both socially and geographically; a mechanical aptitude fostered by an education in engineering and certainly not least, an abundance of pluck and drive.

Forceful, yet never aggressive, his reticence to talk or publicise his own efforts increased year by year. So that the more he accomplished first in flying, then in design and finally in business, to make the name Sopwith famous, his own image receded and the success of his ventures were projected by others to whom he gave fame. He had the intelligence to pick the right men. Frederick Sigrist, was a typical example. Engaged in 1909 to tend to the yacht *Neva* owned jointly by V. W. Eyre and Sopwith, his status was raised from a £2/15/0 per week mechanic to that of manager, so placed that he made a fortune.

From a love of motorcycles as a schoolboy, progressing to motorcars and motorboats in his late 'teens, Sopwith turned to aviation in his 'twenties. But before finally finding the aeroplane as his *métier*, he did have a brief try at ballooning before he came of age. An application for an aeronaut's certificate was, in fact, refused him on the grounds of his youth.

One balloon he had owned in partnership with a Mr. Paddon, became known as the 'Padsop'. On one occasion, ascending in a hot windless day, the balloon was practically becalmed over London. Tiring of its immobility, the two partners decided to descend on the green stretch beneath them—Regent's Park. Some of his friends in London— Frank Butler the motoring pioneer and balloonist, the Hon. C. S. Rolls the first Englishman to cross the Channel by aeroplane and Vere Ker-Seymer the motorist—received a garbled story that he was down in the Zoo! Immediately they dashed around in a car—more correctly an automobile —to render assistance.

The balloon had landed on the opposite side to the Zoological Gardens and as a crowd gathered it was deflated and carried into a large shed. It was here that sculptors were at work on the memorial statue to Queen Victoria that now adorns the approach to Buckingham Palace. Sopwith and Paddon were among the privileged few to see the statue before its official unveiling.

After the ballooning phase, Sopwith turned to motorboats, but did not neglect the thrill of sail and did, in fact, crew the winning yacht in the 1909 Royal Aero Club race.

On September 18th 1910 it could be said Sopwith was infused with his love of aeroplanes. That day, after a slow Channel crossing in the yacht with his friend William Eyre as companion and Fred Sigrist tending the engines, the three put in at Dover to learn that John B. Moisant had landed some seven miles distant, at Willows Wood, Tilmanstone, having made the first flight across the Channel with a passenger. Moisant's companion, who had been his hefty mechanic Albert Filieux, so impressed Sopwith with the capability of the frail-looking Bleriot that he too, resolved to take up powered flying. Moisant had actually attempted a Paris-London flight, but not until October 6th

T. O. M. Sopwith, motorist, yachtsman and aviator. In 1909 he owned a 40 h.p. Napier, was half-owner of a 166-ton schooner, the "Neva", and had just bought a Howard Wright monoplane.

did he reach London, due to the rough English summer weather and a series of mishaps.

By that time Sopwith had visited Brooklands and paid to fly two circuits in the Henry Farman operated by Hewlett and Blondeau—Mrs. Maurice Hewlett and M. Gustav Blondeau who had just formed their successful business partnership. So impressed with this taste of the air, Sopwith sought a machine of his own. At that time he lived at No. 3, Draycott Place, a select area of London mid-way between Chelsea Embankment and Hyde Park. At the other side of the Park, at No. 110 Marylebone High Street, the pioneer designer Howard T. Wright (no relation of the Wright Bros) had an office. There, Sopwith made some enquiries. One evening shortly afterwards he visited the Wright workshop under one of the Battersea railway arches and bought the latest Howard T. Wright monoplane for £630.

Although he had never flown before he felt confident that he could do so and on October 22nd he found he could —but only just. As the machine left the ground after going forward some 300 yards, he pulled the control stick back much too sharply. The machine shot up about forty feet into the air, stalled and crashed, breaking the propeller and undercarriage and crumpling a wing.

Undeterred, Sopwith waited impatiently for the monoplane to be repaired and for weather suitable for flying.

Sopwith takes the air—an ascent in the 'Padsop' from Battersea circa 1908. In the basket, left to right, T. O. M. Sopwith, Frank Hedges Butler and the Hon. C. S. Rolls.

He was daily at the sheds at Brooklands, the famous racing track of the day which, with Hendon, became a centre of British aviation.

Early in November, repairs and some modifications were complete. It was now up to the weather. Friday, November 4th, was probably the best flying day of the year 1910; fine with no breeze. In the morning Sopwith was out rolling, or taxiing as it would now be called, and making short hops. As the day wore on the hops developed into straight flights and then circuits in the air. Finally, a large circuit out over the sewage farm and round the back of the sheds, evoked from the *Aero* magazine's reporter the statement that this was the quickest progress that anyone had made at Brooklands. The following Wednesday he was again in the air, but a burst cylinder head of the 40 h.p. E.N.V. engine put his machine temporarily out of action.

The following weekend was wet and stormy and on Sunday, the most popular flying day, no one attempted take-off from Brooklands. But on the Monday, in spite of a strong breeze which kept others on the ground, Sopwith the novice alone ventured out. He experienced some difficulty in getting off as for some reason no mechanics were available, but Howard Wright himself placed the drags—weights to keep the machine back while the engine was revved up—which were used before chocks.

Already he was impressing observers, indeed the *Observer* on November 13th 1910 reported—'But Mr. Sopwith is a phenomenon. His progress has been incredibly swift; no sooner had he got off the ground for the first time than he essayed and accomplished the most daring turns in the air, wheeling with all the assurance and effect of a Morane.'

Sopwith then made a switch, and turned his interest from the monoplane to the new biplane by Howard Wright, featuring the more powerful 60 h.p. E.N.V. engine. He visited the Howard Wright workshop again in November to arrange the new purchase. At this time he had a pet monkey Oolie which he took around with him. It was unfortunate that on this occasion Grahame Gilmour was also visiting Howard Wright bringing with him his Alsatian. The dog made for the monkey, which jumped up steps the workshop foreman, Harold Snelling, was using to dope the wings of the new biplane, by the application of sago. Covered in the sticky mess it ran from one end of the upper wing to the other causing successive panels of tautened fabric to sag, while Sopwith tried to coax it down and Gilmour hung on to his dog. Howard Wright put on a good face to a prospective customer. By mid-November the aircraft was delivered to Brooklands.

On November 21st, within a month of learning to fly, and within a few hours of first flying in a biplane, Sopwith took his pilot's certificate. Spending the morning taxiing and hopping, the early afternoon doing circuits, he notified the examiners of his readiness before evening.

At that time a candidate for the Royal Aero Club Certificate had to satisfy examiners by three flights, each of three miles round a circular course without landing; they need not all be made on the same day. On completion of each flight the engine had to be cut and the pilot land within 150 yards of a spot previously indicated to the examiners.

Fred Sigrist, the mechanic hired to tend to Sopwith's boat, who became a famous aeronautical engineer.

In 1910, Tom Sopwith's interest turned from lighter-than-air to heavier-than-air craft. This view of Sopwith on the Howard Wright Monoplane, powered by a 40 h.p. E.N.V. engine, was taken October/November 1910.

The test that Sopwith passed that day documented him as one of the British pioneers of the air with Certificate No. 31. By the end of that year, 1910, less than fifty men in the United Kingdom had achieved this distinction.

Flying was the most exhilarating sport Sopwith had experienced and he seized every opportunity to practice. No sooner had his certificate been granted than he went all out for a record. The Michelin Tyre Company had offered a cup, known as the British Empire Michelin Cup, together with a £500 prize, for the longest flight made up until December 31st 1910. Conditions stipulated that contestants must be British subjects using an aeroplane of which all finished or manufactured parts must be of British make.

Currently the record was held by Cody who on Laffan's Plain in an aircraft of his own design, known as the Cody Michelin Cup Biplane, had covered $94\frac{1}{2}$ miles in 2 hours, 24 minutes, setting up a new British record and qualifying for the prize—unless beaten by the end of the year.

The Royal Aero Club were notified by Sopwith that he intended competing for the cup on Saturday, November 10th. Accordingly the club secretary, Harold Perrin, official observers and timekeepers, set out early for Brooklands, which they reached at 8.30 a.m. and erected eight poles to mark out the circuit. This method was deemed fairest, lest a long-distance point-to-point flight give rise to suspicions of an intermediate landing. The Wright had been thoroughly overhauled by the time Sopwith arrived, fortified by a good breakfast and extra warm clothing.

It is said that behind every famous man there was a woman. When Sopwith reached fame in British aviation that woman was neither wife, nor sweetheart, but a sister. No adventurous man could have wished for a more devoted, yet level-headed companion than Sopwith had in his elder sister May. While cautioning him on taking risks and solicitous of his health, she nevertheless backed his ventures to the full. Moreover she was willing to accept the risks of flying with her brother—and did so on many occasions. May accompanied her brother to Brooklands. No doubt it was at her instigation that he wore two flannel shirts, two sweaters, a suit of overalls and a reefer jacket. It was May who gave him his final briefing when he took the air at 10.5 a.m.

The circuit around the poles was estimated at 1 mile, 5 furlongs and 13 yards, so that Sopwith completed well over a mile per lap. His progress was plotted by May who with the mechanics worked out his mileage and placed panels on a board so that he could note his progress and not have to keep count himself. The hours ticked by while the Howard Wright, its E.N.V. engine purring reassuringly, traced the line of poles and banked round the flagged poles at each end. A light breeze, evident mid-morning, had stilled towards midday, and the aircraft flew steadily. For Sopwith, perched in an open cockpit, rushing through the cold winter air, it must have been agony. His right hand was numbed after the first half-hour, but grimly he kept on. At last he noted the figure 95 on the board—he had beaten Cody—now it was on to the hundred miles.

Shortly after 1 p.m., 100 was signalled, it was then a case of continuing as long as possible, in case someone else tried for the record. At 1.16 p.m. he landed. The flight was documented as $107\frac{3}{4}$ miles in 3 hours 12 minutes and 40 seconds; the British record for both endurance and distance. In actual fact, to make sure of passing round the pylons and thereby swinging wide, the actual flight path was in the region of 120 miles.

T. O. M. Sopwith, leaning on the kingpost of his Howard Wright monoplane, watches whilst Howard T. Wright prepares the engine. The streamlined fuel tanks are placed well above the engine to permit gravity petrol feed.

FIRST RECORD

Sopwith breaks the British Record for endurance by the longest flight of 1910 made on December 10th 1910, in his 60 h.p. Howard Wright Biplane. He is seen flying over his car stationed by a lap pylon at the 90th lap and May Sopwith is seen putting out the 100th lap panels. Other views are of Sopwith in his Howard Wright and of the aircraft itself. The full-length photo of Sopwith is contemporary with this period.

CHAPTER TWO

A Record by a Novice

Bleriot's epoch-making first cross-Channel flight in an aeroplane during 1909 not only fired the imagination of the British people, rapidly becoming conscious of aerial events, but acted as a spur for a greater achievement from a British subject. In those days of fervent patriotism, it appeared logical that if 1909 be marked as significant for a French achievement, then British prestige must be restored for 1910.

With the Government of the time as indifferent to aviation as to the Football Association, Britain's place in the air devolved upon three parties—sponsors, press and pioneers. The sponsor in the case was the Baron de Forest who offered a prize of £4,000 for the longest flight from England to the Continent. Conditions were that it must be by a British aviator, in a machine of British construction, and by December 31st 1910. *The Daily Mail,* always ready to champion Britain's place in the air, promoted a competition for this prize and among the entrants was Sopwith.

On Monday, December 5th, Sopwith arrived at Eastchurch with his Howard-Wright biplane to prepare for the record attempt. Rough weather kept him on the ground until Sunday, December 11th, when he ascended for an hour, circling the airfield, finishing with a neat glide into land. His sister May spent much time with him, acting as a kind of training manager. Having flown herself, braced behind her brother in the Howard-Wright, she appreciated that sitting exposed in a slipstream, for hours on end as demanded by a record attempt, would entail good physical fitness. More than once, Tom enjoying a smoke in the hangar had his pipe removed from his mouth by his sister.

For this prize there was keen competition. Other competitors had taken up station closer to the Continent —at Dover. On the cliffs overlooking St. Margaret's Bay, hangars had been erected to house the machines of Claude Grahame-White, Robert Loraine and G. C. Colmore. Among others making attempts were Alec Ogilvie, Lieutenant H. E. Watkins and C. H. Cresswell.

At 8.16 a.m. on the Sunday morning of December 18th 1910, Sopwith set out from Eastchurch with twenty gallons of petrol in the fuel tanks and a thermos flask containing—evidently at sister May's instigation—meat extract. After heading for Canterbury, over which he passed at 950 feet, he set course for Dover, climbing slightly. A report from the Dover Wireless Station at 9 a.m., gave his speed at 50 m.p.h. at 1,200 feet with a following Northerly breeze of 20 m.p.h.

Visibility was poor, and his compass, although carefully slung in rubbers, stuck at N.W. Not until reaching mid-Channel could the coast of France be discerned. His intention was to fly to Châlons, near Paris, a distance of 240 miles. However, dense cloud soon blotted out the sun, making navigation difficult.

To quote his own words, as reported by *The Times*— 'I had nothing to direct myself by, so I just kept flying on. Towns and villages passed below; I knew none of their names. Then the wind began to get more gusty. The machine swayed and lurched and the arm with which I moved the controlling lever began to ache.

Tom Sopwith with his sister May at the time of his record flight to Belgium.

'Just as I was flying over a village at about 800 feet a very ugly gust caught my machine on one side and tilted it partly over. To my consternation the aeroplane refused to regain its normal position even when I exerted the full pressure of the small balancing planes fixed to the rear ends of the main planes. It was a moment I am not likely to forget. Changing hands quickly on my steering lever I leaned over as far as I could from my driving seat so as to be able to throw the weight of my body against the rising wing of my machine. Just when I thought I should slide hopelessly down through the air the machine slowly righted itself, but another gust assailed me and I had to look out for a landing place, although I had 11 gallons of petrol left in my tank and the engine had not misfired once. A field near a village presented itself. I planed down and sat still quite exhausted.'

A Belgian peasant working on the road nearby, stopped working and gazed stolidly. As he appeared disinterested Sopwith walked across to a cottage and enquired of two old ladies his whereabouts. Only then did he realise he was in Belgium, at Beaumont some 9 miles from the French frontier. The ladies, apparently knowing something of English customs, asked what the weather was like in England. Leaving his machine he set out for the nearest telegraph office, but being Sunday, post offices were closed.

Tom Sopwith, who had crewed the winning yacht in the 1909 Royal Aero Club race and had now turned to aviation, is seen with the pipe that Miss May Sopwith discouraged him from smoking when training for record flights.

Meanwhile, his sister May was growing anxious. She had watched him take off from Eastchurch and had immediately set out by car for Dover which she reached an hour after her brother had passed over flying seawards. Putting up at the Lord Warden hotel, she made a series of cross-Channel calls to France, first to Châlons and then to towns on the presumed route. Sopwith had apparently disappeared! Then at 2.30 p.m. a telegram arrived at the Lord Warden from Sopwith in Belgium.

Chilled by the cold and damp Sopwith had tramped to a railway station. After a reassuring cable to his sister, a message was left for Sigrist asking him to retrieve the machine left in the field and to collect his gauntlets and coat left with the cottagers. He impressed upon Sigrist that the biplane must be sent off quickly to England because, if his record was broken by others, he would set out again immediately. He stayed the night in a hotel in Brussels.

Up early next morning he caught a boat train for Ostend and met up at Dover with his sister who was full of news of the other competitors. The windswept ground at Dover had proved an unfortunate site. A gale on the previous Friday had wrought havoc causing hangars to collapse. Loraine's machine was completely wrecked and Greswell's machine had two longerons broken. Grahame-White's Bristol had escaped serious damage and was soon repaired. Noting the fine, but breezy weather on the Sunday, that had enticed Sopwith to make the attempt, and then the sight of Sopwith's machine passing serenely over, Grahame-White was influenced to make his attempt that same day. He ordered his Bristol out soon after 11 a.m. and at 11.30 the propeller was swung.

Taking-off, Grahame-White rose some forty feet before heading out over Dover. Gusts of wind buffeted the machine causing it to pitch and roll and the pilot realised the elements were set against him. Sopwith at this time was experiencing similar deteriorating weather on the Continent. Grahame-White turned back, intending to land on the cliffs, but at this stage the machine appeared to get out of control. Attempting to bank round into the wind again to gain lift, the machine side-slipped into the ground from about 40 feet. From under the tangled mess of wreckage, Graham-White crawled out, blood running down his face. He rose slowly to his feet, walked twenty yards and then fell into a faint. A car took him to the Lord Warden Hotel where he received medical aid for his injuries which fortunately proved to be superficial.

Sopwith went immediately to see Grahame-White and took him back to London in his car. Meanwhile Sigrist, as soon as he knew where his master had landed, set out to arrange for the return of the Howard-Wright biplane lest some other competitor beat Sopwith's record. By Wednesday, a suitable vehicle had been found to convey the dismantled machine for shipment back to England.

As it was, Sopwith's cross-Channel flight of 177 miles in 3 hours 40 minutes was not seriously challenged. Rough weather wrecking machines, had eliminated most of his competitors and tragedy came to another. Cecil Grace had discarded the flotation bags that Oswald Short had designed for his biplane and had abandoned a compass

Tom Sopwith in the 60 h.p. E.N.V.-engined Howard Wright biplane, modified with wind shield and increased tankage, for his attempt to win the Baron de Forest prize for the longest non-stop flight from any point in England to anywhere on the Continent.

in favour of steering by the sun. Taking off shortly after Sopwith's return, he was forced to land near Calais and took off to fly back and try again. He was last observed flying out to sea by the East Goodwin lightshipmen.

As the year 1910 came to a close Sopwith stood by to make flights in case his record should be broken. On December 28th his Michelin Cup record was beaten by Alec Ogilvie who flew a 130 miles flight at Rye. The very next morning, at 9 a.m. Sopwith took off to better Ogilvie, but gusty wind made the effort so tiring that he was forced down after enduring 70 miles. On December 30th, the penultimate day of the year and of the competition, he started an hour earlier, but again was forced down after circuiting the stakes at Brooklands for 70 miles.

Saturday, the last day of the year, Sopwith again started early, but after only seventeen miles he was forced to land with ignition trouble. Sigrist quickly tended to the engine and by 9.40 a.m. Sopwith was again in the air —and this time, his final bid for the Michelin Cup, he kept going well, surpassing first his own, and then Ogilvie's record, finally descending after 150 miles in 4 hours, 7 minutes. But the Cup was not his. That same day, on Laffan's Plain, Cody accomplished 195 miles in 4 hours, 50 minutes. Sopwith lost the Cup to Cody.

The competition for the Baron de Forest prize too, would be decided that day. There was still a chance that one of the competitors at Dover and Eastchurch might make a final bid. However, they were beset by misfortunes. Although Robert Loraine's Bristol Boxkite had been repaired from its earlier accident it was smashed again at Eastchurch; Alec Ogilvie, trying to fly over the Channel

Return from Belgium after his record-breaking flight. Sopwith and his sister May at Victoria Station, London, after driving Montague Grahame White and Miss Pauline Chase (the original "Peter Pan") back from Dover.

from Rye, was forced down on Camber Sands; Grahame-White with a new Bristol presented to him by Sir George White, had it destroyed by fire and Lieutenant Watkins had his machine damaged in a crash.

Only Greswell remained in the running and on this last day announced his intention to wrest the record from Sopwith. Elaborate preparations were made. The tugs *Gnat* and *Lady Curzon* left Dover and the channel boats *Calaiserre* and *Champion* left the French side on patrol. But as the fog persisted over the sea, Greswell abandoned the attempt. Next morning the Secretary of the Royal Aero Club rang Sopwith—'It's yours,' Perrin said.

Ready to go. Tom Sopwith and his record-breaking Howard Wright biplane. Unlike his competitors who started from the coast at Dover, Sopwith started from Eastchurch on Fred Sigrist's advice, that an engine failure is more likely to occur in the first few minutes at full throttle; in which case, if this should happen, it would be better to be over land than sea. Sopwith passed over Dover half-an-hour after take-off from Eastchurch.

CHAPTER THREE

Command Performance

T. O. M. Sopwith at Brooklands with his pet bear. This was before the days of proper clothing for aviators who, because of their exposed position, bound their legs with puttees tied with tape, making sure the boots were overlapped, in an attempt to keep out the chill which had a numbing effect after a few minutes.

On every suitable day in the new year of 1911, Sopwith took to the air. On Saturday, January 18th he made six or seven flights and then set off on a cross-country trip. Over Hurst Park, his appearance coincided with the finish of the three o'clock race, causing eyes to be raised from the field as he circled the track and going down, according to reports, to a mere fifty feet above the crowd. On leaving the race-course he followed along the Thames, intending to land at Datchet where another of his sisters, Mrs. Raikes, lived, but a mist was rising and he wisely turned back to Brooklands.

Next day Sopwith flew direct to Datchet. Landing on the golf course, to the amazement of a solitary player, he left his machine on the links while he went into Datchet for coffee with his sister.

A villager has left an interesting account of the impression that Sopwith made—'Seated on Sunday afternoon enjoying my usual rest, I was suddenly awakened by someone violently ringing the bell, which seemed to carry with it a suspicion of unusual importance. On opening the door my visitor in gasping breath exclaimed "An aeroplane! An aeroplane! Mr. Sopwith has alighted on the golf links, and is returning very shortly." His excitement infected me, and off I started. On my way, I met Mr. Raikes, who kindly informed me that his brother-in-law was in the Manor Hotel partaking coffee and that he intended to start in about ten minutes.

'Breathing quickly, perspiring perceptibly, and with legs stretched to their utmost, I reached May's Crossing, where the signalman informed me that I should be in time as Mr. Sopwith had not as yet gone by. My marathon was telling on me, and when I arrived at the links I found that another half-a-mile had to be traversed. The fields and roads seemed alive with people, and never have I seen such a gathering on the fields usually reserved for the wielders of "drivers, irons and putters".

'But my tiredness magically vanished when I came to the now-to-be-famous spot, and found the bird-machine quietly resting. I was not a moment too soon, for almost at once the intrepid aviator took his seat, and awe and wonder pervaded the field! Such was the solemn stillness that one could hear the beating of the heart and the ticking of a watch, and it seemed as if the dropping of pins could be heard even on the grass. "See," said one, "the engine is starting." Instantly the motor buzzed, the forward plane was elevated, and the aeroplane ran along the grass on its wheels for about 150 yards till it was very close to the "Thistle", when it began to mount into the air. Returning, it gradually ascended higher, and no words could describe the gracefulness of its movement, and the excitement of the occasion. "It seems a dream," said one, whilst another could hardly believe his eyes when the airship [*sic*] sped away from sight amidst applause from the onlookers.

'When all but lost to sight it was noticed that it was

The exposed position of T. O. M. Sopwith in his Howard Wright biplane can be appreciated from this view. In the machine the fuel tank had been placed high between the wings; while the oil tank can be seen, just above the engine, between the radiator tubes.

turning, and to the delight of the still wondering crowd it came back, and with the ease, confidence, and the beautiful motion of a hovering bird, it alighted almost on the very spot from which it started. The crowd was electrified. The youth of the aviator, the placidity of his countenance as he manoeuvred his machine, the charm of his smile at seeing so many viewing him with interest. After waiting a few minutes the whizz of a motor once more announced his departure, and all eyes watched the aeroplane until it was lost to view, when those fortunate enough to have been present went home to relate for many a day, and to remember to the end of their allotted span the first descent of the epoch-making aeroplane on the golf links at Datchet.' Such was the impact of aviation in those days.

On 31st January, Sopwith received his prize at the Royal Aero Club annual dinner. The Duke of Atholl, in the chair, spoke of aviation being a matter of imperial interest and that a prize should also be offered for a machine combining stability and safety. He spoke of a device which, in the event of the machine collapsing would detach the driver [*sic*] and allow him to come down by parachute.

One speech, by Major Sir Alexandra Bannerman, Commandant of the Balloon Corps, went much against the grain. Not only did he state that the aeroplane would not, as had been suggested, revolutionise warfare, but only change tactics; he hinted that the aeroplane had not advanced much since the Wright had first flown in France. At this there were cries of 'Oh!' He said, no doubt recounting the official view, that 'he wanted to see an aeroplane on which one could let go of the controls.' Sopwith was not impressed by these views.

This quest for inherent stability in aircraft was to bedevil the designers of the Royal Aircraft Factory, and the persistence with which they pursued this seeming ideal, allowed such manufacturers as Sopwith, a few years later, to gain success with designs that met the wartime requirement for manoeuvrability in fighting aircraft.

It was up to Sopwith to make a short speech, his first on aeronautical matters, in giving thanks on receipt of his prize money. He modestly gave the misfortunes of his colleagues as the reason for his own success. He thought that many aviators, and certainly he was speaking for

Sopwith in his hanger at Brooklands about the time of his flight to Windsor Castle early in 1911. As he explained, it was not so much of a Royal Command, as King George V himself well appreciated the temperamental state of aircraft and their engines at the time. Nevertheless, Sopwith successfully coaxed his biplane to Windsor.

himself, would be pleased to help the Army and Navy if they were given a chance. He said that what was needed was a lighter British engine, for the current in-line engines, compared to a French rotary engine, carried extra weight equivalent to a passenger.

The Baron de Forest, in complimenting Sopwith on winning the prize, said it was one of the pluckiest efforts ever seen in any sport or industry. His effort had not gone unnoticed in other quarters. King George V expressed a wish for Sopwith to fly to Windsor.

An anxious Sopwith reached Brooklands on the cold foggy morning of Wednesday, February 1st 1911, to fly to his Sovereign. The frost was so keen that the water froze in the radiator pipes as they were being filled. Realising that this might cause a burst, he nevertheless resolved to fly if the fog dispersed. But at mid-day, Brooklands was still enveloped in thick fog. After ringing Windsor, he was surprised to learn that it had bright sunlight. With this news he decided to take-off in the mist.

The same Howard Wright biplane in the form in which T. O. M. Sopwith flew it to Windsor after Fred Sigrist had carried out modifications to Sopwith's suggestions. The fuel tank has been re-positioned behind the pilot, wing extensions have been fitted and detail improvements made.

Sopwith in his Howard Wright biplane as used for his flight to Belgium.

At exactly 1.0 p.m. Sopwith took off and found the lower atmosphere a thick woolly veil. He climbed steadily to 1,000 feet, took direction by compass and soon emerged in bright sunlight. It was bitterly cold, but the engine functioned well and so warmed the radiator that some of the tubes, which had been split by icing, started leaking.

Twenty minutes later Datchet church came in view and as Sopwith circled the links, he could see from the black specks on the ground that a crowd was gathering. He landed to cheers and was welcomed by his sister and brother-in-law who took him by car to their house for lunch. The aircraft was left in the charge of the Police.

After lunch Sopwith motored to Windsor and met Sir

Sopwith in his Howard Wright biplane showing the revised fuel arrangement by Sigrist.

Charles Cust of the Royal Household, who accompanied him over the golf course below the East Terrace, in order that a suitable landing spot could be selected. He decided upon the East Lawn as allowing the best landing and take-off run.

News soon spread that Sopwith was flying to the King. Hundreds of people, including many Eton boys, hurried to the castle, passing Sopwith on his way back to Datchet Golf Links. The E.N.V. engine gave a little trouble before starting and it was ten minutes after the scheduled time that Sopwith hove in sight flying over the Datchet Road. After gliding down across the East Lawn at about 400 feet he turned over the towers, along the East and North Fronts, and circled the Round Tower while the crowds cheered. Built on a mound, the massive tower dominated the flat countryside and so Sopwith's circuit was witnessed by many people in the district brought to their doors by the unfamiliar sound of an aeroplane.

It was by then a fine winter afternoon, with a blue-grey sky and a glinting sun making the aircraft look whitish. On landing Sopwith was met again by Sir Charles and was presented by him to his Majesty, who congratulated him on his flight. His machine was then inspected by the King and Princes Henry, George and John. King George V, a sailor himself, asked some searching questions and noted the leaking radiator.

Arriving by motor car, Mr. and Mrs. Edward Raikes and sister May were presented to the King who also had a chat with Sigrist who was standing by in case anything on the aircraft should need attention and to assist on take-off. At 3.20 p.m. Sopwith took the air again, but before setting course for Datchet he flew round at a low altitude. At the Datchet golf course, mist was already rising again and the machine was left overnight, where it had landed, near the fourth hole. In clear weather next day Sopwith returned his aircraft to Brooklands where Sigrist immediately repaired the radiator.

Brooklands at that time was also the Mecca of the motorist as well as the aviator. Sopwith retained an interest in cars, but in 1911 chiefly as a means of getting to his aircraft. There is a story of him betting several of his motoring friends a fiver each that he could lie on his back in Piccadilly for five minutes without being moved on by the police. At that time it was possible to park cars down the centre of Piccadilly, and Sopwith calmly drove his car to the centre, climbed out and lay on his back under the car ostensibly tinkering with the engine. He won his bets!

While Sopwith had been flying at Brooklands, he was keen to note what others were flying. Impressed by the Martin & Handasyde (to combine later to become Martinsyde) Type 3 monoplane, he ordered a model for himself. This new monoplane, the Type 4B, powered by a 50 h.p. Gnome rotary engine, cost £1,100.

Certainly Sopwith had a bias towards the monoplane at this time. With Gustav Hamel, the famous aerobatic pilot, he visited the Bleriot School at Pau in France and flew the new 70 h.p. Gnome-engined Bleriot monoplane. This too, he bought. He was seeking fresh fields to conquer and his imagination had been fired by a new challenge, issued by the American newspaper magnate, William R. Hearst, for the first New York to Pacific Coast flight.

CHAPTER FOUR

American Adventures

Sopwith did not become head of one of the largest aircraft companies in the World by chance; it was because he had the intelligence to seize opportunities, plan meticulously and had necessary capital to promote his ventures. For his American tour he appointed a manager to make arrangements, J. Dudley Sturrock who had served Claude Grahame-White on an earlier tour, and took a team of three mechanics out—Fred Sigrist, Jack Pollard and Harry England. A new 70 h.p. Bleriot monoplane was shipped out and within hours of it being unloaded at New York from the S.S. *Provence,* Sopwith, accompanied by his sister May, was disembarking from the S.S. *Amerika* of the Hamburg-America Line.

They were met by another of their sisters, Violet, now wife of an American officer, General E. Burd Grubb, who had recently been appointed Director of the Old Soldiers' Home at Kearny. It was here they stayed.

Early in May Sopwith had assembled his machine at the Hempstead Plains aviation field on Long Island. Next morning he made a solo test flight and was evidently so pleased with the monoplane's functioning that he stayed an hour in the air. Shortly after landing he invited Phillip W. Wilcox, an American aeronaut, for a trip. For a while all went well, the machine rose quickly then, at about 100 feet, a sudden gust of wind appeared to upset the aircraft and the Bleriot turned partly on its side and fell rapidly. Sopwith's impression was of being in a vacuum. A mere ten feet from the ground he regained control—but too late. With another ten feet of altitude all would have been well. As it was, the machine hit the ground at Mineola Park and crumpled into a mass of wreckage.

As Sopwith explained later to enquirers, he had met 'what aeroplanists call a hole in the air'—in fact what is now called a down draught. Although he had felt the controls begin to 'bite' again near the ground, the aircraft had 'mushed' on downwards and was wrecked. Neither pilot nor passenger were thrown from their seats, but Sopwith had been jerked forward and he caught the side of his face on the *cabane.* Apart from this and a slight leg injury and bruises he was unhurt, and similarly Wilcox for whom this was his third crash in six flights!

A cable was sent off to France straight away for a new 70 h.p. Gnome-engined Bleriot monoplane, costing some $12,000. Meanwhile, his Howard Wright biplane had arrived from England and was being assembled at the Philadelphia Driving Park, Point Breeze. There, after testing the machine, he took up his sister Violet, Mrs. Burd Grubb, for her first flight, staying about six minutes in the air.

Another passenger that day was Charles M. Clark, a photographer who took several pictures of the Navy Yard as the aircraft passed over League Island, and May also made her first flight on the American continent.

Next day the Howard Wright biplane barely escaped being wrecked. The wind was gusty and Sopwith waited patiently for conditions to improve before leaving and then was faced with an unexpected gust that seemed to cause more of a down-draught than uplift. Wrestling with

T. O. M. Sopwith ready to take his sister May for a flight in the Howard Wright biplane.

the controls, a few feet above the ground, he approached a four-foot fence across the park, clearing it with only inches to spare.

Conscious that it was the custom of the French pioneers—and France was leading the world in aviation at this time—to display their prowess by flights around the Eiffel Tower, Philadelphians seemed to expect an attempt to encircle their magnificent City Hall. Lincoln Beachy, a balloonist, had encircled it in an airship in the summer of 1907, and in 1910 J. Armstrong Drexel, who made a

T. O. M. Sopwith with his sister at Kearny in America during 1911, before setting out on his flying tour. The Americans, confident in the achievements of their pioneers since the Wright Brothers first flew at Kittyhawk in 1903, were amazed at the performances of the British and French machines introduced by T. O. M. Sopwith and Claude Grahame White.

The Bleriot Monoplane, that Tom Sopwith bought in France and had shipped to America. Left to right, Harry England, Jack Pollard, M. Pannier (mechanic to the American pilot Ovington) and Fred Sigrist. This was the basic Type XI Monoplane which differed little from the earlier model used by Louis Bleriot in his epic Channel crossing in 1909 and it became one of the most used of pre-1914-1918 War aircraft. This two-seat model, powered by a 70 h.p. Gnome engine, relied on a wing-warping device for lateral control.

cross-country flight over Philadelphia, essayed several times to make the attempt, but never succeeded. Sopwith wisely confided in no-one his attempt. After taking off on one of his flights from Point Breeze on his own, he set course for the city. Quickly disappearing from view of the spectators, it was realised he was 'off on a tour'.

The noise of his engine over the built-up area of the town brought people out of doors. From factories and stores people flocked into the street. Over an aptly-named Broad Street, his machine was plainly visible and many spectators started running in its direction as if to keep pace with it. America had not then brought in laws restricting the activities of airmen in built-up areas. Sopwith was showing Philadelphia what he could not have done in his own country.

From his highest point above Chestnut Street—if you know Philadelphia—he descended to 500 feet to circle the City Hall. As the local paper put it—'he circled the tower as gracefully as any of the pigeons which make their nests under the William Penn statue. Leaving the tower behind him, Sopwith headed his machine in a southerly direction for the return trip to the track. Hundreds stood on roof tops and craned their necks until the machine had disappeared.'

Sopwith was evidently quite low as a paper said that he turned west at Snyder Avenue! Passing over his race track starting point, he continued until he had reached the Schuylkill River, where a large grain elevator had evidently caught his eye. Making this his turning point he returned to Point Breeze.

On one occasion Sopwith accompanied by a passenger, Mr. Harry Doughton, dropped a message into the Navy Yard addressed to the Commandant, Captain Grant. According to reports, it was a card inscribed 'Thomas Sopwith presents his compliments to the Commandant of the Navy Yard by aeroplane delivery'. Later, he paid a personal visit. His sister May, Sturrock his manager and a mechanic left by car for the Yard while Sopwith, with a passenger, set out by air and landed on a baseball field after players had scattered. He was met by Captain Ritterhouse of the U.S. Marines, the acting commandant in the absence of Captain Grant. As the baseball field was rather small for take-off, his pupil passenger had to return by car, to lighten the craft.

On his last day at Point Breeze, shortly before flying to the Navy Yard, a big touring limousine drove on to the field and stopped before him. Inside was Miss Margaret Dunlap of 2017 Walnut Street, a prominent member of

The Sopwith team in America in front of the Howard Wright Biplane. Left to right (standing), Fred Sigrist, J. Dudley Sturrock, May Sopwith, Tom Sopwith and a hired American mechanic. Those seated are Harry England and Jack Pollard.

Making the best of a bad job after a gale that had sprung up an hour before flying was due to start at the Columbus, Ohio, Aviation Week in late May 1911. Some of the Sopwith team are seen resting amongst the debris after their initial salvage efforts; they are, left to right, Sturrock the Sopwith team manager, Pannier a French mechanic from the Ovington Team, T. O. M. Sopwith and Fred Sigrist. In spite of this initial mishap, Sopwith gained several prizes at this display.

the younger set in Philadelphian Society and a great beauty to wit. In the fashion of the day, she wore an enormous hat. 'Are you going to take me up?', she demanded. 'Not unless you take your hat off', Sopwith laughed in reply. Her hat came off and she was escorted to the aircraft and given a trip of about seven minutes.

It was time to move on to Columbus, Ohio, for their Aviation week, May 29th to June 3rd. Sopwith had no less than four mechanics brought over from England, headed by Sigrist. This led to speculation that he intended breaking another record and when two of the mechanics at Ohio visited a Third Street engineering firm for a 35-gallon tank, it was rumoured that he planned to beat Henri Farman's endurance record of eight hours twelve minutes made in France the previous year.

In fact, this was so. The new tank would permit nine hours in the air and the existing ten-gallon tank on the Howard Wright was being converted as an oil tank. The attempt was revealed when Dr. John C. Eberhardt, a member of the Dayton Aero Club, was appointed official representative of the club to observe Sopwith's attempt.

On Sunday May 29th, Tom Sopwith in common with the other aviators at Columbus checked over his machine. The Howard Wright was wheeled out onto the grass and the engine tested to the full while six men hung on to keep it back. At almost the last minute it was pointed out to the authorities that the long grass would impede take-off and a mower was set to work.

At 2.30 p.m. next day the show opened with the firing of a mortar followed by displays by the aviators advertised. Two events showed that if the military authorities were slow to appreciate the potential of the aeroplane, it was certainly not so by the pioneers. At 4.30 p.m. a bomb-throwing contest was scheduled at imaginary battleships marked out on the field; there was also target firing from pistols carried by the pilots. On May 30th there was even 'wireless telegraphy from ships [*sic*] in the air' scheduled.

The weather was not kind to the meet. On the first day a gale sprang up an hour before flying was due to start. One tent was blown down and Sopwith's Howard Wright was first hit by a flapping tent wall and then a collapse of part of the tented hangar damaged the rudder. Display flying was cancelled, but Sopwith soon had his machine repaired and made a test flight. On an apparent impulse, he flew out six miles to the Columbus Club golf course where he landed. According to reports he made his own take-off unaided. Swinging the propeller himself, he ran with the machine hurdling part of the framework into the seat to grasp the controls!

In spite of this practice, Sopwith did not do well in

A general view of the Sopwith camp after the gale had abated, showing the same four persons seen above sitting in the lee of the partly collapsed tent in which the Howard Wright biplane was damaged. The aircraft on the right is shown packed for road transport.

23

Pannier, Ovington's French mechanic who assisted in the Sopwith camp, after a sleep amongst the wreckage.

the competitions for quick starting. He took 207 feet to get into the air compared with Baldwin's 162·6 feet. However, when it came to the bomb-dropping competition, Sopwith for a time led.

On the last day of the meet there were two mishaps, one involving Sopwith. First the all-metal framed 'Red Devil' of Captain Thomas S. Baldwin, dropped 25 feet, smashed through a fence and fell in a ditch. Then Sopwith in his Howard Wright, skimmed over to the spot and saw that all was well with Baldwin. He turned to come back, half rolling, half flying. As he passed the judge's stand, the end of the starboard skid dug into a small hillock. As the local paper put it—'there was a crashing of wood for a second and a half, and when it was over Mr. Sopwith had no aeroplane to speak of'.

Sopwith returned to Long Island, and when his aircraft was repaired used the Nassau Boulevard aerodrome for a series of passenger-carrying flights over the Island. Ex-Lieutenant Governor Timothy L. Woodruff, who had stood by and watched his wife being carried into the air, himself made a flight on June 23rd. On landing he said, 'Well, you know that was the most wonderful experience I ever had. I was never more surprised in my life with the security and ease of it. When we came down it was at an angle of 45 degrees. Now, on one of the Coney Island pleasure machines your heart is in your mouth when you make such a dip, and they call them scenic railways, too. Why, you can't see a thing on them. Up in the air you get the most wonderful panorama.' Such recommendation, reported verbatim in the press, brought much custom.

There was the occasional stunt. At 7.25 p.m. the Oyster Bay express from Oyster Bay to New York passed Nassau Boulevard. Tom Sopwith, with sister May, met the train one evening and paced it along the track. May waved her handkerchief in reply to passengers who thronged the windows and the engineer tooted the whistle in salute.

Another stunt brought wider acclaim—this time a ship was involved. The giant four-funnelled White Star liner *Olympic* had just completed her maiden voyage to New York. Sopwith had intended flying out to meet her on arrival, but fog had prevented this. However, his former passenger, Mr. Woodruff, commissioned Sopwith to take a message to the ship as it passed The Narrows between Staten Island and Brooklyn, outward bound, in the mid-afternoon of July 28th. A coincidence brought much publicity to this attempt. A Mr. W. Atlee Burpee, seed-merchant of Philadelphia, had embarked for Europe on the *Olympic* which sailed at 3 p.m. A few minutes after the ship had left its berth, Mr. Burpee realised his pince-nez glasses were useless. He went to the signal cabin and had the following message transmitted:

'On Board S.S. *Olympic*
To: John Wanamaker, N.Y. Spring in eyeglass broken. Mail by next steamer, Morley's Hotel, London.
W. ATLEE BURPEE'

This message was received by the Marconi station on the roof of the famous Wanamaker Store, an installation allied to a new slogan of 'Mail, Wire or Wireless orders taken'. Wanamaker's publicity manager was not slow on the uptake. He telephoned Sopwith to fly out a replacement spring.

Sopwith was already leaving Nassau Boulevard in his Howard Wright, taking with him Richard R. Sinclair, Woodruff's secretary—who was finding that secretarial work could have some surprising ups and downs—to carry and drop a small package with messages on a ship.

The Howard Wright rose at 3.27 p.m. and in the slight haze went back and forth over Brooklyn. Finally the *Olympic* was sighted approaching The Narrows. Descending slightly from 1,000 feet to fly over the giant liner, Sopwith found good sighting impossible by the four clouds of smoke that rose from her stacks, and felt too, the rising warm air. The package was dropped, but its landing could not be seen through the smoke. Alas, it fell in the water.

Passengers had crowded excitedly onto the decks and the *Olympic* gave three shrill blasts on its siren. Burpee, having seen the package fall, hurried to the signal cabin and sent another message:

'On board S.S. *Olympic*
To: John Wanamaker, Broadway and Tenth Street, N.Y. Just now, four-fifteen p.m., within one hour of sending you my order by wireless an aeroplane has passed over the Olympic, causing considerable excitement and dropping a package which, unfortunately, just missed the deck. Is it possible that this was an endeavour of yours to execute my order without waiting for steamer. If so, your advertisement in today's Herald fails to show the Wanamaker spirit of enterprise.
W. ATLEE BURPEE'

The advertisement referred to was an invitation for passengers departing in the *Olympic* to use the Wanamaker wireless station to send messages to relatives.

Sopwith landed, as intended, to meet his sister on a baseball field behind the clubhouse of the Crescent Athletic Club. By that time the *Olympic* was much too far out to attempt dropping the actual package from Wanamaker —but nevertheless this attempt brought much publicity. For the message-dropping venture, Sopwith received a valuable silver cup that evening, presented in the Garden City Hotel by the Aero Club of New York. It was inscribed—'To Thomas Sopwith in commemoration of his flight to the Olympic'.

Such publicity, allied to such important personalities as Henry W. Taft, brother of the President of the United States, being taken up for a flight, brought much custom.

So well did Sopwith do for passengers that in some quarters it was resented. 'Those exuders of coin who have exchanged a nice yellow fifty dollar bill for a five-minute flight in Mr. T. O. M. Sopwith's taxiplane will be gratified to know that the price constitutes a delicate compliment to American social exclusiveness. Mr. Sopwith, who is regarded as England's premier wealthy aerial sportsman, tactfully placed the minimum fare limit at $50 when he began his omnibus service at Nassau Boulevard. He will readily take as much money as you may care to spend and accommodate you with a flight of corresponding duration and distance'. Thus reported one paper which apparently resented business acumen by one not of their country.

This paper quoted that Americans returning from the Coronation of King George V found that an aerial trip could be taken from one of the aerodromes around London (Hendon presumably) for $20. It went so far as to suggest that a person of moderate means who wanted a flight for $50 could take a steerage passage to England, take a flight and still have change! What the paper resented most of all was that Americans were not making similar money in Boston. It concluded 'Mr. Sopwith's American enterprise contains a hint for active American sportsmen of wealth. Mr. Alexander Smith Cochran, who owns the cup-winning yacht *Windward,* might send her over to England and make a thrifty fortune by operating it for a season in some British seaside resort as a taxi-yacht. Social England would undoubtedly lionise him and gladly pay $100 for a five-minute ride in this famous racing craft'. Strangely, Sopwith was destined to become as famous a yachtsman as Cochran and, in truth, the few flights per day would not cover the outlay of the new Bleriot in place of the one wrecked, let alone the upkeep of his manager and mechanics.

Next came an attempt at legal restraint on Sopwith's activities. A Bill of Complaint was filed in the Circuit Court by Wilbur and Orville Wright, representing the Wright Company, to restrain a Mr. T. O. M. Sopwith of England for conducting passenger-carrying flights at the Garden City Estates.

The Bill, in legal phraseology, alleged 'that prior to a date in 1903, the Wright brothers were the original joint inventors of a new flying machine of great value and ability, and constituting the first instance in the history of the countless attempts to produce flying machines wherein the heavier than air machine ever made aerial flights, and wherein the machine was within the control and under the will of the operator.'

Concluding, the Bill stated that the machine imported by Sopwith from England was covered by the Wright Patent of May 22nd 1906, and that Sopwith should be immediately restrained from his flights in the U.S.A.

For the next few days, a strong wind making flying unsuitable, decided the issue and by the time of the next aviation meeting, at Chicago, the new Bleriot had arrived. The Wrights were also somewhat mollified by Sopwith negotiating to buy one of their aircraft.

The nine-day aviation meet from August 12th to the 20th was an important event and was exceedingly well attended. One local paper after the opening day wrote that—'before the wondering eyes of 306,000 spectators, more than a score of daring pilots of the air drove their mechanical birds in the greatest aviation meet since man began to out-distance the birds'!

As soon as the opening gun had fired, Sopwith was taking off with two passengers for the $3000 prize offered for the first aviator of the meet to stay up in the air for over an hour with 2 passengers. There was considerable competition for the prizes, most of America's pioneer airmen and some from other countries participated. To name them, with their aircraft; Lincoln Beachey (Curtiss), G. W. Beatty (Wright), Captain Paul W. Beck (Curtiss), Frank T. Coffyn (Wright), Andrew Drew (Wright), Eugene Ely (Curtiss), Earle L. Ovington (Curtiss and Bleriot), John J. Frisbie (Curtiss), Howard Gill (Wright), Lee Hammond (Baldwin), James V. Martin (Grahame-White), J. A. D. McCurdy (McCurdy), George Mestach (Morane), Oscar A. Brindley (Wright), Phillip O. Parmelee (Wright), C. P. Rodgers (Wright), Rene Simon (Moisant-Bleriot), T. O. M. Sopwith (Bleriot and Wright), Arthur B. Stone (Queen), J. C. Turpin (Wright), James Ward (Curtiss), A. L. Welsh (Wright), and Witner (Curtiss).

Competing against such men in a variety of craft added greatly to Sopwith's experience in judging the merits of an aircraft. He is quoted as saying, 'Speaking from a mechanical standpoint, I think the Wright machine is a monstrosity. I don't see how it could be any worse and still it seems to fly reliably. As you say in this country it seems to get there, but that chain from the motor to the propellers is a very bad arrangement. The chain is the worst feature of the Wright machine, but the slow motion of the propeller makes up for it at least partially.

'If you can get the power from a propeller making 500 revolutions a minute you will get about 75 per cent of efficiency out of your machine. If the propeller makes 1,200 revolutions a minute you can get only about 50 per cent efficiency. That is what gives the Wright machine an advantage. With its two propellers it gets the power with-

Jack Pollard and Harry England photographed during the Sopwith team's American tour.

out making as many revolutions as where a single propeller is used, and I think the advantage in efficiency more than makes up for what it loses by having that horrid looking chain.

'If you want speed the monoplane has it. The Nieuport is the fastest in the world, and the Morane comes next. Both are faster than a Bleriot, but they are all much alike in construction.'

Sopwith was well satisfied with the general handling of his new Bleriot on the first day of the Chicago meet. There was no Nieuport there to challenge his Bleriot. In the initial competition, a speed race for monoplanes, he was closely run by Ovington's Bleriot. But Sopwith's time of 13·52 minutes was a record for the twelve-mile course of nine laps. In the alighting contest, in which the aviators had to make their touch-down closest to a marked spot, his 8 feet 8 inches was the best of the day. On a race to the local Country Club, Sopwith's time was 15·58 minutes; running him up was Ely on his Curtiss at 17·58 minutes, and this time Ovington did not finish. In the 500 metre climb, Sopwith was thought to have tied with Rene Simon for the American record, but he was later credited with this record. When the meet ended Sopwith had won more prizes than any other aviator, although in actual cash, he came second with $14,000 to C. P. Rodgers' $15,000.

The meet had been marred by fatal accidents to St. Croix Johnstone and William R. Badger. There had been talk of cancelling the last few days lest more accidents happen. Among those advocating a continuation was Miss May Sopwith, who wrote to the local paper, in the true manner of our pioneers of the air, 'I am sure there isn't any reason for the meet stopping now. The people have come here to see the machines demonstrated, and there isn't any reason for stopping the programme.

'Someone says that the machines should not be flown until they are perfected, so that the accidents will not be so frequent. How on earth do we know when they will be perfect? We may not live to see it. We want to know what is being done, and what has been done, and the men want to show us their machines and how to manage them. It's all fair and honest, and even if accidents do happen, they are expected and should be accepted as the inevitable.

'People who understand aeroplaning and what it means aren't foolish enough to want to call the meet off now. I want it to go on and I am willing my brother shall fly and I want him to. We may not have another chance for some time to have the world's greatest aviators, and the world's greatest machines, before us. It may be the last chance for us, and it's the greatest wonder of the age.'

While Sopwith directed policy, he did not interfere with details. All arrangements were left to his manager Sturrock and Sopwith wisely left press matters in his hands. In fact, his reluctance to talk to the Press became something of a joke. All questions were referred to Sturrock. One persistent reporter, who invaded his sanctum, said the dialogue ran something like this:

'Do you intend to fly again today?' asked the reporter.

'See my manager,' responded Sopwith.

'Did you experience trouble with your motor in the flight this afternoon?' persisted the reporter.

'Better see my manager,' replied Sopwith.

After similar questions with the same response, the reporter, to break the imperturbable aviator, said 'Mr. Sopwith, it is reported that you are going to attempt a flight to the moon this evening in an attempt to ascertain if it is really made of green cheese'.

'Now, you really *must* see my manager,' retorted Sopwith.

Sopwith had not lost sight of the Atlantic to Pacific flight and had bought a despatch car to use in this project. He had not lost his love of cars and reported critically on his new Velie Torpedo—'The car is doing bully good work. I am very pleased with the Velie. Don't you know that one thing that has struck me in the American motor cars in comparison with the English machines is the fact that they are geared so low. The Velie is a smart car and responds beautifully. While I do not care how fast I drive my Bleriot in the air, I do not like to speed in my motor car. I shall use my Velie at all of the aviation meets throughout the country, and expect to take it back to London with me when I return. What appealed to me particularly in the Velie was the flexibility of the motor; its ease of handling and its clean cut, accessible construction.'

September brought another aviation meet, the Harvard-Boston meet at Quantum. By this time a famous compatriot of Sopwith's had joined the lists, Claude Grahame-White, who promised keen competition since he was bringing a Nieuport monoplane. The papers tipped Grahame-White as favourite with Stone's Queen monoplane and Ovington's Bleriot as runners-up. Ovington's mechanic thought that if the wind was high his master's 70 h.p. Bleriot would beat Grahame-White's 75 h.p. Nieuport; he appreciated the merits of a Nieuport, but was well aware it yawed in the wind. Sigrist was not so confident of his master. The Bleriot had cambered wings for maximum lift in passenger carrying and although racing wings had been ordered from France it was doubtful if they would arrive in time.

The highlight of the meet was the Boston Lighthouse race, involving a flight from the Squantum field to circle the Boston lighthouse, and return. The previous year, Grahame-White had won the contest with 34 minutes 11 seconds elapsed time for the thirty-mile round trip; he had every chance of winning it again. Only four other entrants challenged him. Beachey and Ely on Curtisses both met with mishaps, the former mistook the Graves lighthouse for the Boston one and the latter alighted on an island after a radiator pipe had burst. This left the three monoplanes, Ovington's Bleriot with the race meeting number 13 on the rudder and a Dragon painted along the fuselage, Grahame-White's white Nieuport with No. 3 on the rudder and No. 7 Sopwith's Bleriot.

Knowing his friend Grahame-White had the faster machine, Sopwith used every artifice to get the maximum performance. Competitors started at two minutes intervals and Sopwith had noted Grahame-White's turn at the first pylon and thought he could manage to make it tighter. He planned to make sharp turns and fly high with the wind, but low on return with the breeze against him. The course was strange to him, but the great white lighthouse stood out from afar. That Beachey had missed it was due to a misunderstanding at the contest briefing. As he drew near, overflying rocks and water, Sopwith came down low, making his turn around the tower at only a few yards above the water. On his way back he passed Ovington, 800 feet higher and the two rivals waved to each other.

As it was, Grahame-White made the best time in 31 minutes 5 seconds—but he was disqualified for not passing over the start pylon. Thus the winner was Sopwith with 31 minutes, 33 seconds, who had a clear win over Ovington's 35 minutes 42 seconds.

Alternating from his Bleriot to his Wright, Sopwith won contest after contest. On one day he came second to Lieutenant Milling on the figure 8, on bomb-dropping he got to within 13½ feet of the target compared with Milling who came second at 77 feet. On quick start he came top, beating Milling by ½ second. On a one-way Boston lighthouse trip that day he came second to Grahame-White.

Back to New York, joined by Grahame-White, Sopwith gave exhibition and passenger flying from Brighton Beach Aerodrome situated on the ocean front between Manhattan Beach and Coney Island. It was here that the Wright Brothers, Wilbur and Orville, called to see Sopwith and here Ernest Newman, who had been tending aircraft in a Central American expedition, joined the Sopwith team. On September 10th perfect weather had brought crowds to the beaches, yet the air was apparently turbulent. Grahame-White was expected to make a circuit of the Statue of Liberty but air currents out in Upper Bay proved tricky and he returned. On the way back he saw Sopwith's aircraft low over the water; horrified he saw it hit the surface, turn over and sink.

Sopwith had left the field with Lee Hammond, another aviator, at 3 p.m. After circling the track around the field several times he headed out to sea and turned to fly along the beach over the heads of bathers in the surf. Opposite the Manhattan Beach Hotel he made a wide turn to return. It was then his engine started spluttering and he made for a clear piece of beach below the Brighton Beach Hotel—but he was too low. The machine glided down steeply, and was skimming just clear of the water when a wave caught a wing. The tail went up in the air and pitched the craft over in the water. Hammond tried to jump, but both men were caught under the machine.

Both Sopwith and Hammond had to fight for their lives after being precipitated into the water and kept under by the wings. It was some seconds before the two men reappeared. Kicking free from the aircraft they surfaced and clung to a wing of the overturned craft, which then sank lower under their weight and they struck out for the shore. A lifeboat put out but fortunately the motor boat *Evelyn* was close by and intercepted their swim. Sopwith was fished out of the water first. Hammond was so exhausted that he was transferred to a rescue craft and taken ashore where a doctor revived him. Later, in a bathing suit, he rejoined Sopwith in salvaging the aircraft.

The tangled wreckage of the Wright was taken in tow by the *Evelyn* to the diving platform in front of Parkway Baths. There, with the help of scores of bathers it was lifted on to a bathing raft and towed ashore. While the planes were broken, the engine seemed intact.

Spectators noting Sopwith's non-return realised that something was wrong, but George W. Beatty was about to take off for an exhibition flight. The engine was run up and the machine was straining when Beatty signalled release. Immediately the machine charged forward towards the spectators' fence. Onlookers jumped back out of the way as the aircraft struck a rail, spun round and crashed full tilt into the fence. Beatty unseated and uninjured sat in the wreckage with the engine still running which mechanics, rushing forward, soon stopped.

Grahame-White had landed to see if he could help Sopwith. Since only he and Eugene Ely were left to entertain the crowds, he returned at 5.30 p.m. to announce that he would fly to Nassau Boulevard. Several thousand watched him start. The Nieuport surged forward like a racing car for a few hundred feet, then the tail rose in the air as the machine buried its nose in the ground. A doctor who raced across to the spot found Grahame-White sitting on his craft swatting mosquitos.

Sopwith was among the first to reach Grahame-White. Both now had their aircraft smashed, but the summer flying season would soon come to a close and the idea of a trans-American flight was shelved. Undaunted by their mishaps, fortified by prize money, they turned to see what America could offer on the ground.

The end of the Howard Wright Biplane. Upside-down in the water, off Manhattan Beach, its wheels can be discerned in front of the boat left centre.

CHAPTER FIVE

Derby Winner — and Trainer

Late in 1911 after his return from America, Sopwith purchased a new Martin-Handasyde monoplane which The Aeroplane announced as "The Magnificent Martinsyde".

Having absorbed much of what was going on in aviation in America, Sopwith returned to Britain in October 1911 and waited impatiently for his aircraft to follow. He was given the opportunity to fly when asked to test the new Martin and Handasyde monoplane, the 'Magnificent Martinsyde' as it became nicknamed, and expressed himself as delighted with its performance. Two days after Christmas 1911, having just returned from Paris, he took a dozen passengers one or two at a time aloft in the Martinsyde including two of his sisters.

On Saturday afternoon, December 30th, his final passenger was Lieutenant Parke and it would appear that Sopwith set out to test the nerve of this naval officer for he did violent banked turns, then cutting the engine he glided down to within a few feet of the floodwater adjacent to the Brooklands ground before opening up and, as one observer put it, 'generally playing tricks calculated to cause more excitement than entertainment to spectators'. He did have a mishap with this aircraft, but it was no fault of his flying. Landing at St. Albans, the undercarriage suddenly dropped down into a hole, quite three feet deep, hidden by thick grass. The tip of the skid took the shock and both the skid and propeller broke. Sopwith had the machine dismantled, and when neatly packed on a trailer, towed it back to Brooklands himself.

Sunday, January 28th, was a great day for Sopwith, his machines having arrived from America and the first, his Bleriot, having been assembled and tuned up by Sigrist, he was able to participate in the Brooklands flying on one of his own proven aeroplanes. Nevertheless, he took the opportunity to size up others. When next day Cody invited him, with two other certificated aviators, Gilmour and Valentine, to fly in his biplane, he accepted with alacrity and sat perched behind Cody while the latter skimmed the ground, jumping fences and telegraph wires, before climbing and going through manoeuvres. Astounded by its climbing ability, he afterwards did a series of competitive trials with his 70 h.p. Bleriot against Cody's biplane. He was more than ever anxious to exhibit his own machines for he now had good reason for it—to advertise his new venture —to be known as The Sopwith School of Flying of Brooklands, Weybridge, Surrey.

The new departure was announced on February 1st. Advertisements proclaimed four 'Entirely Different Types of Aeroplanes' and claimed this was a variety greater than that of any other flying school. Terms for tuition were £75 per person including all breakages and third party risks. Special terms were offered to officers of either Service.

Sopwith had an advocate in that famous aeronautical journalist, C. G. Grey, who wrote, 'The latest move at Brooklands, and a welcome one, is the opening of an aviation school by Mr. T. O. M. Sopwith. He has now got a whole fleet of various machines, and it has struck him that with so great a variety he ought to be able to turn out much better pilots than the man who is confined to one type of machine. His school machines will comprise his old Howard Wright biplane, familiarly known as "the family tank", his American Wright (Burgess Wright) biplane, a school Bleriot, and a Howard Wright monoplane. The two biplanes will be fitted with dual controls, so that the pupil can learn the controls in the air. Mr. Sopwith will also have his 70 h.p. tandem Bleriot for his own use and will, probably, have the Martin-Handasyde to fly as well. With Mr. Sopwith himself as chief instructor, and that prince of engine tuners, Mr. Sigrist, to keep the machines in order, the pupils at such a school ought never to lack either machines to fly or the best of tuition. To the man who wishes to become a really good all-round flier, Mr. Sopwith's school should appeal at once'.

By February 7th, Sigrist had the American Burgess Wright in flying trim and the performance of this 'outsider' astounded the veterans. Among those that Sopwith took up in it that month was Captain F. H. Sykes who, six years later, became Chief of the Air Staff.

Hailed as the winner of the Aerial Derby—Tom Sopwith's Bleriot about to land at Hendon after completing an 81-mile circuit of London, June 8th, 1912.

In this photograph Sopwith is seen seated (front) in his 70 h.p. Bleriot Monoplane. His passenger wears his cap back-to-front in the approved "racing" style! The fully sprung undercarriage, and the pylons supporting the flying and landing wires are clearly seen.

Having established the school on a sound basis, Sopwith was already seeking new fields to conquer. His old friend Howard Wright having amalgamated with the Coventry Ordnance Works had become their new aeroplane department manager. The firm intended to compete in the military trials with a new revolutionary aircraft designed by W. O. Manning; it had already been agreed that Sopwith would be the test and demonstration pilot.

So as not to be tied to his School when opportunities came, Sopwith appointed F. P. Raynham his chief pilot, manager and instructor at the School. While Raynham conducted the routine instruction, Sopwith gave air experience flights. On February 25th he took up Mrs. R. F. Scott, the sculptress wife of the famous explorer and mother of Peter Scott, in the Burgess Wright, then Admiral Buckman, finishing the day skimming over the Brooklands track in the Wright with Mrs. C. G. Grey.

By March 7th the two-year-old Howard Wright biplane was made ready together with the monoplane, which had been fitted with a 40 h.p. A.B.C. engine, so that all four school machines were in commission. Howard Wright, interested in the innovation to his designs, the engine in the monoplane and dual control in the biplane, was a frequent visitor and that March he took a course of flying instruction under Sopwith.

April 16th, the first pupil to pass out from the school, Mr. D. G. Young, took his R.A.C. certificate using the Burgess Wright in the tests. Five days later this machine was nearly lost. Raynham, flying in gusty wind, was blipping the engine as he descended. When turning over the new bridge across the Brooklands racing track and making for the gap between the Sopwith sheds and the main block, there was an explosion and pieces from the machine—including Raynham's cap—flew through the air. The machine still at about 150 feet levelled out and for a moment spectators held their breath wondering if the controls had been affected, then the nose dropped again and Raynham glided in to a perfect landing. A cylinder had blown off the Gnôme, knocked off the engine shield and plunged through the lower plane. One piece of metal had caught Raynham's cap, worn back to front in the motoring style of the day, and sent it spinning.

The Gnôme engine was completely wrecked. In its place Sigrist installed a 35 h.p. Green driving geared-down propellers. On May 2nd Sopwith tried it out and it flew perfectly. It was a busy time for Sopwith, both he and Raynham were flying all through the week taking passengers on flights or on instruction. Sunday, May 5th was the busiest day of all. A series of competitions were run, the most important being a cross-country race to Chertsey and back for the Second Malcolm Cup. Two biplanes and four monoplanes were entered; it was won by Sopwith in his 70 h.p. Bleriot, who wrested the lead from C. P. Pizey by a matter of seconds.

An even more important event for Sopwith that Sunday was the first flight of the first Coventry Ordnance Biplane, built by Howard Wright to the design of W. O. Manning. After some preliminary running of the 100 h.p. Gnôme which drove at half speed the $11\frac{1}{2}$ feet diameter propeller, Sopwith and Manning took their side-by-side seats and set out for taxiing runs. After some 30 yards, when a speed of almost as many miles per hour had been attained, the machine began to lift and Sopwith brought up the tail and did a series of short hops. Turning at the end of the field he came back and flew some twenty feet off the ground. As dusk was falling, only two more runs were possible before it was put back in the sheds.

Next morning Sopwith continued on the C.O.W. biplane and took up three passengers, two sitting one each side of the fuselage on the lower plane. This was followed by circuits with Manning and an unintended spiral descent was made through his passenger mistaking an instruction and switching off the petrol during a turn. Next day there was trouble with the chain drive, delaying further tests.

During May, Sopwith visited France. As a keen sailor it was inevitable that he would attempt a marriage of his two loves, the sea and air. He made his first flight in a seaplane, a Donnet Lévêque, near Paris, flying both under and over one of the Seine bridges at Juvisy. His visit was short—an important event claimed his attention in Britain.

The two photographs on this page are of the Burgess-Wright aeroplane bought by Sopwith in America, and brought to England, where Sigrist modified it for School (instruction) work at Brooklands.

Saturday, June 8th 1912, was the day of the First Aerial Derby, sponsored by *The Daily Mail*, to which 45,000 spectators paid admission to Hendon and many thousands more witnessed along the route. The 'track' was an 81-mile circle around London, starting and finishing at Hendon, with a 230-foot chimney at Kempton Park, the Grandstand at Sandown Park, Russell Hill at Purley, a tower at Epping, Government Buildings at Purfleet and Barnet waterworks towers as the turning points.

There were fifteen entrants, including S. F. Cody, Gustav Hamel and B. C. Hucks. Sopwith had entered his 70 h.p. Gnôme-engined Bleriot. The competitors started at 4 minute 15 second intervals, the winner being the pilot who completed the course in the quickest time—starting all together involved too many risks.

A burst of cheering rose from the crowd as the first aircraft returned—Sopwith on his Bleriot—but his time had yet to be assessed in relation to the others. Guillaux on a Caudron was actually seen approaching by some of the crowd at Hendon, when he was forced to descent short, out of fuel. After all had landed, the best times were announced as follows:

T. O. M. Sopwith (Bleriot) 1 hour 23 minutes, 8 seconds
G. W. Hamel (Bleriot Monoplane) 1 hour 38 minutes, 46 seconds
W. Moorhouse (R.M. Monoplane) 2 hours 0 minutes, 22 seconds
J. Valentine (Bristol Monoplane) 2 hours 26 minutes, 39 seconds

Hamel was then declared the winner, for Sopwith was disqualified for allegedly failing to go round one of the check points. An appeal to the judges, considered some weeks later, established that the observer had not been in a position to judge Sopwith's turn and from other witnesses there was no doubt that Sopwith was indeed the real winner.

In other respects Sopwith enterprises were flourishing. A Henri Farman biplane had been acquired in May for general school work to add to what was by the standards of 1912 an 'air fleet'. The Bleriot had been fitted with a covered-in fuselage and at the same time the wings and tail were re-covered and doped; Sopwith tested the machine in its new configuration on June 7th.

A minor mishap to the Farman—a wheel coming off causing a broken propeller—occurred on June 19th but it was soon repaired. During July 1912, Captain T. I. Webb-Bowen (five years later a Brigadier-General in the Royal Flying Corps), Mr. V. H. N. Wadham, Lt. P. L. W. Herbert, Mr. H. Sweetman-Powell and Captain R. C. W. Alston took their official R.A.C. Certificates after passing through Sopwith's school.

In late June, Raynham left the school to go to L. Howard Flanders Ltd. and on July 9th Mr. Copland Perry joined to take over his place. One of his first pupils was to become known as 'The Father of the Royal Air Force'—Trenchard, then a Major. Having to pass out as a pilot within four weeks or lose his chance for a Central Flying School course by being over-age, he went to see Sopwith to explain his predicament. Sopwith, impressed by his manner, promised to do all he could, scheduling his first lesson early morning on July 18th. By the last day of the month Trenchard had passed his tests and was awarded his certificate, No. 270, on August 13th. At this time another pupil, also destined to become a Marshal of the Royal Air Force, trained at the school, Captain (later Sir Edward) L. Ellington; his certificate No. 305 was dated October 1st. Another success that month after tuition starting with Sopwith himself, was the designer Howard T. Wright. And, a few days earlier, a pupil has passed out whose influence would in the future be profound, not only to the Sopwith organisation but to British aviation—H. G. Hawker.

Additional cooling for the engine was obtained by placing the pilot's cockpit alongside, instead of in front of it. The crossed chain drive to the starboard propeller, to contra-rotate and so balance out torque, is clearly seen in both photo-photographs.

CHAPTER SIX

A Round of Records

The amazing thing is that in spite of test, school, competition and passenger flying, Sopwith found the time to enter an entirely new field—aeronautical design. A tractor biplane built by the school staff with wings of Wright camber and form, was tested on Saturday evening, July 4th 1912. Powered by a 70 h.p. Gnôme engine, it rose quickly. After several straight flights, Sopwith took two of his mechanics up and the machine appeared to rise with equal facility. The only criticism was that it could do with a larger rudder and this was soon effected.

This first complete Sopwith-built aircraft had an interrupted career. On July 12th Messrs. Gordon Bell and J. Charteris set out to fly the aircraft to Cowes where Sopwith was practising for a speedboat race in America. After take-off from Brooklands, the aircraft side-slipped on its first bank and came to rest in pieces across a dyke at the

The first of the Sopwith aeroplanes, the Sopwith-Wright hybrid, was first tested on Saturday evening July 4th 1912. The sweep of part of the banked motor racing track at Brooklands can be seen above the lower wing.

sewage farm adjacent to the flying field. Both occupants escaped unhurt. Five days previously, Captain Alston practising taxiing on the Henri Farman had hit a Bristol and the Burgess Wright had shed its starboard wheels on landing. Such mishaps frequently occurred to airmen in these early flying days.

Apart from retaining his love of boats, Sopwith's aeronautical interests were not confined to aeroplanes. His attempts to marry his nautical and aeronautical interests were reaching fruition by an agreement he had made with the boat-builders, S. E. Saunders Ltd. of Cowes. Saunders built a hull while wings were constructed in the Sopwith sheds under the direction of Sigrist to effect a flying boat. Before it was ready, the Military Aeroplane Competition was under way and Sopwith's attention was occupied on Salisbury Plain as a test pilot.

The Coventry Ordnance Works now had a second machine fitted with a 120 h.p. Chenu engine. It arrived without incident, but a magneto coupling sheared when the Chenu was started, putting it out of action, while the earlier machine, towed there by a steam traction engine, was delayed. Eventually, Sopwith did get a few flights in with the first (Gnôme-engined) aeroplane, but it failed to climb satisfactorily. In mid-August he handed over its testing to Copland-Perry and set out again for America. He had been asked to drive the hydroplane *Maple Leaf IV*, built by Saunders for Mr. E. Mackay Edgar, and had been practising at Cowes on *Minimum*, a Saunders-hulled boat capable of 30 knots, belonging to Norman Clark Neill.

With *Maple Leaf IV* representing Britain in the British-American Cup Races—originally called the Harmsworth Trophy Races, Sopwith retained the Cup for Britain.

Back in September, Sopwith was ready on the 18th to test the Sopwith biplane, rebuilt after its mishap. On October 8th he flew it to Farnborough in $14\frac{1}{2}$ minutes, where he demonstrated it at the Royal Aircraft Factory, Carrying a passenger it climbed to 1,000 feet in a little under three minutes.

Sundays were seemingly red-letter days for Sopwith. Sunday, October 13th his new Wright made its début. This was a modification of his old Burgess Wright biplane, but with innovations to Sopwith's specification incorporated by Sigrist. The shielded cockpit was a decided improvement, and the 40 h.p. A.B.C. engine promised continuing reliability. It was intended to enter this machine for the Michelin Duration prize of £500 and Sopwith no doubt remembered his numbed hands in his attempts two years earlier, when he made provision for the pilot's seat to be protected. But this time he did not fly; an employee recently trained at his school flew in his interests. Thus started the working association between Thomas O. M. Sopwith and Harry G. Hawker the young Australian; the latter did his first tests in the new Sopwith-Wright on October 15th 1912.

The very next day Hawker made a record attempt, staying in the air $3\frac{1}{4}$ hours before being forced to descend with a fractured valve spring. On the 21st he tried again, but was so buffeted by the wind, after 160 minutes, that he realised he could not stand the strain of the eight hours needed to break the record. Next day he flew $3\frac{1}{2}$ hours before rain came, shorting the magneto and so forcing him down with engine failure. Perseverence had its reward on the 24th, when he remained aloft for 8 hours 23 minutes, landing in darkness, to achieve a new British Duration Record. During the flight over Brooklands, the Sopwith Wright had been joined several times by Sopwith up in his rebuilt tractor biplane.

With this success, and mindful of the recent mishaps to his aircraft in the hands of pupils, Sopwith made a decision that proved vital to British aviation. At Brooklands there was already a Flanders School to where Raynham had moved, flourishing Bristol and Vickers schools as well as a Ducrocq school; his friend Grahame-White had a large school at Hendon where there were also Bleriot, Black-

The Sopwith-Wright hybrid, of 44 feet wing span, could carry two passengers in front of the pilot, this led to the 3-seat design shown on the opposite page.

burn, Deperdussin and Ewen schools. Sopwith reasoned that there was little point in continuing in competition on flying instruction now rapidly becoming a routine matter, when so much could be accomplished in the design field. The school work was gradually given up. Sigrist concentrated on building aircraft to Sopwith's design and Hawker, already proving his ability as a pilot, was to be the test and demonstration pilot. And this was how the Sopwith Company came to set out to produce aircraft.

The first draughtsman was appointed on October 21st. He was R. J. Ashfield, a master at Tiffin's and an air enthusiast. Employed at times by newspapers, he had collected the draft of an employment advertisement from Sopwith for a draughtsman—and applied for the job himself. His £3 per week was considerably higher than his teaching salary.

The first business transaction in this new field came in November, by the sale of a Sopwith tractor biplane to the Admiralty. Hawker, with Harry Kauper, a fellow Australian, as passenger, set out on the first delivery flight of a Sopwith aircraft to the Services on Friday, November 22nd 1912. Leaving Brooklands at 11 a.m. they rose to about 2,000 feet in a couple of circuits and set out for Eastchurch. After passing over Croydon, thick fog blanketed out both landscape and sun. Moreover, the compass proved useless, and revolved slowly. After flying in the mist for thirty minutes, Hawker deemed it advisable to risk a descent to try and discern their whereabouts. Suddenly Kauper sighted the deck of a ship! After banking round to change direction they saw at an indicated 200 feet by their altimeter the top of a tree, but without sight of the ground! Circling this area, the outline of a field was eventually made out and Hawker brought the biplane down. Walking along they discerned a farmhouse a mere 200 yards away that had escaped their notice from the air.

As the fog persisted they could not continue until 10.30 next morning, when they made the remaining twenty-five miles to Eastchurch in half-an-hour, and handed over the machine. Shortly afterwards it was allotted the serial No. 27 in the service number system that continues to this day. A second machine already in building was also delivered to Eastchurch later in the year, becoming No. 33 in Royal Navy service.

Having founded a factory, a quick return was needed on the first sale. Sopwith approached Commander Oliver Schwann (appointed at that very time to the newly formed Admiralty Air Department as Assistant Director) for an early payment. Schwann sportingly offered to pay £900 from his own account, claiming it from the Admiralty in due course; however, the Navy paid fairly promptly.

With this money Sopwith was able to take possession of a recently closed skating rink, close to the station at Kingston-on-Thames, as works while retaining the Brooklands sheds for testing. This presaged the pattern for the future when Kingston became the production centre and Brooklands the testing and experimental shops.

The new Company was registered as The Sopwith Aviation Company. They announced in advertisements early in 1913 that the company had established 'large works' and were prepared to undertake 'Aeroplane construction in any Branch'. In hand on the shop floors was a new tractor biplane and the Bat Boat flying boat hull built by Saunders

The first Bat Boat under construction in the newly-acquired disused skating rink at Kingston. Powered with a 6-cylinder engine, and with the highest possible quality of construction, both for the Saunders-built hull, and Sopwith-built wings and airframe, this machine created great interest at the 1913 Olympia Aero Exhibition.

The first true Sopwith design, the 3-seat biplane with non-inflammable celluloid windows. While there appears to be a tailskid, it seems that with the use of small outrigged front wheels, and extended skids behind the main wheels, landing was intended to be on these alone in the manner of the earlier Sopwith-Wright hybrid.

was having its wing superstructure erected. The latter was now looked upon in a new light, for in January 1913 the Royal Aero Club had received a letter from Mr. A. Mortimer Singer, containing an offer of a £500 prize in competition for a machine capable of rising from both sea and land. The Bat Boat could well be adapted.

It was a case of all hands to the factory. Some six fitters and carpenters were engaged, including a lad—the former 'Buttons' at the rink. Hawker was given charge of the fitters, while Sigrist did design and general supervision, under Sopwith's general direction. Realising the danger of being bogged down in administrative work, Sopwith appointed a general manager. This was R. O. Cary, a member of the Royal Aero Club, who had arranged and managed a number of notable exhibition flights for Gustav Hamel and had been with L. D. Gibbs & Co. which became the Universal Aviation Company. Another man with proven experience in aviation had been brought into the rapidly expanding organisation.

Cary arrived at a time of feverish activity. The Olympia Aero Exhibition opened on February 14th. On stand No. 22, booked for the Company, Sopwith proposed exhibiting the Bat Boat together with the new tractor biplane, one of two ordered by the Admiralty, specified as to be similar to the two aircraft already delivered, but speedier. This aircraft was the first to be fully designed by the works staff—it was a true Sopwith.

The new biplane, completed a bare week before the exhibition, was transported to Brooklands for test, where Sigrist installed a reliable air speed indicator. Sopwith then took it up with Hawker. At 40 m.p.h. it rose easily and Sopwith went up to about 300 feet and opened up the Gnôme, to find that he could get the indicator up to 73 m.p.h. It landed at about 35 m.p.h.; altogether a most satisfactory first finding. Dismantled, the biplane was taken to Olympia.

When it came to sending the Bat Boat to Olympia, now fully erected on the floor of the skating rink, there came a hitch—it was too big to go under the wide rink doorways! A builder had to be called to remove bricks to permit it to be drawn out into Canbury Park Road, for loading on to the waiting lorry.

Both the constructional methods and superb finish of the Bat Boat brought favourable comment in the aeronautical press. From Olympia it went for trials at Cowes, where both Hawker and Sopwith took turns in trying to coax it off the water without success. Working late one evening, Sopwith finally appeared to get it to rise—but only momentarily as it flopped back on to the water damaging the hull. Beaching it and retiring exhausted to a hotel, Sopwith and Hawker returned in the morning to find it wrecked by a wind that had sprung up in the night. Immediately they planned a new model, this time with ailerons in place of its warping wings.

Meanwhile the biplane that had been exhibited was taken to Brooklands and re-assembled ready for delivery to the Royal Navy. This time the Navy arranged to collect after an acceptance test at Brooklands. Lieutenants Spenser Grey and L'Estrange Malone reported to the Sopwith shed on Saturday March 1st 1913. Hawker demonstrated it both in the air and on the ground and the two naval pilots formally accepted it as H.M. Aircraft No. 103 and flew it away to Hendon.

The second machine was completed that spring and Hawker flew it to Farnborough on May 8th where it underwent War Office tests for climbing and ascents from the test ground. With one passenger and fuel for four hours it climbed 1,000 feet in 2 minutes, 22 seconds. Next morning, Whit Saturday, it was put through the speed variation test and in gusty conditions flew steadily as slow as 35 m.p.h. and then reached 75 m.p.h. Straight from test it was flown back to Brooklands, then on to Hendon and entered in the altitude contest on the programme that day for the Fifth London Aviation Meeting, held over the Whitsun holiday.

The new Sopwith quickly out-climbed the four other contestants and at 7,000 feet disappeared into the overcast. Hawker, after finding it impossible to break out above the cloud, descended, but could not recognise the countryside. He landed in a field at Ponders End to enquire the way, and then set off to reach Hendon in time to collect the Contest trophy.

Whit-Sunday, at Brooklands, the biplane did passenger-carrying flights and on the Bank Holiday Monday flew in the Brooklands meeting, winning the cross-country race.

Press interest in the second 3-seat Sopwith biplane. One of the cameramen in the centre is believed to be Alan Fenn. Note that the fuselage depth has been increased, the window panels refined by having curved, instead of square corners, and the tyre section increased. Although designed as a 3-seater it was used as a 2-seater in service.

On May 31st, Hawker took the aircraft up to 11,450 feet, which was confirmed later as a British height record.

Following these successes, new orders were promised and another administrative post was created; Sidney F. Burgoine was appointed assistant works manager. He had worked for Thornycroft's and then in his brother's business, Burgoine the boat-builders of Hampton Wick; more recently he had accompanied the Sopwith team to America in 1911. The experience of a boat-builder was now added to the staff at a time when the Bat Boat was ready for sea trials. Furthermore, such were the possibilities with the new aeroplane that further models were being put into production for fitting with floats as hydro-aeroplanes (the term seaplane or floatplane was only then coming into use).

The floorspace was re-organised with fitters' and carpenters' benches along one wall and offices on the other. The former learners' rink, known as 'Mug's Alley', was used as a drawing office, and nearby was the general office in charge of a Mr. Frank Spriggs—who as Sir Frank Spriggs K.B.E. was to become Managing Director of the Hawker Siddeley Group. There was little room for a proper design office and a hutted one was later built for Ashfield on the roof, to which an access staircase was built from the rink.

During May the Bat Boat had been fitted with wheels to make it amphibious. Hawker tested it at Brooklands on Monday the 25th. However, it proved unstable in the air and on landing a strut broke damaging the port aileron. The tail and elevator areas were increased before being sent to Cowes for trials prior to competing for the Mortimer Singer prize. The rules decreed that both entrant and pilot must be British subjects and the aircraft constructed in the U.K. Between sunrise and sunset on one day, a series of land to sea and *vice versa* flights had to be made, with a passenger, of not less than five miles, reaching a minimum altitude of 750 feet on each flight and of 1,500 feet on at least one of them.

Early in June, Hawker was flying the Bat Boat from Cowes. On one occasion he was half-way to winning the prize, having completed three flights from land to sea and return, but on the third time a wheel buckled on landing. There was no suitable spare and further trials were delayed. It mattered little, for such was the Sopwith lead in amphibious craft that there were no competitors running close.

Hawker did not wait for Cowes, but returned to Brooklands to push a few more records around in other directions on the now famous biplane. On June 16th he raised the existing height record with two passengers from 8,400 to 10,600 feet. Later, with one passenger, he rose to 12,900 feet to beat his own solo British height record of 11,450 ft.

Hawker returned to Cowes with his fellow Australian, Harry Kauper, another Sopwith employee. With Kauper as a passenger, to comply with competition regulations, Hawker set out in July to win the prize. One attempt nearly ended in disaster. A site on top of the cliffs near Cowes had been selected. Taking off over the cliff edge the tall grass so retarded the speed of the machine that it fell over the cliff edge before flying speed had been reached. It picked up the necessary speed in the ensuing dive, but only when within ten feet of the water.

A new site was chosen on the Portsmouth side of the Hamble River. From there on July 8th 1913, observed by Messrs. R. Savage, J. Herbert Spottiswoode and T. Howard Wright on behalf of the Royal Aero Club, Hawker made the necessary alightings and won for Sopwith Aviation the Mortimer Singer Prize of £500.

With a record breaking aeroplane and its adaptation as a floatplane, a prize-winning amphibian in flying boat form, the new Sopwith Company had remarkable success and versatility. *The Aeroplane* summed up their achievement at this stage—'The all-round success of so many different types of Sopwith machines is quite one of the most remarkable things in the history of aviation either in this country or elsewhere, and does credit to the perspicacity of Mr. Sopwith and practical knowledge of Mr. Sigrist'.

CHAPTER SEVEN

Prizes and Enterprises

The Admiralty already had the good faith expressed by *The Aeroplane*. An example of the Bat Boat was purchased and three floatplane versions of the successful tractor biplane had been ordered, fitted with 100 h.p. Anzani engines. The first of the Sopwith floatplanes to reach the Navy had been accepted at Calshot in June 1913. It had completed its handing over trials satisfactorily, but shortly after Lieutenant Spenser Grey, who commanded the newly formed Calshot Naval Air Station, had taken possession, the propeller, breaking up after hitting a buoy, punctured a float and put the aircraft out of action for a few days.

While the naval authorities made ready to try out their first Sopwith floatplane, the military authorities were showing a marked interest in the original landplane version, now that ailerons replaced the original warping control. On June 29th, another Sunday, Captain A. G. Fox, of the Royal Flying Corps tried out the new version in a semi-official capacity, but he evidently reported favourably as orders followed for four. By August, Sopwith Aviation was advertising for fitters, erectors and woodworkers.

Sopwith's other interests, motoring and boating, were certainly not neglected. In fact, in September 1913 he won for the second time the International Motor Boat Trophy for Great Britain. In spite of losing at the start of every day's run he achieved a maximum speed of 48 knots (approximately 56 m.p.h.) on his Saunders-built motor boat *Maple Leaf IV*.

Sundays continued to be red-letter days for Sopwith aircraft. On Sunday, July 27th 1913, Hawker took up three passengers in the tractor biplane, Messrs. Bellew, Dukinfield-Jones and King, reaching a height of 8,420 ft. which constituted a world altitude record for four persons in the one machine.

By that time a new project was interesting the new Company; the *Daily Mail* had offered in May a £5,000 prize for a 'Circuit of Britain Race'. This sum would be awarded to the entrants of the aircraft completing a prescribed circuit around Great Britain within a 72-hour period; the contest opening on August 16th 1913. Both entrants and pilots had to be British subjects and the aircraft entirely of British manufacture.

By mid-July there were four who had paid the £100 entry fee—the Sopwith Aviation Company, Mr. F. K. McClean, Mr. S. F. Cody and Messrs. Radley & England. Sopwith was expected to enter his Bat Boat and it evinced some surprise when his entry was known—a new floatplane. Based on the 100 h.p. Anzani floatplane recently delivered to the Navy, this model was to be powered by a 100 h.p. Green engine to comply with the regulations concerning an all-British product.

Once again, Sopwith found himself left alone in the field. Cody was killed in his seaplane on August Bank Holiday, the Radley-England waterplane was without a suitable engine and Short's could not get their entry ready in time. Then, after crowds had waited at dawn on Saturday, August 16th, expecting to see the Sopwith seaplane start at 6 a.m., it failed to materialise. In spite of the time limit, not until 11.47 a.m. did Hawker take off for Ramsgate, 144 miles distant, which he made in 144 minutes. Leaving at 3.2 p.m. for Yarmouth, 96 miles away, again 60 m.p.h. was averaged, but on arrival—Hawker collapsed.

With so many able helpers in the Sopwith organisation, there had not been the need for the solicitous aid of sister May—until an emergency like this. Sigrist was at Yarmouth in case all was not well with the machine and May had been there to watch progress and cheer Hawker on. It was May who had Hawker taken to a nursing home where he passed a comfortable night and although better, was unable to get up. It was thought he might have suffered from inhaling fumes from the engine but Kauper, his passenger, was unaffected.

Sopwith, on being told, arranged for Mr. Sydney Pickles to take on the aircraft. He arrived at Yarmouth on the Sunday and attempted an early start on the Monday. He found the sea too choppy to set off and that water had

Typical of pre-1914 - 1918 War week-end scenes at Brooklands and at Hendon when large crowds came by foot, bicycle, train and car to watch flying. The Sopwith 3-seat Tractor is coming in to land, and as may now be seen—the aircraft had a tail-skid of an unusual twin-strut design.

The first Sopwith floatplane, a version of the 3-seat Tractor biplane built to Admiralty orders. (The trailing edge of the starboard top wing was not touched by 'Air Review')

got into an elevator. There was nothing for it but to have the machine brought back to Cowes where, as a precautionary measure, longer exhaust pipes were fitted.

The *Daily Mail* decreed another date, August 25th, for the contest and there was a possibility that at least the Short might be ready in time and Hawker well enough to take the Sopwith. As it was, the Short was withdrawn and the Sopwith taking Hawker and Kauper set out alone—this time at dawn. At first all went well, the itinerary was as follows:

Monday, August 25th, 1913

Took off from the Solent	5.30 a.m.
Checked past Dover Castle	6.55 a.m.
Alighted at Ramsgate (144 mile leg)	8.08 a.m.
Took off from Ramsgate	9.08 a.m.
Checked passing Aldeburgh	10.11 a.m.
Alighted at Yarmouth (96 mile leg)	10.30 a.m.
Took off from Yarmouth	11.44 a.m.
Checked passing Cromer	12.27 p.m.
Checked passing Filey	2.34 p.m.
Alighted at Scarborough (150 mile leg)	2.45 p.m.

(Here Hawker and Kauper went aboard the yacht *Naidia* to have a meal and rest; Hawker looked quite well but Kauper had a slight headache. Take-off was delayed a few minutes as difficulty was experienced in getting the aircraft to head into the wind.)

Took off from Scarborough	4.22 p.m.
Alighted at Seaham	5.35 p.m.

(This was an unscheduled alighting due to a burst exhaust pipe heating water connectors and boiling away the water. The radiator was re-filled with sea-water at Seaham)

Took off from Seaham	6.40 p.m.
Alighted at Beadnell (105 mile leg)	7.40 p.m.

(This was a world record for a day's over-sea flying)

Tuesday, August 26th 1913

Took off from Beadnell	8.05 a.m.
Alighted at Montrose	9.48 a.m.
Took off from Montrose	10.25 a.m.
Alighted at Aberdeen (113 mile leg)	10.55 a.m.
Took off from Aberdeen	11.55 a.m.
Checked past Banff	1.0 p.m.
Alighted at Cromarty (134 mile leg)	2.05 p.m.
Took off from Cromarty	3.10 p.m.
Checked past Inverness	3.45 p.m.
Alighted at Oban (94 mile leg)	6.55 p.m.

Wednesday, August 27th 1913

(Set out from Oban 5.42 a.m., but due to water-logged float could not take-off and returned for repair)

Took off from Oban	6.48 a.m.
Alighted at Kiells (30 mile leg)	7.55 a.m.
Took off from Kiells	8.25 a.m.
Alighted at Larne (81 mile leg)	9.30 a.m.
Took-off from Larne	11.00 a.m.
Fell into sea off Loughshinny (96 mile leg)	1.15 p.m.

Fifteen miles north of Dublin, Hawker decided to alight on Loughshinny, near Skerries, to adjust engine

The Sopwith Daily Mail *'Circuit of Britain' Race entrant at Scarborough, August 25th 1913. This photograph was reproduced and sold later that season, labelled 'Mr. Hawker's Waterplane', by one of the local businessmen.*

valve springs. As he descended and banked round for a landing run, his foot slipped from the rudder-bar. The machine crashed out of control into the water. Hawker got out without hurt while Kauper broke an arm. There was no evasion in taking responsibility for the accident. Sopwith said that all his machines had an end clip on rudder bars to prevent foot-slip—except this machine. Hawker was equally frank, when asked the cause, he just said, 'Carelessness'.

It was a failure, but one that gained considerable merit, for only a Sopwith had made the attempt that was open to all. The *Daily Mail* very sportingly gave £1,000 to Hawker for a gallant attempt in covering 1,043 miles.

The engine of the competition seaplane was salvaged —an engine often cost as much as an airframe—and after overhaul was fitted to the latest tractor biplane from the works in order to constitute an all-British landplane eligible for competitions by the patriotic sponsors. 'Backing Britain' was the tenor of the times.

Hawker first tried out the new machine on October 4th 1913 after which minor adjustments were made. Four days later, taking off from the grass in front of the Sopwith sheds at Brooklands, Hawker set out in the direction of Cobham. As he rose over the trees bordering the Weybridge-Byfleet road, a down-current caught the aircraft. Hawker, realising that a crash was inevitable, took his feet off the rudder bar and braced his body. The machine hit the ground in a side slip so that a crumpling wing took the main force of the impact. Apart from straining a back muscle and bruising his forehead, Hawker escaped injury. Again, he was equally frank when asked the cause—'Overconfidence', he said.

Rebuilt, the machine was ready for Hawker to use to try for the Michelin Cross-Country Competition on November 19th. He was not successful, but it mattered little, for a new compact single-seater for racing was under construction at the skating rink works at Kingston upon Thames— the Tabloid. When asked in 1961 who actually designed the first of the line of Sopwith Scouts, Sir Thomas Sopwith answered that it was a family affair, Harry Hawker had 'a crack at it', as did Sigrist, himself and the boys in the shop, but he thought that Hawker was the chief instigator.

The new biplane was demonstrated at Farnborough on November 29th. It climbed at 120 feet a minute, could fly steadily at as low a speed as 36·9 m.p.h. and reach 92 m.p.h. with its 80 h.p. Gnôme engine. With the flying season over with the onset of winter, Sopwith permitted Hawker to take the Tabloid to Australia to give a fillip to aviation in the Commonwealth by a Sopwith. Meanwhile, Sopwith engaged Howard Pixton, formerly with Bristol's, to be the chief test pilot and Victor Mahl, who was both an engineer and pilot, was also engaged.

The fateful new year, 1914, was faced with confidence. On the penultimate day of 1913 the sixth of the 3-seat Tractor Biplanes had been delivered to the Army at Farnborough. Howard Pixton flew it there accompanied by Lieutenant Mapplebeck, R.F.C. Over Guildford the 80 h.p. Gnôme failed and they were compelled to land at Sutton Court near Lord Northcliffe's residence. Lady Northcliffe kindly invited the aviators to lunch, but they had to refuse and work on the engine to reach Farnborough before dusk.

The first of the Batboats with a 100 h.p. Green engine, shown flying over the Solent.

Two days later Pixton flew it on its acceptance test. In the new year, on January 2nd, Pixton and Mahl flew the seventh and on Sunday the 18th the ninth and last of the Army order for Sopwith 3-seat Tractor Biplanes was tested and flown to Farnborough two days later. After that the factory concentrated on a new production order for Tabloid biplanes and a variety of aircraft for the Navy—a new Bat Boat, floatplane version of the successful biplane, a torpedo-carrying seaplane and their first foreign order—a 'pusher'-type trainer seaplane for the Greek Navy.

One of the minor jobs that came the way of the factory, recalled Sopwith's earliest invitation to fly before the King. This time it was the famous aerobatic pilot Gustav Hamel who had received the royal invitation. However, when testing his Morane-Saulnier the evening before, to make sure all was well—the crankshaft of its engine fractured. Frantically Hamel sought help. Telephoning Kingston 1777 at 7 p.m. he received an answer—as the Sopwith works were on overtime. The works manager, Cary, gave every assistance and had two of his mechanics strip down a Gnôme in order to provide a replacement for the broken shaft.

In view of the Admiralty orders, the First Lord of the Admiralty, Mr. Winston Churchill, visited the Sopwith works in February where a new machine was being constructed for the Admiralty. This was named the Sociable as it was a side-by-side two-seater, but it came to be known as *Tweenie* in service. It was, in effect, a compromise between the standard 3-seat biplane and the single-seat Tabloid. Pixton tested the machine and handed it over to the Navy on February 19th. Lieutenant Spenser Grey set

The sole Sopwith exhibit, at the 1914 Aero & Marine Exhibition at Olympia, was the new Batboat with a 14-cylinder 200 h.p. Canton Unne (Salmson) engine. This radial engine was cooled by a centrifugal pump, driven by toothed gearing from the engine timing wheel, which circulated water throughout the radiator.

out with it straight away for Hendon, but owing to mist and rain lost his way and descended in succession at Wembley, Edgware, Barnet and then in a field north of Hendon, to ask the way. Churchill was at Hendon next day to make his first flight in the machine, but owing to the strong wind, Lt. Spenser Grey made it a short flight and did not go higher than 600 feet. Visiting Calshot on February 23rd, Churchill had a chance to see the new Sopwith floatplane in service. Two days later he was back at Hendon for a flight in the Sociable.

On Saturday, February 28th, Lt. Spenser Grey went up to 10,600 feet in the Sociable and then got lost in the London fog and landed at Barking. Returning to Hendon, where the air was clearer, he found Churchill yet again ready for an airing, and this time set off on a fairly long flight. In this way Sopwith aircraft were getting publicity at the two main centres of aviation, Hendon, where thousands paid a shilling at the weekend to enter the airfield public enclosures, as well as at their Brooklands flying base. Meanwhile, at the other side of the world, a Sopwith aircraft was winning acclaim. Hawker, in the original Tabloid, was delighting the population of Sydney, Australia, at the Randwick racecourse on February 22nd.

In March, the Olympia show came round again. It was one of the major aeronautical events of the year. Open from 10 a.m. to 10 p.m., March 16th to 25th (Admission 1/-), the 1914 Aero & Marine Olympia Exhibition had fifteen manufacturers exhibiting 26 different types of aircraft; yet strangely Sopwith, at Stand 44, exhibited only one aircraft—the Bat Boat. This was a new model, powered by a 200 h.p. Salmson engine with a compressed air self-starter. The hull was fitted with wireless—Sopwith having learned that all naval aircraft, except trainers, were to be fitted for wireless telegraphy. *The Aeroplane* remarked that the design and construction was of such high class that the exhibit deserved special attention.

During the spring of 1914 the first serious accidents occurred with Sopwith aircraft in naval and military service, but neither could be ascribed to any constructional failing. The first concerned the Churchill aircraft—the Sociable—on Wednesday, March 25th. It had been flown that morning from Hendon to Eastchurch by Lieutenant Spenser Grey accompanied by Engineer-Lieutenant Aldwell. The two officers set out to return during the afternoon. Barely had they reached 250 feet when the machine appeared to turn without banking and so commenced a flat spin to the ground. Just before impact, it appeared to flatten out a little, but too late to avoid the crash which completely wrecked the machine, trapping the occupants. It took twenty minutes to extricate the unconscious Lt. Grey who,

The new Salmson-powered Batboat on trials in the Solent. The hull of the original was built by S. E. Saunders Ltd., of East Cowes, Isle of Wight, using the patented 'Consuta' method of construction. An example of this new model was sold to the German Navy.

This Sopwith Type C floatplane No. 138, built to Admiralty Contract C.P. No. 30775/14, fitted with the Sueter patent torpedo-dropping gear, is seen at Cowes in early 1914. The 200 h.p. (Salmson) engine was fitted with an extended propeller shaft to permit streamlined fairing. It was delivered to the R.N.A.S. at Calshot.

although suffering from concussion, fortunately broke no bones; Lt. Aldwell with a fractured skull and broken thigh was fully conscious. Both were taken to the sick bay at Eastchurch and then transferred to the Naval Hospital at Chatham where, in time, they made a full recovery.

The second accident was also the first collision to occur with aircraft of the Royal Flying Corps. Lieutenant C. W. Wilson, R.F.C. took off from Farnborough around 4 p.m. on May 12th for Brooklands on one of No. 5 Squadron's Sopwith 3-seat Tractor Biplanes. He returned to Farnborough about an hour later, just after Captain E. V. Anderson, Black Watch, who was endeavouring to familiarise himself with the same type, took off with Air Mechanic Carter. Evidently neither pilot saw the other and the two aircraft, Nos. 324 and 325 collided in mid-air and fell to earth. Only Lt. Wilson survived and he was able to walk away, albeit with a broken jaw and extensive bruising.

The Company in March 1914 had been formally reconstituted as a limited company and registered as The Sopwith Aviation Co., Ltd., with a capital of £26,000 in £1 shares. This followed the acquisition of premises in Canbury Park Road, about a hundred yards from the skating rink (now the Kingston School of Art). Company objects were stated as follows: *To acquire the business now carried on at Kingston-on-Thames and at Brooklands, under the style of 'The Sopwith Aviation Company' and to carry on the business indicated by its title. First directors: Thomas O. M. Sopwith, Reginald O. Cary and Gertrude May Sopwith. Qualification: £1,000. Remuneration: £50 each, per annum. Registered office: Canbury Park Road, Kingston-on-Thames. Private Company.*

At the close of March the first of the seaplanes for the Greek Naval Air Service was ready for test, so that the Company had prospects of an export market as well as supplying both the Admiralty and War Office with aircraft. A new naval requirement was for shipborne aircraft which meant incorporating folding wings. With this in mind Short Bros. had taken out a patent wing pivoting gear and were thereby in a favourable position to acquire a lion's share of the orders. However, to compete in this field, Sopwith asked Sigrist to use his ingenuity on this task and he soon came up with a more compact variation of a folding gear, using a cable and drum mechanism, actuated by a wheel in the cockpit. This idea was patented (No. 22,440/14) in the name of the Sopwith Company Ltd. and F. Sigrist.

At this time those concerned with the Company had no idea of the World War that was to break out later in the year. In the spring of 1914, thoughts from the Kingston factory were centred on the 1914 summer flying season and the competitions in which they could participate.

Although the original 1913 'Circuit of Britain' Sopwith floatplane entry was wrecked in the contest, the Admiralty were sufficiently impressed as to order a rebuilt example. This was taken on Royal Naval Air Service strength as No. 151. It later served in No. 4 Wing, R.N.A.S.

CHAPTER EIGHT

Sopwith Schneider Success

The Sopwith entry in the 1914 Schneider Trophy Contest at Monaco; Howard Pixton was at the controls and he was the only entrant to complete the course—at a speed of some 85 m.p.h. He then went straight on and lapped at a world record speed of 92 m.p.h.

There had been no British competitor in the first of the Schneider Trophy contests in 1913, but in the second contest of 1914, held at Monaco, T. O. M. Sopwith represented Britain with one of his aircraft.

The Sopwith entry was virtually a Tabloid on floats but with a 100 h.p. Monosoupape Gnôme engine. Trials of this floatplane at Hamble on April 1st did not go well. A central float placed too far back caused the machine to somersault during preliminary taxiing, throwing Pixton out who, although unhurt, found it difficult to keep above water in his flying clothes. It was then decided to scrap the single-float arrangement and the machine was dismantled and taken back to Kingston. There the floats were modified and a new chassis was built. But time was running short, as the contest was due to start the following week.

Five days later, company employees were asked to report in at dawn on April 7th. The new machine had been completed the previous afternoon and it was to be taken down to the Thames for test. Although a trailer was used for most of the way, manhandling was called for to get it on to the water just below Kingston Bridge. Its flotation test was completed satisfactorily and Pixton intended to fly it on the stretch above Teddington Lock. However, crowds had gathered to watch the unusual spectacle of an aircraft on the Thames, including officials of the Thames Conservancy Board who objected to an aircraft being flown from water under their jurisdiction.

Sopwith had more resource than to argue with officials or act against their wishes. Next morning he asked for another early attendance and at 5 a.m. the Schneider Cup entry floatplane was trundled to Richmond which, being down-river, came not under the Conservancy Board, but the Port of London Authority. The floatplane was pushed into the water near Glover's Island. After a little taxiing on the water to get the feel of the aircraft's response, Pixton tried to take-off towards Eel Pie Island. As the engine was mis-firing and the reach of this part of the river was short, little more than water-skimming was achieved, but it was sufficient to prove the efficiency of the new floats.

Before Pixton could conduct further trials he was required to switch to the original landplane version, to make his first flight in a true Tabloid. As related, the prototype Tabloid had been taken to Australia by Hawker, and the next, fitted with floats, he had just tried on the Thames. Now the first of production orders for the Services awaited test at Brooklands. Pixton took his first up on Friday April 16th. Since his services were required at Monaco, Sopwith arranged for Mr. R. H. Barnwell of the Bristol Flying School to take over at Brooklands temporarily as the Sopwith test pilot. The following day he expressed his confidence in the Sopwith product by doing four loops in succession.

Pixton then left for Monaco. Victor Mahl, who had tuned up the engine before the machine was crated up, also went and Sopwith himself, already installed in the Bristol Hotel, was arranging for the reception of the aircraft and his team.

The crated Tabloid floatplane had arrived at Monaco during April 16th, the day Pixton had tested the landplane version in Britain. Next day it was erected outside the Hervieu tent rented by Sopwith. Some worry was caused by rust in the engine—evidently the effects of the immersion at Hamble. By Saturday evening all was ready, but as the sea was rough, a practice run could not be made. The rules allowed attempts at any time between 8 a.m. and sunset. In this period the competitor had to make a preliminary lap of the course marked out on the sea between Monaco and Cap Martin, alighting twice and then flying a series of laps around markers—the pilot achieving the fastest running being the winner.

On the Sunday the Sopwith contingent, supervised by Burgoine, were up at 5 a.m. and loaded the floatplane with two hours fuel and oil. It floated with its tail well down and with the elevators drooping almost in the water. In towing it out, it was found advisable to have two men standing on the front of the floats to lift up the tail clear of the water. From the moment the engine was opened up, the machine surged forward, and it was a matter of a mere 100 yards before it lifted from the water. After practising banked turns and speed runs, Pixton brought it in satisfied with the handling. However, the engineers were not satisfied.

Mahl considered that the engine was running too fast

at its 1,350 r.p.m., and that in its two-hour contest flight the engine would overheat. In consequence, the Lang propeller fitted was exchanged for an Integral model of similar diameter, but coarser pitch. Since full trials had been precluded by the short time available, and engine consumption in this configuration was not known, an additional fuel tank was fitted. This supplementary tank of some 5½ gallons was lashed in position in the cockpit and connected to the main 24 gallon tank.

Next morning, April 20th, was the day of the contest. As soon as the signal went at 8 a.m. the French team sent out their two Nieuport monoplane seaplanes. They were followed by M. Burri, the Swiss entrant, piloting the only flying boat in the contest—an F.B.A. Next, after an interval of about fifteen minutes, Pixton set out and after making his prescribed alightings, flew steadily round the course, banking steeply at the turns.

The only other British contestant, Lord Carbery on a Deperdussin monoplane floatplane, gave up after one lap due to trouble with his 160 h.p. Gnôme engine. Such was the speed of the Sopwith and its superb cornering in Pixton's hands that it drew ahead of the earlier starting Nieuports. Both French pilots opened their throttles to keep their position, but this apparently over-taxed their engines; first M. Espanet dropped out, followed by Pierre Levasseur on his 17th lap. Pixton continued, but on his fifteenth lap the engine started misfiring and for the rest of the time one of the nine cylinders failed to fire. Burri was still in the race, but ran short of fuel on his 23rd lap and had to land for refuelling. While Burri was so engaged, Pixton crossed the finishing line, having completed the necessary 28 laps. He covered the 280 kilometres in 2 hours 13 seconds, averaging 85.5 m.p.h. and won the contest.

Instead of landing, Pixton opened up his engine and went all out for another two laps, risking over-taxing his engine now that the contest flights had been made, in order to attempt a 300 kilometre speed record. This he did at 92 m.p.h.—a world record.

Pixton made a good landing in spite of the sea having become very choppy. While waiting for Mahl to set out in a motorboat to escort him in, a wave damaged the elevator. There was still time for the other contestants' floatplanes to try—Roland Garros on his Morane-Saulnier monoplane, the remaining French entry, or the Americans, Weymann with a Nieuport monoplane and Thaw with a Curtiss biplane—but they knew they now had no chance of winning. It was acknowledged that Sopwith had the fastest seaplane in the world.

The prestige value of this victory was enormous. It had far-reaching consequences. Hitherto the French, undisputed leaders in aviation, had viewed with amused contempt the British entrants in international aviation competitions. They had also regarded a biplane form as greatly inferior to the monoplane in speed events. The British victory had come as a blow to French national pride and marked the high tide of the French as leaders in world aviation. From this time, for the next four years, the lead swung to Britain. Dallas Brett summed it up in his *History of British Aviation* as 'the most important event which had ever happened in the history of British aviation'.

Pixton was a man of simple tastes. He was greeted on landing by Jacques Schneider himself who invited him to call for whatever drink he liked to celebrate the event. No doubt he expected something like a call for champagne of specified vintage from a particular great House; as it was, Pixton said he'd have a Bass!

That evening Pixton and Sopwith were dined at the invitation of Jacques Schneider in the Hotel de Paris; among the guests were Harry Delacombe and Harold Perrin representing the Royal Aero Club. The prizes were presented next day by Prince Albert I of Monaco.

Meanwhile, at Brooklands, Barnwell was still testing Tabloids. On Wednesday, April 22nd, he flew one to Farnborough where it was recorded as having a 39.6 to 94.9 speed range. Another Tabloid left the works the following Friday for Brooklands where it was given engine tests. Pixton, back on routine work, was there to test it on the Sunday. When delivering it during the ensuing week he had to make a cross-wind landing at Farnborough. A buckling wheel turned the machine over, but with little damage to the Tabloid. Pixton was evidently unaffected, as the following Sunday, when it rained nearly all day, he took up another Tabloid and was, in fact, the only pilot to fly from Brooklands that day.

On Tuesday May 12th the Royal Aero Club entertained Sopwith and Pixton to lunch at the Royal Automobile Club in honour of their Schneider Cup victory. The chairman, the Marquess of Tullibardine, referred to Sopwith's brilliance as a constructor and pilot and said that he was now one of the foremost constructors in the world. Pixton, a man of few words, said modestly that any pilot could have won on the machine he was given to fly.

The winning Sopwith flying over the course at Monaco, April 20th 1914.

CHAPTER NINE

Harry Hawker at Home

Harry Hawker, the Australian genius, who had a great influence on Sopwith designs, and whose name is perpetuated in the great aeronautical complex, Hawker Siddeley Group, of which Sir Thomas Sopwith is Founder and currently President.

Meanwhile, on the other side of the World, Hawker was showing his compatriots not only what a Sopwith aircraft could do but, to many, what an aeroplane was. This was a triumphant return for the young Australian, who had made his name in the Mother country.

Harry Hawker had been born in Victoria, at Moorabbin, a small village then known as South Brighton, on January 22nd 1889. Like many other men who were to make their mark, he was a poor scholar and after trying four different educational establishments he left school altogether at twelve and went to work for a motor firm, Messrs. Hall & Warden, for five shillings a week, having given his age as fourteen.

A mechanical aptitude had been inherited from his father, a blacksmith, and he soon became something of an expert with cars. He moved on to Melbourne to supervise a fleet of cars owned by a Mr. de Little at the attractive salary of £200 per annum. This enabled Harry Hawker to save, so that when he met by chance young Harry Busteed, inspired by the sight of a Bleriot monoplane and set on going to Europe, he could join him without difficulty. Yet a third Harry, and an aircraft enthusiast to wit, joined them on the boat—a young mechanic called Kauper. They reached London in May 1911.

Busteed went straight to the Bristol School at Larkhill where he learnt to fly and joined the British & Colonial Aeroplane Company. Entering the Royal Naval Air Service in 1914 he retired from the R.A.F. in 1930 as Air Commodore Henry Richard Busteed, O.B.E., A.F.C. It is said that but for his outspoken ways and disregard for protocol, colloquially known in the Service as his 'Aussie attitude', he might well have climbed even higher in the Service.

While Busteed made his name in service aviation, the other two Harrys were to make theirs in design and engineering. Their start was not auspicious. Lack of references for work in England proved a handicap, and employment was difficult to find. With savings dwindling they were forced to seek cheaper lodgings. On July 29th Hawker's luck changed when he managed to get a job with Commer Motors at 7d per hour. This job he changed in January 1912 for one with Mercedes for a $2\frac{1}{2}$d. per hour increase and on March 18th an even better chance with the Austro-Daimler Company making aero-engines.

Kauper, meanwhile, did not find employment for some months. Eventually, he successfully answered an advertisement by T. O. M. Sopwith for a mechanic. Once installed at Brooklands he contrived to bring his friend into the organisation and the rising success of the company soon brought that opportunity. After being interviewed by Sigrist, Harry Hawker started work on July 12th 1912.

Sigrist was quick to appreciate Hawker's mechanical aptitude and made good use of him; for some months there was a fifteen-hour day of a seven-day working week. Hawker had aimed at saving £40, the price of a fare home if needed, but re-assured and happy in his work, he offered the sum to Sopwith as fees for flying tuition.

Soon he was put in charge of the Sopwith sheds, representing Sopwith's in flying competitions and conducting the firm's test flying. With the Round Britain Race attempt, his record and exhibition flights, Hawker was becoming a household word. When he arrived in Australia at Fremantle on January 13th 1914 aboard the R.M.S. *Maloja*, he met with public acclaim. Visiting his parents next day at St. Kilda, he was welcomed by a Mayoral reception.

Arrangements were made with the C.L.C. Motor and Engineering Works at Melbourne to provide workshop facilities. On January 26th the Tabloid, which had been carried crated aboard the *Maloja*, was assembled and exhibited in the engineering firm's showrooms. A trial flight was delayed through red tape—the special castor oil for engine lubrication was held up by the Customs.

His first Australian flight was on Tuesday, January 27th 1914, at Melbourne. It was not the first aeroplane to be flown in Australia, but certainly the fastest and most manoeuvrable ever seen on that Continent up to that time.

It could not be said that Hawker was typical of the air pioneers at the time. He did not walk nonchalantly to his craft, wave to the crowd and rev the engine in a series of impressive bursts. Instead, he walked to his aircraft with toolkit in hand; went round tightening screws and nuts, strumming wires to test their tautness, checking controls and, when he ran up the engine, he listened to its tune.

Once in the air, his actions were more polished than flamboyant, but nevertheless as impressive as the most intrepid aviator. He was never reckless and knew the limitations of his machine. He regarded mishaps as human failings in the tending of the machine, never pure accident. Local papers reported him as climbing steeply, banking at 45°, switchbacking with engine alternately starting and stopping, and steeplechasing over the countryside. At this stage he had not yet ventured his first loop.

On landing, Hawker's characteristics were again manifest. He did not respond to acclaim from the crowd like others by standing up in his cockpit, or by leaning against his machine. His first thought was to get his aircraft safely away from the crowd cordoned off, then he would respond.

In rather the same way as Sopwith had visited the King by landing on the East Lawn, Windsor, so Harry Hawker landed on a lawn at Government House when visiting the Governor-General of Australia, Lord Denham. As it happened, Lord Denham was no stranger to aviation, having flown as a passenger in a Grahame-White biplane.

The Australian tour was self-supporting. Australians clamoured for a flip and offered £20 a time. An exhibition on Caulfield Racecourse, Melbourne, drew a crowd estimated at 25,000 to see Hawker fly. Unfortunately they became over-enthusiastic and thronged the course. After one landing Hawker was obliged to veer and ran into railings striking spectators. One, struck by the propeller, was found to be a Mr. G. K. Francis, an uncle of Harry Kauper's. No one was seriously hurt and flying continued. Two of his passengers, a Miss Dixon and a Mrs. Clive Daniels, are believed to have been the first Australian women to take to the air. While landing with Miss Dixon on Elsternwick golf links, the aircraft was 'bunkered'. Swinging round as it touched the ground, the undercarriage buckled and the propeller splintered. Unruffled, Miss Dixon said she would take another trip.

On February 12th Hawker left for Sydney and eleven days later gave a flying exhibition to 20,000 people assembled at Randwich Racecourse. Among his passengers this time were Lord Denham, and Miss Strickland, a daughter of the Governor of New South Wales. At Albany, in New South Wales, after an exhibition flight he raised the Australian height record to 7,800 feet, during which he was for

Hawker with his contemporaries early 1914; he is linking arms with his fellow Sopwith test pilot, Victor Mahl.

a time hidden by a blanket of cloud. The cold, however, affected the Gnôme engine which was missing and it finally gave out at 300 feet. As he attempted to glide back into the field, he was confronted by tall railings and realised he would not clear them. Rather than risk flying into them and overturning, he deliberately put his nose down to bring the machine in short. Again the undercarriage buckled and the propeller splintered, as the Tabloid went up on its nose. By the cloud of dust that rose, Hawker was obscured and the crowd feared the worst. But Harry was walking away. His comment to the press was—'Blame me, it was my fault, not the machine's, I let the engine get cold'. Hawker hoped to popularise aviation in Australia and was anxious lest the aeroplane be condemned by the journalists as a dangerous machine. He had wisely brought out two spare propellers and spare undercarriage and wing struts. Within four days the machine was mended and he was dealing with thirty contracts for flights at £20 a time.

During this time he was formulating ideas on how his aircraft could be improved, but they could only be carried out in workshops in England, and this was an incentive to return. In any case, the intention had been to fly for the Australian summer season only and then return to the Sopwith factory. Late in April 1914, Harry arranged for the Tabloid to be dismantled, re-crated and shipped, while he took passage back to England. As soon as his ship had berthed at Tilbury, he made straight for Brooklands.

The prototype Sopwith Tabloid two-seater which Harry Hawker took to Australia in 1914. T. O. M. Sopwith left no-one in doubt as to whose aeroplane was in the air!

CHAPTER TEN

The Weeks before the War

The end of an era. A flying scene at Hendon in 1914 before war was declared. The Sopwith Tabloid is in the air, and the 3-seat Biplane is in the foreground. To the left of the ladder to the top of the pylon can be seen two members of the St. John's Ambulance Corps whose personnel tended to Hawker, June 20th 1914.

Sopwith's first export order was now under way. The Greek Navy was establishing a Naval Air Arm and Sopwith aircraft were chosen—not the standard proven type, but an entirely new design for a pusher floatplane.

The Greek Government had appointed in 1913 a British Naval Advisor, Rear-Admiral Mark Kerr, who, being a forward-looking naval officer, took cognisance of the value of aircraft. He pointed out that, in the narrow waters surrounding Greece, a capital ship would be the prey of destroyers or submarines. He advocated torpedo boats, an air establishment, a chain of wireless reporting stations and anti-submarine nets. King Constantine, brought up to believe in the capital ship—and the bigger the better, took some time to convince, but was eventually converted to Kerr's views and asked that his Prime Minister, Mr. Venizelos, and his own officers be similarly converted. Greek naval officers, noting the rapid building of British Dreadnoughts, did not share Kerr's view, but as a compromise, and perhaps to keep Kerr quiet as he had the King's ear, they agreed to allot money to buy three seaplanes to establish a Naval Air Arm.

Kerr contacted Captain Murray Sueter of the Admiralty's Air Department at a time when Sopwith had been called in to discuss a torpedo-carrying seaplane. As Sueter himself wrote of this time 'We called on the services of Mr. Tom Sopwith, the "stand-by" of the Royal Naval Air Service, in many of the difficult air matters that came up for solution'. Sopwith was asked to provide a spotting aircraft, initially fitted as a trainer, and capable of being fitted for bomb or gun-carrying. This he did.

Building had commenced late in 1913 and by early March 1914, testing was under way at Hamble. Testing was delayed a week or two by the pinion at the end of the Anzani engine's crankshaft, which drove the magneto, oil pump and cam gear, becoming loose. Also, larger elevators and rudder area were found to be necessary. In the new configuration the first was tested on Wednesday, April 22nd, and rose 3,000 feet in 17 minutes with full load. The first three floatplanes, similarly modified, were shipped to Greece later that spring and Admiral Kerr, to try and impress the Greeks how easy flying was, decided to learn to fly himself. His mission had included the pilot Collyns P. Pizey and four British mechanics, Messrs. Gaskell, Lapray, Radley and Simms, who were given Warrant rank in the Greek Navy. Pizey, who had been instructing at the Bristol Flying School in 1913, had often competed with Sopwith in races. A variety of aircraft were now being handled by the Sopwith organisation as evinced from the log of movements in a week of 1914:

May 14th. Fourth Tabloid of Army order left Canbury Road works for Brooklands. At Calshot, a new seaplane, built as a reconnaissance gun-carrying floatplane, was under flying tests. This was the SPGn, a large pusher with a 90 h.p. Austro-Daimler engine. It was taken on Royal Naval Air Service strength as No. 93.

May 15th. A landplane version of the Circuit of Britain seaplane, with a 100 h.p. Green engine, left the Brooklands sheds to fly to Portsmouth where, at nearby Calshot, another of the 100 h.p. Anzani-engined seaplanes was on test for the Admiralty.

May 16th. A wet Saturday. No flying.

May 17th. A new Bat Boat, similar to that exhibited at Olympia, underwent trials in the evening calm.

May 18th. While Victor Mahl did a few passenger flights with a standard 3-seat biplane from Brooklands, Pixton stood by at Calshot waiting for the choppy sea to abate. Again the wind dropped towards evening and Pixton ran final trials on the 200 h.p. Bat Boat before handing it over to the representatives of its purchaser—the German Admiralty.

May 19th. Passenger flights on the 3-seat biplane at Brooklands. At Calshot the 100 h.p. Green biplane giving demonstration flights to the Navy.

May 20th. The fifth of the Tabloids for the Army delivered to Brooklands from the works, followed by the Schneider Cup winner which the factory had modified to have a landscape chassis making it into a Tabloid.

May 21st. Pixton testing the 100 h.p. Gnôme-engined Tabloid (ex-Schneider Cup floatplane) with a view to retaining it as a Sopwith demonstration aircraft.

Probably one of the finest (for the period) ground-to-air photographs ever taken, this one shows a Tabloid on a test flight. Note the white panel—as yet unnumbered—on the rudder. This aircraft was scheduled for the Navy.

It will be seen that the Sopwith works were a hive of activity and supplying aircraft to the Army, Navy and foreign government months before the First World War commenced. At times test pilots were flying different types of aircraft from bases miles apart on the same day, such as Victor Mahl on the penultimate day of May who flew the 100 h.p. Tabloid at Brooklands in the morning and was testing one of the Greek pusher seaplanes from Calshot in the evening. That same Saturday the eighth Tabloid for the Army arrived at Brooklands from the works showing evidence of the first major modification in this initial production order, in that it had a 'Vee' type undercarriage.

The 1914 Aerial Derby, originally arranged for Saturday, May 23rd, had been cancelled due to bad weather and postponed until Saturday, June 6th. Two days before the event Sopwith himself brought the 100 h.p. Tabloid to Brooklands, after overhaul at Kingston, by towing it behind his car. Two Sopwith aircraft were entered, and flown to Hendon, both Tabloids—an 80 h.p. standard machine, temporarily offset from the Army order, to be flown by Pixton, and the 100 h.p. special, ex-Schneider Cup floatplane, marked boldly with SOPWITH in large letters along the fuselage, to be flown by Barnwell. Temporarily bearing the event numbers 18 and 21 respectively, neither achieved success on this occasion.

A ship from Australia, docking at Tilbury early the following morning, made another Sunday an eventful day for the Sopwith organisation. Harry Hawker returned. It might be thought that attention to his baggage and settling into his new accommodation might have occupied all his attention on that day of arrival. Not so with Hawker. The prospect of taking to the air after six weeks at sea overrode any idea of arranging for the essentials of existence on the ground. He made a bee-line for Brooklands and was up in the 100 h.p. Tabloid that same afternoon, demonstrating that he had lost none of his skill.

It became usual to enter a Sopwith in any event going and the 'First Race from London to Manchester and Back' on Saturday, June 20th 1914 was no exception. This event, like the famous London-Manchester Race, was sponsored by the *Daily Mail*, this time for a gold trophy and a £750 prize offered by Pratts—the petrol people. Hawker was entered with the 100 h.p. Tabloid on which he had practised. Two days before the race, after three laps in succession, there was an excess of petrol in the carburettor jet which fired and burnt through a longeron. Hawker landed safely and the aircraft was sent off to Kingston for repair. It was returned next day and in spite of rain a test was made to ensure all was well for the morrow.

The day of the race dawned bright and the sun, drawing up the damp, caused mists. Hawker en route from Brooklands to Hendon for the start, lost his way, but eventually picked up the Welsh Harp and made his way from there. His Tabloid was given the race No. 14 in place of the No. 21 of the Aerial Derby and he was made scratch man. Last to take-off, he was first back. But, as with the first Circuit of Britain attempt, so again, Hawker had retired ill. He came down at Coventry and shortly afterwards flew straight back from there. For some time after landing back at Hendon he was unable to speak. The cause could not be ascertained but Dr. Leaky, chief of the Hendon St. John's Ambulance Corps, had a theory that it was due to air pressure on the ear drums.

The following Saturday, Hawker had a discussion with a naval officer who advocated a larger rudder area on small types such as the Tabloid. Following this, at 7 p.m., he

One of the military Sopwith 3-seat biplanes used by a flight of No. 5 Squadron, Royal Flying Corps, on Salisbury Plain from mid-1913 to mid-1914.

went up in the 100 h.p. Tabloid and at just above 1,000 feet, over the Byfleet Road, did a dead engine loop, by diving steeply, shutting off the engine and pulling back the stick. He completed the loop perfectly, but immediately following this manoeuvre the machine spun down. Disappearing from the view of the Sopwith sheds crew, behind the trees on St. George's Hill, it spun into the top of a tall tree and then fell vertically, bringing down several large boughs, folding back the wings which closed like a lid on the fuselage and telescoping the landing gear into the fuselage. Hawker got out completely unhurt and rode back on the pillion of a motor cycle. Sunday morning he was up on an 80 h.p. Tabloid! In those days Hawker suffered more from long straight flying than aerobatics and crashes!

However, the machine was smashed and could not be repaired in time for the London-Paris-London Race—as C. G. Grey put it 'The Sopwith entry was scratched, owing to Mr. Hawker's failure to alight successfully on a tree-top'.

Hawker had another chance to indulge his ideas. The Tabloid prototype which he had taken to Australia had arrived back at the Kingston works. He now had this modified to his specification. It was fitted with a new rudder, was rigged with dihedral on the lower plane and the rear fuselage section was left uncovered. On July 11th he tested it at Brooklands.

Five days later another new Sopwith aircraft, a floatplane temporarily with a wheeled undercarriage, arrived at Brooklands, an entrant for the 1914 'Circuit of Britain' arranged for mid-August. This was a contest for seaplanes, sponsored by the *Daily Mail* who were offering a £5,000 prize. This time it appeared that the firm would be warmly contested. The entrants by the end of July were:

Contest Number, Aircraft and Pilot
1. British Curtiss Flying Boat with 2 × 100 h.p. Curtiss-Austin engines piloted by Mr. A. Loftus Bryan.
2. British Curtiss Flying Boat with 120 h.p. Beardmore Austro-Daimler engine piloted by Captain E. C. Bass.
3. Sopwith Bat Boat with 200 h.p. Sunbeam engine piloted by Mr. Howard Pixton.
4. Sopwith Tractor Floatplane with 100 h.p. British-built Monosoupape Gnôme engine piloted by Mr. V. Mahl.
5. Grahame-White Floatplane with 100 h.p. British-built Monosoupape Gnôme engine (pilot to be announced).
6. Avro Tractor Floatplane with 150 h.p. Sunbeam engine piloted by Mr. F. P. Raynham.
7. Eastchurch Aviation Company Floatplane with 120 h.p. Green engine piloted by Mr. F. B. Fowler.
8. Blackburn Hydro-Biplane with 130 h.p. Salmson engine piloted by Mr. Sydney Pickles.
9. Beardmore (D. F. W.) Tractor Floatplane with 120 h.p. Beardmore Austro-Daimler engine (pilot to be announced).

Sopwith aircraft were beginning to face fierce competition. Already the original tractor biplane was being replaced in Army service. No. 5 Squadron operated two flights of Henri Farmans and one of Sopwiths. While they found the Sopwiths pleasant to fly, they were still having teething troubles and when in July Avros were received, also with the 80 h.p. Gnôme engine, the Sopwiths were relegated to training roles. Lord Carbery appearing in the Hendon-Manchester-Hendon race with a Bristol Scout, portended a rival to the Tabloid.

The Sopwith Company strove to produce new designs. Simultaneous with preparations to design a seaplane for the 'Circuit of Britain', the Company were planning to enter an international event for landplanes—the Gordon Bennett trophy, sponsored by the publisher of the *New York Herald*, which also carried a purse of £1,000. For this, a 'slim-line' Tabloid fuselage was prepared, a new fin and rudder was devised, while the original Tabloid wing form was retained. But, before the racer was completed for the Gordon Bennett competition or the Circuit of Britain run, the United Kingdom had declared war on Germany on August 4th 1914.

This Sopwith seaplane built in 1914, embodied the short folding-wing principle, for which Sopwith had acquired a licence from the firm. Later Sigrist devised a wing-folding system of his own which was patented. In all, twelve of these Admiralty Type 807 floatplanes were built, Nos. 807 - 810 and 919-926.

CHAPTER ELEVEN

Sopwiths at War 1914

War meant a hastening of deliveries from the shop floors and a number of new orders. So that the War Office and Admiralty would not vie with one another, it was tacitly agreed that certain manufacturers would be regarded as exclusively Army suppliers and others Navy contractors. Sopwith's, then contracting to both services, became an Admiralty contractor.

Apart from an increased tempo of work and enlarging the premises, the atmosphere in the shops along Canbury Road changed little, but at the Brooklands sheds it was greatly different. Brooklands had become a military camp under the command of Captain F. V. Holt, of the Oxfordshire and Buckinghamshire Light Infantry. No. 2 Reserve Squadron R.F.C. was forming there and the whole of the Bristol School's aircraft, plus two Vickers School Boxkites, had been taken over. The Bleriot School continued, with the proviso that pupils had to be suitable candidates for the R.F.C. training and passing into the Royal Flying Corps.

tragedies in which duty officers making their rounds were shot in error and at least one deaf civilian was shot, but for an intruder to fire back suggests that something sinister was intended. But to this day the identity of the intruder is still a mystery.

At the two Kingston factories, workmen toiled at the spate of orders in from the Admiralty. Already engaged on orders for ten Admiralty Type 860 and six Type 880 floatplanes the new requirements were for 24 2-seat tractor landplanes (a type that became known later as the Spinning Jenny) and 13 Tabloids. New staff were taken on, but a few of the old hands, being Territorials, were called up and a few others joined the rush to get into the Services before the war was over. The greatest loss was the test pilot Pixton who, now that Hawker had returned, felt need of a change and joined the Aeronautical Inspection Department; he was gazetted a 2nd Lieutenant in the Royal Flying Corps in April the following year. Mahl took his

The employees and their products. This photograph was taken outside the Sopwith works early in the 1914-1918 War. The airframes on the lorry are of Sopwith Schneiders. They show the construction typical of the period, a wire-braced, box-girder structure, with a rounded top decking. Later the structure would be fabric covered and doped, except the forward portion which would be covered in aluminium sheet.

As with most military establishments at this time, when spy scares were rife, the airfield was under guard. A detachment of the Royal West Surrey Regiment (Special Reserve) were detailed to guard the sheds and before the month was out the first alarm had occurred. In the early hours of an August morn, a Private Robertson on sentry duty not far away from the Sopwith sheds heard someone prowling round. Three times he challenged in the traditional way—'Halt! Who Goes There?' After the third challenge without a responding 'Friend', Pte. Robertson fired. The intruder, whoever he was, fired back, hitting Robertson in the arm with a revolver bullet. The guard turned out at the first shot and the sheds including the Sopwith premises were searched, but without result.

Most such scares came to nought. There were several

place as test pilot of the Sopwith floatplanes, which were tried on Southampton Water using sheds at Woolston as a base. Another loss to the firm was an engineer, Mr. L. Moore Lilley who, being on the Reserve, was called up in October 1914 as a 2nd Lieutenant in the Army Service Corps (Mechanical Transport Section). So that employees remaining could show themselves to be on work of national importance, a special badge was issued to the staff.

News, in spite of the rigid censorship on happenings on the Western Front, was gradually reaching the factory. Sopwith Aircraft had been in action with both the Army and the Navy.

The British Expeditionary Force commenced embarking for France on August 9th 1914, five days after the declaration of war. Their supporting Royal Flying Corps

Used on home service at Eastchurch in the first years of the war was an example of the original Sopwith biplane.

element was composed of Nos. 2, 3, 4 and 5 Squadrons and an Aircraft Park. None of the R.F.C. squadrons had Sopwith aircraft, but the Aircraft Park had; in addition to its assembled ten B.E.2s, three B.E.8s and three Henri Farmans, there were four Sopwith Tabloids in crates. The park arrived at Boulogne on the 19th.

Two of the Tabloids were assembled before the end of August and were flown during the Battle of Le Cateau by 2nd Lieutenants C. G. Bell and N. C. Spratt while still on the strength of the Park. Later the Tabloids were distributed to Nos. 3 and 4 Squadrons.

While No. 4 Squadron were at Compiegne, on Saturday, August 29th, a German machine came over their encampment and dropped three bombs. Spratt, attached to the unit, immediately took off in a Tabloid. As a squadron officer wrote in his diary, Spratt 'made twenty rings round it and fairly put the wind up the German and then came back'. At Seris, the following Wednesday, where No. 4 Squadron had arrived only that day, a German machine came over towards evening. Spratt again gave chase in the Tabloid; arriving back in the dark he did a complete somersault in the Tabloid without hurting himself. The Tabloid was dismantled and placed on a trailer and does not appear to have been used again. Repairs were difficult to effect until the retreat from Mons was over and the position had stabilised.

The Navy in 1914 was making a wider use of its own Sopwiths in action—and on the Western Front. The Admiralty had considered it important to deny the use of territory within 100 miles radius to German Zeppelins and to attack all airships operating in that radius. Explaining this, in these words, to the French Marine Headquarters at Bordeaux our Naval Attaché used a surprisingly modern phraseology and actually referred to the Admiralty's wish to take measures necessary to 'maintain aerial command of this region'.

With French assent, plans were made to place between 40 to 50 aeroplanes at Dunkirk or other convenient places on the French coast, supported by over fifty armoured cars and 200 marines. Under Wing Commander C. R. Samson, R.N., three squadrons were formed, Nos. 1 to 3 under Squadron Commanders E. L. Gerrard, S. D. A. Grey and R. B. Davies respectively. Eventually it was planned to have twelve machines to each squadron, but initially it was a case of collecting what they could.

A naval board of survey visited the Sopwith sheds and reviewed the aircraft available. To supplement two standard Tabloids, they made an offer for Hawker's runabout, the Tabloid prototype. These three aircraft became Nos. 167-169, and a very famous trio they became. Also in the Brooklands sheds was a standard Tractor Biplane, Sopwith's general utility and passenger-carrying aircraft, a familiar sight at pre-war meetings at Hendon and Brooklands. This, too, was impressed by the Admiralty who soon had the large lettering SOPWITH painted over and its R.N.A.S. No. 906 was marked on the rudder.

All four aircraft were sent to Eastchurch, where Samson's force was concentrating, to join the heterogeneous array of aircraft that mechanics were busily engaged in making serviceable. Among them were two other Sopwith Tractor Biplanes, No. 33, the second aircraft that Sopwith had delivered to the R.N.A.S. in 1913 and No. 103, another pre-war delivery. The first unit, including the two Sopwiths, Nos. 33 and 906, flew to Ostend on August 27th but were later withdrawn to Dunkirk where reinforcements arrived. Both Sopwiths flew on reconnaissance on September 9th; No. 906, flying over Doullens, Fierent and Blanzy, was out two hours, while No. 33, reconnoitring Lille, Tournai and Douai, made a three-hour flight. For the days following the weather was bad, with gales and rain, and Samson grounded the aircraft on the assumption that it was not worth risking the few machines available and that it was unlikely that Zeppelins—the bogey of the Navy at this time—would be out.

On September 15th, a windy and overcast day, Samson decided that at least one reconnaissance should be made in view of the reports of small bands of Uhlans probing forward and sent out Flight Lieutenant Collet on Sopwith No. 906 to patrol Lille, Valenciennes, Condé and Tournai. Collet landed after three hours without having sighted any enemy troops.

At this time Antwerp, in the path of the German advance, was being reinforced by British Marines. Considered thereby to be a safe base, Samson re-organised his squadron and established a forward base at Antwerp for striking against German bases, in particular the German Zeppelin sheds at Düsseldorf and Cologne. Antwerp soon became a rallying point for a selection of famous pre-war Sopwith aircraft making their first strike in the Great War.

First to be sent was No. 906, the impressed Tractor Biplane; Lt. Collet left Dunkirk in it at 11 a.m. September 16th, accompanied by Major Gerrard flying Samson's famous B.E.2 No. 50. Two days later they were joined by the two Tabloids, Nos. 167 and 168 that had staged through Dunkirk, where another Sopwith, No. 103 in which both

Winston Churchill and his wife had once flown, was being prepared for the flight. The Tabloids were joined next day by the Sopwith Churchill itself, No. 149 which, repaired since its crash, had been hurriedly fitted at Dunkirk with an extra fuel tank and bomb-dropping gear.

Originally, an attack was scheduled for the 19th, but the Tabloid No. 167 having landed in soft sand was in need of repair and the attack was finally made on September 23rd. Sopwith aircraft have come to be regarded as fighters yet this first strategic bombing attack of the 1914–1918 War was by a force, small admittedly, but composed predominantly of Sopwiths. The aircraft engaged and their allotted targets were as follows:

B.E.2a (No. 50) flown by the Force Commander, Major C. L. Gerrard, R.M.L.I. (Düsseldorf)
Sopwith Tractor Biplane (No. 906) flown by Flight Lieutenant C. H. Collet (Düsseldorf)
Sopwith Churchill (No. 149) flown by Lieutenant-Commander Spenser Grey with Lt. Newton Clare (Cologne)
Sopwith Tabloid (No. 168) flown by Lt. Marix (Cologne)

The attacks were timed to be simultaneous and the aircraft started at daybreak. It was clear as far as the River Roer, but from there a thick mist extended to the Rhine, blanketing the ground. Only Collet managed to locate his objective. Gliding down from 6,000 feet, the last 1,500 through mist, he sighted the giant Zeppelin shed at Düsseldorf from 400 feet while about a quarter of a mile distant. Bombing from that height, his first bomb fell short and exploded but the other two, although appearing to hit the shed, apparently failed to explode. Collet returned direct to Antwerp. The other two pilots flying Sopwiths, after flying round vainly trying to locate their target, landed short and were refuelled. The commander's B.E. had a fuel system failure near the River Meuse and landed by some Belgian troops. Happily, all four aircraft were back at Antwerp by 1 p.m. that same day. While little material damage had been done, the raid had considerable moral effect. Grey reported that he had lost a bomb and didn't know where it had dropped.

Gerrard's report of the operation advised that further attacks should not be made until a properly organised flight of fast machines was available and recommended that the Sopwith Gordon Bennett machines should be sent out in boxes.

The Sopwith Tractor No. 103, too late for the venture from Antwerp, was used from Dunkirk as soon as it was ready. It was first sent out on September 24th, flown by Lt. R. B. Davies with Bombardier Wyer as observer, to attack the German airfield observed at Roeulx. No enemy aircraft were seen, but four sheds near the railway line south of Roeulx were assumed to be hangars and four grenades were dropped on them. Goods trains were seen moving on the line between Denrain and Cambrai and a 20 lb. Hales bomb was hopefully dropped on the line to disrupt this communication.

In the course of this reconnaissance No. 103 became the first Sopwith aircraft to be fired upon and next day, attacking the same two targets, in the hands of Lt. Davies with Capt. Courtney observing, it received the first hits by a Sopwith—two rifle bullets went through the wings. No. 103 was then moved to Morbecque where Bristol T.B.8s Nos. 916 and 917 constituted No. 3 Squadron, but Samson considered the machine unsuitable for war purposes as it would not climb well and it was relegated to coast patrol work at Dunkirk. However, more reinforcement had arrived, including Sopwiths.

Towards the end of September, the Royal Naval Air Service on the Western Front was predominantly a force of Sopwiths. On September 27th it was as follows:

No. 1 Aeroplane Squadron, Antwerp
B.E.2A (No. 49) B.E.2A (No. 50)
Sopwith Tractor Biplane (80 h.p. Gnôme) (No. 906)
No. 2 Aeroplane Squadron, Antwerp
Sopwith Tabloids (Nos. 167, 168 and 169)
Sopwith Churchill (No. 149)
No. 3 Aeroplane Squadron, Morbecque
Sopwith Gunbus (converted seaplane) (No. 901)
Bristol T.B.8s (Nos. 916 and 917)
Dunkirk Depot
Deperdussin Monoplane (No. 7)
Sopwith Seaplane, converted landplane (No. 58)
Sopwith Tractor Biplane (No. 103)

The prototype Tabloid. This compact Sopwith biplane was aptly nick-named the Tabloid at a time when aircraft in general were not given type names. Moreover, a threatened lawsuit over the use of the name by a manufacturer of a medical compound of that name, apparently only endorsed it in the public's mind. Although never officially bestowed, 'Tabloid' it became by popular acclaim!

Bleriot Monoplane (70 h.p.)
Bleriot Monoplane (Lord Edward Grosvenor's own)
Headquarters Flight, Morbecque
B.E.2a (No. 46) (damaged)
Bleriot Monoplane (No. 39) At Dunkirk for write-off
Short Seaplane (converted landplane)—wrecked next day

Every clear day, the Sopwiths at Antwerp were out reconnoitring Brussels which had fallen to the enemy. Their pilots brought back the intelligence that the old Belgian airship shed at Berchem had part of the roof removed—suggesting the Germans would enlarge it for Zeppelins.

Gerrard intended using the Sopwiths for a further attack on Zeppelin sheds in Germany, but it was proving difficult to keep the force serviceable. On September 25th Flight Sub-Lieutenant Lord Carbery had started out with Prince de Ligne on Tabloid No. 169. Just after take-off the engine started missing and the pilot, trying to turn back on to the airfield, stalled from 100 feet. The Prince's foot was crushed and Lord Carbery broke the patella of his knee. Gerrard decided to have this prototype Tabloid re-built to standard form—converted to a single-seater, the fuselage covered in and a small fin fitted.

Next day the Churchill was out of commission; Flight Lieutenant Newton Claire broke an axle on take-off and as a result, when landing an hour later, it capsized and damaged the top plane. Fortunately neither pilot nor his passenger were hurt. That evening, Lt. Marix on Tabloid No. 168 gave chase to a German monoplane and was overtaken by darkness; he landed at Geel and returned to Antwerp where the engine was found to need overhaul.

The situation at Antwerp soon became precarious. On September 28th the Germans commenced their bombardment of the outer forts and advanced on the city in the days that followed. Both the Sopwith Tractor No. 906 and the Tabloid No. 168, piloted by Lts. Collet and Marix respectively, made at least one bombing attack on railway junctions at Herenthals and Aerschot in a vain effort to stem German reinforcements. Truly, the world's first strike aircraft were Sopwiths.

Now came another Sopwith on the scene—of yet another type—the Gun-bus. A further order of six Sopwith pusher floatplanes for the Greek Government had been comandeered by the Admiralty and the last two, Nos. 900 and 901, were fitted with a landplane chassis and sent to Dunkirk. On October 5th, Flt. Cdr. Bigsworth flew No. 901 up to Antwerp, in the course of which it was fired upon over Ghent. Next day Bigsworth, scouting around Antwerp, sighted an enemy kite balloon, but was chased away by enemy anti-aircraft fire when he approached.

At 11.30 a.m. on October 8th the Germans commenced bombarding Antwerp from the south. Shells were passing over the airfield which lay between the town and the enemy. It seemed doubtful if the screen of forts could hold out much longer and plans were made for evacuating the British Marines if the town fell. The machines, all Sopwiths except for a single B.E., were trundled out their sheds and the engines were tuned ready for take-off.

Major Gerrard, realising the forts could not hold out, and that in falling back a valuable vantage point for striking at Germany would be lost, made a final effort to launch the Sopwiths on another attack against Zeppelin bases. October 9th dawned dull and misty, but towards midday it cleared. Lt. Cdr. Spenser Grey and Flt. Lt. Marix were alerted to attack airship sheds at Cologne and Düsseldorf 112 and 103 miles distant respectively.

Spenser Grey in Tabloid No. 167 left the field at 1.20 p.m. Finding a thick mist over Cologne, he flew continuously around the district looking for the sheds—Intelligence had given him two different positions for these, one to the north-west and the other to the south of the town. Flying at 600 feet, he remained over the city some 12 minutes drawing fire from the ground. Having failed to locate any sheds, he made for the station in the city centre, where many trains were seen to be drawn up and there he dropped his two bombs. He landed back at 4.45 p.m.

Marix left ten minutes after Grey on Tabloid No. 168 and flew at above 1,000 feet to Düsseldorf, where the sheds were clearly visible. Attacking the target in an easterly

The improved Batboat at Cowes, Isle of Wight, on a beaching trolley. An increased dihedral on the lower wing of this model allowed better sea clearance for the wings. The Batboat was not only the first successful British flying boat, but almost certainly the first British flying boat to fly at night.

Tabloid close-up. Note that each wheel is fitted with its own half-axle, in a way that became a characteristic of Sopwith aircraft. Application to patent this was made in 1914 and was granted November 1st 1916 (No. 109146) in the name of the Company and T. O. M. Sopwith.

direction, he dived down to 600 feet and dropped his bombs. He failed to see if they had penetrated the shed but within thirty seconds he could see the roof fall in and flames shoot up to 500 feet. Under fire, the Tabloid was badly damaged and fuel ran out when still twenty miles from Antwerp. Landing, he first borrowed a bicycle and later a car to return to his base.

At 8.30 p.m. that evening, the airfield itself was subjected to bombardment, damaging the aircraft. It was intended to take a tender out at first light and retrieve Marix's Tabloid, but at 11.30 Gerrard became aware, unofficially, that a general evacuation had been ordered. As there were Germans in the woods bounding the airfield, and already mechanics were being fired upon, the R.N.A.S. detachment crowded into two cars and drove to Ostend.

This was the end of the service of several famous aircraft. The Sopwith Tabloid prototype (No. 169) which Hawker had taken to Australia had been left in the Bollekens factory at Antwerp for conversion; with it was No. 149, the one and only Sopwith Churchill, under repair. It seems likely that the Belgians destroyed them to prevent their falling into enemy hands. The Tabloid No. 167 was wrecked by enemy fire and Belgian patriots had to abandon No. 168.

The depleted force staging at Ostend was now ordered to concentrate at Ghent where three of the nine aircraft were Sopwiths, the Tractor Biplanes Nos. 103 and 906 and the Gun-bus No. 900. Soon Ostend fell and the force withdrew further and were tasked with reporting on German use of Ostend Harbour. On these reconnaissances No. 103 was used and came under fire again. Unfortunately the machine, now old and having been exposed to the elements for two months, would no longer climb satisfactorily and both this and No. 906 were scheduled for return to England; No. 103 continued in service at Eastchurch until 1916.

Only two of the Gunbuses, Nos. 900 and 901, had arrived in France and since Samson considered them inferior to the Maurice Farmans then arriving, he decided to return them to England. Meanwhile Sopwith mechanics at Brooklands were testing another Gunbus with a 150 h.p. Sunbeam engine in place of the original 100 h.p. Gnôme. During the same month, October, Hawker was testing the two Gordon Bennett racers, in two different forms, as the Admiralty had now called these forward for completion and purchase. While the second had a slightly better performance than the Tabloid, the Admiralty were not influenced. The reasons—the Navy wished to standardise on types and the Bristol Scout was showing promise.

On November 1st came the firm's first fatality. An engineer, Reginald Alston who was doing design work, set

The new patent Sopwith split axle, incorporated on late production Tabloids, did much to prevent accidents such as this. However, by its very lightness, mishaps were bound to occur on some of the improvised airfields in France; and anyway it is the wheel that collapsed first!

Lieutenant Spenser Grey, R.N. (left) and a Petty Officer Artificer (right) at Hendon, with the Sopwith Gordon Bennett racer impressed for war duty with the Royal Naval Air Service.

out with Mahl at the controls of a new seaplane. Barely had it left the water, than it suddenly dropped and dived back into the water. Mahl was thrown clear and picked up. The wreckage drifted down Southampton Water and was brought ashore later off Netley Hospital where Alston's body was recovered.

Daily Mahl continued testing seaplanes from Woolston and before the end of the year there came the first of the production Schneider Cup floatplanes.

Yet another Sopwith aircraft type had been in France. This was No. 58 originally delivered to the R.N.A.S. as a Sopwith Tractor Floatplane in 1913. In the need to supplement Samson's force it had been converted to a landplane and arrived at Dunkirk at 4 p.m. on September 27th. While there it was fitted for bomb-dropping and retained as a reserve. It had flown to Cassel on October 1st piloted by Lt. Chambers and carrying Lt. Dyott, but was returned later to the Dunkirk depot and then to England where it was re-allotted to Yarmouth Naval Air Station. On November 19th, a few miles north of Yarmouth, No. 58 stalled from 150 feet and when hitting the water the explosion of a practice bomb stunned the pilot who would have drowned but for the devotion of Petty Officer James C. S. Hendry, the observer, who extricated the pilot from the wreckage and kept him above water until rescued. Hendry was awarded the Albert Medal.

A close liaison was maintained between the Sopwith factory and the R.N.A.S. When Squadron Commander C. I. Randall of the naval engineering branch went to Paris in November to purchase aeronautical stores, two 135 h.p. Canton Unne engines were purchased from M. Donnet for the Sopwith works.

As the weather deteriorated and thoughts turned to the hardships of men looking after the aircraft—handling parties for floatplanes were often up to their waist in water in an icy sea—a fund for Royal Naval Air Service comforts was launched by Mrs. Sueter, wife of the Director of the R.N.A.S., who lived at The Howe, Watlington, Oxfordshire. Of all the organisations that responded, probably the best response of all came from Sopwith's—from top to bottom. The Company Directors contributed £12/10/0 and a collection among employees raised initially £13/3/4 and a further £9/18/7 and £7/0/8 during subsequent collections in 1914. It was logical that the firm should support funds for naval airmen, since they were the men using the bulk of their production; however, the Royal Flying Corps Air Fund also benefited by £16/19/3 through a collection in the factory.

In supporting the men who flew his machines, Sopwith gave unstintingly. The Royal Aero Club had instituted, with the approval of the Lords Commissioners of the Admiralty and the Army Council, a Flying Services Fund that had been originated by M. Andre Michelin. The fund provided benefit for officers and men of the Royal Naval Air Service and Royal Flying Corps that were incapacitated on active service and for the widows and dependents of those killed. Andre Michelin, Chairman of the tyre company of that name, donated an initial £1,000 to get the fund going. This was immediately matched by another £1,000, donated personally by T. O. M. Sopwith.

But in 1914 the most important day for Sopwith was the penultimate day of the year. On that day he married the Hon. Beatrix Mary Leslie Hore-Ruthven, daughter of the 8th Baron of Ruthven.

TAILPIECE !
with apologies to W. Boddy Esq. Editor—Motor Sport

CHAPTER TWELVE

Sopwiths go East—1915

A Sopwith 860 under test after building late 1914. The pilot is probably Victor Mahl. The floatplane shown in this photograph survived until March 1917.

By the end of 1914 not a serviceable Sopwith aircraft appeared to remain in the Western Front. The Tractor Biplanes were too old and warped after the exigencies of the campaign, the Gunbuses were considered unsuitable and the Tabloids, in spite of initial successes and acclaim, were not sufficiently rugged for improvised airfields and were replaced by Martinsyde and Bristol Scouts. Sopwith aeroplanes were discarded by both Army and Navy. But Sopwith seaplanes still offered some promise, although the major seaplane event in the war so far had been the attack on the German naval base at Cuxhaven, Christmas 1914—in which all participating seaplanes had been Shorts.

The extent of the rejection by the Services of the Sopwith for operations had not yet been fully realised by the works. Although the liaison between the Admiralty and the factory was good, particularly the relationship between T. O. M. Sopwith and the Director of the Air Department, the actual deployment of the aircraft was a naval secret. Nevertheless, the cessation of orders for aeroplanes in favour of new orders for seaplanes gave evidence of a shift of requirements. The Kingston works were as busy as ever; in fact working to capacity as sub-contracting of Sopwiths commenced, but the new trend meant that in test-flying, Mahl was busier at Woolston than Hawker at Brooklands.

The year 1915 opened on a wet Friday, curtailing flying, but next day Mahl had one of the new Sopwith 860 floatplanes on endurance test (actually No. 854) over the Solent. On January 5th, the first of the production Sopwith Schneiders, so called as they were virtually the same as the model used for the firm's Schneider Trophy race success in 1914, appeared on the Solent.

Later in the month, one of the Greek seaplanes, impressed for R.N.A.S. use, had a more powerful engine fitted in lieu of the earlier 100 h.p. Gnôme. Hitherto, Sopwith aircraft had almost standardised on 80 and 100 h.p. Gnômes. The need for more powerful engines became apparent if service specifications were to be met and the Gunbus was being tried with a 150 h.p. Sunbeam, while the 860 Type seaplane had 225 h.p. Sunbeam engines. One of the early deliveries of the latter type collided with a pier and was towed to Woolston for repair.

Meanwhile, the lower-powered 807 Type seaplanes, built during the latter part of 1914, were in service—and about to enter a new theatre of operations.

When H.M.S. *Ark Royal*, the first ship to be fitted exclusively as a seaplane carrier, sailed from Sheerness at midnight on Monday, February 1st 1915 for the Dardanelles, more Sopwith aircraft were carried than any other type. Aboard were ten aircraft: Sopwith Folder Seaplanes Nos. 807, 808 and 922, Wights Nos. 172 and 173, Short No. 136 and four crated Sopwith Tabloids.

Heavy gales in the Channel and Biscay obstructed work on the aircraft lashed in the holds, but on reaching Malta trials of the Short and one of the Sopwiths were carried out from Valetta harbour on February 14th—with unfortunate results for the Sopwith. Flight Lieutenant W. H. S. Garnett, with Petty Officer Marchant as his passenger, clambered into Sopwith Folder Seaplane No. 808. The 100 h.p. Monosoupape Gnôme started without trouble, and the machine gathered speed. While skimming the water at the point of take-off, a float, hitting the swell, broke off and the machine turned over in the water.

The wreckage was hauled aboard and repairs were put in hand while the ship sailed on to Tenedos, reached on the 17th. That same day of arrival a reconnaissance was ordered of the Dardanelles entrance defences and this time No. 807 was hoisted out, with Lieutenants R. H. Kershaw and A. G. Brown as pilot and observer respectively. Several runs were made on the surface but the seaplane failed to leave the water so, in desperation, some of the fuel was drawn off to reduce weight. Lightened, the machine rose, but failed to gain sufficient height for reconnaissance.

The first important task for a Sopwith aircraft in the Dardanelles campaign was carried out on February 19th. Flt. Lts. N. S. Douglas and E. H. Dunning, detailed as pilot and observer respectively, flew in Sopwith Folder No. 922 with orders given as follows:

The wreck and salvage of No. 920, a Sopwith type 807 floatplane. Three such floatplanes were sent out to Bombay in the S.S. Persian *for No. 4 Expeditionary Squadron, R.N.A.S. with the task of destroying the German cruiser* Koenigsberg *hiding in the Rufigi Delta. No. 920 was erected and flown on February 11th 1915 and No. 921 the next day; the third machine was kept as a spare. The three machines were embarked in the auxiliary cruiser* Kinfauns Castle *for Neororo Island to carry out practice flights in secret, with 50 lb. and 16 lb. bombs.*

Sopwith type 807 No. 921, believed the subject of the photo below crashed a week after landing on Neororo Island. The bombing plans were later abandoned with the other two Sopwiths, due to the limitations of their Gnôme engines in a tropical climate and the effect of the heat on the floats which had to be filled with water when the machines were not in use as a protective measure. Nevertheless No. 920 was repaired and later operated from Mafia Island; subsequently it was sent to Mesopotamia where it was finally written off in 1917.

1. Reply by W/T 'yes' or 'no' to question 'Are two guns of No. 1 Battery nearer to barracks than to cliff'.
2. Then spot on to No. 1 Fort for H.M.S. *Inflexible*.
3. Be 4,000 feet if possible, but 3,000 further off if necessary.

As it was, the wireless set short-circuited and the mission could not be carried out to the letter. Thus all three Sopwith Folders had now been used and each had been found wanting in some respect. The swell at Malta that wrecked No. 808 was unfortunate—but the Short No. 136 tested the same day had remained seaworthy. The Sopwith was soon repaired and ready for use the same day as the first reconnaissance was made.

Cmd. R. H. Clarke Hall, the *Ark Royal*'s captain, stated in one of his first despatches home that while the Canton-Unne engines of the Short and the Wight floatplanes had given little trouble, the 100 h.p. Gnôme Monosoupapes of the Sopwiths, apart from a small defect on the pressure system, were difficult to coax up to the required revolutions to develop their power. The wireless failure was unfortunate, probably spray getting in the set during take-off caused the shorting. However, the other machines were having their difficulties.

No further flights were attempted until February 25th when the Short, hoisted out for reconnaissance, was found to have a cracked cylinder, and Wight 173 found the sea too heavy to get off and, in any case, water shipped over the engine shorted the distributor. Next day the Sopwiths Nos. 807 and 922 flew for 25 and 32 minutes respectively on attempted scouting flights.

Meanwhile reconnaisances both by air and on foot were made off Tenedos to find a suitable landing ground for the four crated Sopwith Tabloids, but without success.

Sopwith Folder No. 922 came in March to prove itself the champion of the three Sopwiths, in spite of occasional setbacks. On the 1st, when Flight Lieutenant N. Sholto Douglas set out with Petty Officer B. J. Brady, the bottom of a float was carried away. However, two days later the same two took the machine on a reconnaissance to Kephez Point and located a minefield, reported on the number of guns in two forts and another battery position. The day following, again on No. 922, they reported Turkish troop movements by wireless. On the 5th the machine came under fire and Sholto Douglas was wounded in the leg; six holes were later counted in the fabric.

The first Sopwith Seaplane to come under fire had been No. 807 in the course of 66 minutes flying to locate field guns on March 4th. This was hit eight times, but neither Flight Lieutenant R. H. Kershaw or Petty Officer Marchant were hit. Patched up hurriedly, it was sent out again that day, but strained its undercarriage and returned lop-sided. But No. 808 had an even worse time, ending in its early demise. On March 1st a broken piston valve necessitated an early return from reconnaissance. Three days later, when its crew, Flt. Lt. W. H. S. Garnett and P. O. Marchant, searched a piece of coast for the enemy to cover a demolition party which had made a landing, carburettor trouble forced their return and on a second attempt that day, it failed to climb through ignition trouble. On the 5th it was taken up by Garnett with Flight Commander H. A. Williamson who was going to spot for the guns of the battleship H.M.S. *Queen Elizabeth*. At 3,000 feet the propeller burst in the air breaking the starboard top wing, causing the machine to spiral down. Both occupants were injured, the machine was wrecked, but the engine was saved. No. 808 was written off.

It was realised that with the limitation of the Gnôme Monosoupape engine, and the poor condition of the airframe, that No. 807 could only be useful as a single-seater. On March 14th it was flown by Flt. Lt. E. H. Dunning to find out if a pilot could control the machine properly and work the W/T set at the same time. The test was cut short after 33 minutes by the engine overheating and when tried again, two days later, a float was torn away getting off the water and in salvaging the wings were damaged.

Not until March 28th was No. 807 ready again, when Dunning, on his own, spotted for H.M.S. *Majestic*. But its days were numbered. In April ignition trouble and tappets out of adjustment led to its withdrawal for overhaul.

It was clear that if Sopwith were to continue as suppliers of seaplanes, higher powered engines were essential. At Woolston Victor Mahl had been busy taking one float-

Nos 920 and 921 at Mafia Island for operations against the German cruiser Konigsberg which had sought refuge in the Rufiji Delta. According to an Official History they 'were soon found wholly unsuited for work in the tropics (for which they had not been designed).

plane out after another from the Sopwith sheds, alternating between the new heavy floatplanes with a 225 h.p. Sunbeam engine and the light Schneider Cup machines. Both types were urgently needed for the Dardanelles campaign and were hurriedly shipped out.

On Monday, March 29th, Mahl did not feel well enough to carry out testing, but on the Tuesday, feeling better, he was in the air all next morning trying to catch up with a backlog of work. Later, he collapsed. Rushed to hospital he was operated on for appendicitis on the Wednesday and died the following day. He was 25.

Sopwith Schneiders No. 1437 and 1438 that Mahl had so recently tested arrived at Tenedos for *Ark Royal* in the S.S. *Moorgate* on April 9 and were erected within two days. As single-seaters, they had limited uses; No. 1437 was allotted to H.M.S. *Doris*. Seemingly of greater potential value were the two Sopwith 860 Type Seaplanes with 225 h.p. Sunbeam engines that arrived aboard the S.S. *Aragaz*. However, with seven machines already aboard the 'Ark' there was no room for their erection. An empty steamer or collier was required and on May 9th, when the S.S. *Penmorrah* was allocated as a store ship, the 225 h.p. Sopwiths were transferred to this ship for erection.

The U-boat had become the bogey of off-shore operations and most flights in late May were anti-submarine patrols. Flights with the large Sopwiths were brief. On the 30th Flt. Lt. Bromet, accompanied by C. P. O. Finbow, kept up nearly an hour before the engine cut and they were forced to descend, being towed back by an M.T.B. (Motor Torpedo Boat). Next day Flt. Lt. Kershaw after a fourteen-minute test flight with Finbow, lasted 45 minutes in the air before the engine cut.

Kershaw had more success with the Schneider No. 1438 which he had taken on detachment to H.MS *Minerva* earlier in the month. He had reconnoitred Smyrna, and on another occasion dropped bombs near the harbour entrance. On return, on the same machine, he had looked for suitable territory for establishing an aerodrome and also dropped pamphlets over the Turkish lines. Schneider No. 1437 had been erected on April 24th but having broken a propeller, was kept unserviceable through lack of spares.

During this time No. 922 had done much useful work, kept in service by spares from the wreck of 808 and the frame of 807, but on May 22nd Sholto-Douglas had failed to coax it over 800 feet. After a test flight next day he found the engine carbonised and it was changed, but its performance was then even worse, 500 feet being its maximum effort on May 24th.

The Sopwiths had been plagued with engine troubles; had their engines performed more satisfactorily it is possible that many more Sopwith 2-seat seaplanes would have been ordered. On June 1st, the *Ark Royal's* commander reported to the Admiralty on his aircraft. The Schneiders he pronounced 'fine weather machines', but qualified this by saying that they had proved useful where detailed information was not required. For detailed reconnaissance two-seaters were necessary and the Sopwith Folder No. 922 was said to have been 'very satisfactory' but the type was not recommended as the spring of the floats tended to make the machine bounce in getting off the water and the floats themselves, being very lightly built, broke up easily. Of the higher-powered Sunbeam-engined Sopwiths, Clarke Hall intimated that they were bedevilled with engine troubles, and not by any aerodynamic failings.

This sealed the fate of the future Sopwith floatplanes, apart from the Schneiders whose redeeming feature was that they could so easily be stowed aboard ships. However, a few of the folder seaplanes already built gave sterling service as will be related. But it is at this point that of the Navy's two-seat reconnaissance seaplanes, Short's took pride of place.

CHAPTER THIRTEEN

Sopwiths at Sea—1915

Messrs R. J. Turk and Sons were well - established boat builders and repairers. They also let out boats on hire. This establishment, with its various facilities and slipways was not far from the Works at Kingston, and must have been a boon to Sopwiths when carrying out flotation tests of Sopwith Schneiders.

While the Sopwith floatplanes floundered in the Dardanelles, and Sopwith aeroplanes were discarded on the Western Front, the Admiralty still had hopes in early 1915 of employing both usefully from home bases.

The Sopwith Two-Seater Scouts and Tabloids were being delivered crated to East Coast Stations for anti-Zepp patrols, and parties of Sopwith employees visited stations to erect them. At the same time, Schneider Cup floatplanes were reaching the same destinations, stored ready for shipborne use. Although called Schneider Cup floatplanes, this was something of a misnomer for the Schneider Trophy was not a cup.

At the beginning of the war, work had been put in hand at Chatham Dockyard to convert the cross-Channel packets *Empress, Engadine* and *Riviera* to seaplane carriers. Now ready, it was planned in March 1915 to take *Empress* to sea, accompanied by the Harwich Force of light cruisers and destroyers, so that her Short floatplanes could reconnoitre the Nordern Zeppelin sheds and bomb the wireless telegraphy station at Norddeich, while Schneiders would be taken to deal with any Zeppelins.

Bad weather delayed the operations and it was May 3rd 1915 before the Harwich Force took to sea. By this time the allocation of carriers had changed; *Engadine* and *Riviera* sailed together with a more recent seaplane carrier convert, the *Ben-my-Chree,* formerly an Isle of Man packet. Choppy seas, however, made launching of seaplanes impossible and when three days later the force set out again, it was thwarted by fog.

On May 11th the Force set forth again under Commodore R. Tyrwhitt. At 5 p.m. a Zeppelin was spotted estimated at some 70 miles distant. Tyrwhitt ordered up a seaplane and an attempt was made to fly a Schneider off the *Ben-my-Chree*. The ship had been fitted with staging forward, and wheels had been fitted on the Schneider's floats. As usual, it was the engine that caused the failure. The backfiring of the Gnôme caused such tremors that the platform collapsed! Meanwhile, *Engadine* was hoisting out her three Schneiders which took-off but failed to find the Zeppelin due to fading visibility—and only one returned. Of the other two, one spun into the sea and both pilot and machine was lost and the other, forced to alight with engine trouble, was wrecked on the sea; the pilot was rescued and the aircraft salved.

Twelve days later the Schneider's landplane counterpart, the Tabloid, rose in defence of the homeland. Three, Nos. 1209-1211, had been allotted to Eastchurch and on May 23rd, when an enemy aircraft was reported over Margate, Nos. 1209 and 1211 rose to intercept. At least they landed safely, but no sighting of the enemy was made.

An even more galling failure came on June 2nd. Part of the Harwich Force at sea, sighted a Zeppelin. The light cruiser H.M.S. *Arethusa* hove to in order to hoist out the Schneider it was carrying for just such an opportunity. The floatplane made a very successful take-off but by the time it had risen to 1,800 feet, thick smoke made by destroyers below was mistaken by the pilot for his pre-arranged recall signal—and so the Zeppelin L5 escaped.

To the public at large, Sopwiths were apparently as successful as ever. Their failings were cloaked by the censorship of war, but their successes could be promulgated and a success on June 6th brought a breath of pre-war activity back to Brooklands.

Inbetween coping with the production of the Schneider floatplanes and the Two-seat Biplanes, a variation of the Tabloid had been designed by Fred Sigrist and became used by Hawker as the company's hack machine. On this Hawker resolved to break the official British height record, then held by Squadron Commander E. F. Briggs, D.S.O. at 14,920 feet, standing from pre-war days.

The 'Sigrist Bus' was tuned up on June 5th and fitted with a new Lang propeller to give greater efficiency at a lower engine speed—1,100 revs instead of the usual 1,300—in view of the long running entailed. Next day, shortly before lunch, Hawker flew the 'Sigrist Bus' from Brooklands to Hendon; in effect from a purely military airfield to one that still had facilities for civil flying. A sealed barograph was fitted by Royal Aero Club officials and Hawker set out.

As later confirmed from the barograph, and given as the official British height record, Hawker took the biplane up to 18,393 feet. The last 100 feet, at the apparent limit of its ceiling took eight minutes to attain; in another seventeen minutes Hawker had brought the machine back to earth. He was none the worse except for feeling the cold and a slight bleeding of the gums.

Around the time of this attempt T. O. M. Sopwith had sought permission to visit the B.E.F. in France. This was arranged for 10th and 11th June. Crossing to Boulogne, he motored to Royal Flying Corps Headquarters at St. Omer for discussions and then visited a series of aerodromes to see for himself aircraft in the Field. The squadrons then had B.E.2cs, Moranes or Vickers Gunbuses, with some having mixed equipment including Maurice or Henry Farmans, Avro 504As, Bristol Scouts and Caudrons. There was not one Sopwith on the Western Front. But it is significant that in the months following this visit to France, the Sopwith designs that came to lead in British aviation, were formulated. A coincidence of this visit was that the authorisation of the pass for Sopwith to visit France was signed by Captain F. Festing, whose son, a former Chief of Imperial General Staff is now a firm friend of the Sopwith family.

On July 4th the Harwich Force tried again with the joint object of reconnoitring the Ems, where transports had been reported, and to bring Zeppelins to action. Both *Engadine* and *Riviera* accompanied the Force, both carrying Sopwith Schneiders. Approaching Ameland, Zeppelins were sighted and the Force reduced speed while *Riviera* stopped down to hoist out three Schneiders, all recently delivered new from the factory. None got off the water. One-by-one, their floats broke off in the choppy sea as their Gnômes vainly strained to lift them from the surface.

The last of a Navy order for 100 Sopwith Babies is here seen outside Turk's boathouses.

These repeated failures did much to create a distrust in the Navy of air operations. Not until 1916 were further attempts made to take Schneiders out with the Fleet. Nevertheless experiments continued; as the Commander-in-Chief of the Home Fleet, Admiral Jellicoe, pointed out, some spotting facility was imperative to match the Germans with their Zeppelins. If the Schneiders were to fail to rise from a sea, then experiments had to be pushed ahead from a steadier surface—a ship's deck.

By this time the new seaplane carried *Campania* was at Scapa Flow, available for trials. A former Cunard liner, the ship had been converted to carry up to twelve seaplanes and had a flying deck built above the poop stretching to the bows. On August 6th 1915, Flight Lieutenant W. L. Welsh

Eleventh of a batch of 100 Sopwith Schneiders, en route for delivery to the Royal Naval Air Service. This floatplane has more elaborate Serial Numbers and a Union Jack painted above them. The lorry, with its solid rubber tyres is interesting, in that it has the letters 'O.H.M.S.' painted on what was then called the scuttle, at the rear end of the bonnet.

climbed into the cockpit of a Schneider held down at the top of this ramp. The engine was run up, allowed to warm, then revved up before Welsh waved his helpers away and the aircraft was released. Gathering speed as it shot down the ramp, with the ship steaming into wind at 17 knots, the Schneider took the air with just a few feet to spare from the end. The precise measurement was no rough estimation, for the deck had been chalked and the wheel marks ended just seven feet from the end of the 120 feet ramp.

If a light Schneider took such a long run, a two-seat Short or Sopwith seaplane would need a longer ramp and *Campania* was modified to increase their run to 210 feet. But the possibilities had been shown. This was not the first time an aeroplane had been launched from a ship, both the British and United States Navy had achieved this pre-war. That this was achieved with a standard service aircraft was significant and it led to the time, in a matter of three years, when most of the capital ships of the Fleet had an aircraft aboard—and of those carried all were Sopwiths. But this was in the future, the trials and tribulations of Sopwith aircraft were not quite over.

The Sopwith Two-Seater Scouts were withdrawn from the coastal stations of Yarmouth and Killingholme in mid-1915 and relegated to training, being issued to stations at Hendon, Chingford and Cranwell. Due to a proclivity for spinning the machine had earned the unhappy nickname of 'Spinning Jenny'.

While the Schneiders had been wrecked by pounding seas, the Spinning Jennies received their pounding at the hands of trainee pilots. Their short service life is illustrated from the record of three consecutively numbered machines in the middle of the production batch of 24. No. 1063, which had only been delivered to Chingford in June, was crashed and completely wrecked by Probationary Flight Sub Lieutenant de Ville on July 2nd. No. 1064, rebuilt at Grain after an earlier crash and sent to Chingford mid-August, was crashed by an experienced pilot, Flight Lieutenant Arnold,

This photograph was taken in the Sopwith Works, about the time of the change-over from Schneiders to Babies.

on September 19th, wiping off the undercarriage and straining fuselage and wings. No. 1065, flown by Smylie another 'Sub' on probation, was crashed a day previous to No. 1063 wrecking the undercarriage, straining the wings and damaging the engine bearers.

R. R. Soar, later a Sopwith Triplane ace, wrote of the Spinning Jenny—'At Eastchurch 1915 there were two, No. 1072 and a wrecked one. I used to fly the serviceable one while waiting to fly the one Tabloid which R. D. M. Campbell eventually crashed after I'd had only a few flips in it. It was generally said that the 1072 type spun in the air from which you couldn't get it out. I had a bet that I'd spin the sod and get it out. Young pilots, including me, were scared stiff of the bugger spinning—and it spun fairly fast but came out easy on pushing joystick on dashboard. It sometimes spun after landing and turned 180° and set off like hell down-wind and I could never stop it.' One of the last Spinning Jennies to survive was No. 1062 at the Navigation School, Portsmouth, for ground instruction.

The Zeppelin attacks on Britain had caused an outcry and the public clamoured for action. Defence of the United Kingdom at that time was vested in the Admiralty and certainly their east coast aerodromes were best placed for meeting this menace. However, it took too long for intercepting aircraft to rise to reach the intruders and the idea was mooted of meeting them halfway—at sea—by taking Schneiders out in trawlers that would ride at anchor. When a Zeppelin was heard or sighted, or reported by W/T to be approaching, the Schneider could rise and be in position to intercept the returning airship.

Kingfisher was the first trawler to be equipped for this work and night after night it was at sea. Sopwiths had effected a series of improvements to the Schneider and with this new use for the single-seat floatplanes and more encouraging reports of their use in the East being received, a further hundred were ordered. This new batch, incorporating the improvements were the first production Sopwith Baby floatplanes. It was specified that they should have detachable fuselages for ease of stowing. Deliveries commenced in September 1915 and the order led to a series of advertisements by the firm for additional labour.

The employees continued to remember those who used their aircraft and around Christmas 1915 they added £6/7/2 to the £9/13/6 they had collected in the summer for the R.N.A.S. Comforts Fund; and to this the Company Directors added £50. Early in 1916 the employees entertained the directors to a dinner and yet another £20/1/9 was collected for the Fund on this occasion. This concern for those who used their aircraft was manifest in other ways, for whatever the present shortcomings, due chiefly to engine troubles, workmanship of Sopwith products was never in question.

While the Schneiders were tested on the Solent, the Sopwith sheds at Brooklands were still used for trials on the modification being carried out on two tractor biplanes, the 'Sigrist Bus' and 'Hawker Runabout', both derived from the Tabloid and powered with 80 h.p. and 50 h.p. Gnômes respectively. Hawker occasionally took his 'Run-about' to Hendon, demonstrating its amazing speed range of 22 to 84·6 m.p.h. and its aerobatic capabilities. It was the forerunner to the famous manoeuvrable fighters to come. The fortunes of Sopwith Aviation Ltd were on the turn.

CHAPTER FOURTEEN

Manifold Expansion—1916

An early production 1½ Strutter used by No. 5 Wing, R.N.A.S. was Flown by Flt. Sub-Lt. Rouse, it was used to escort a bombing raid on Bruges, February 7th 1917 and to bomb Ghistelles two days later. It served until October 23rd 1917 when it was crash-landed at Dunkirk.

WANTED. First class engineer's draughtsman; must have had charge of fair size drawing office; good organiser essential. No person engaged on Government work need apply. Write, stating age, particulars of experience, and salary required, to the Sopwith Aviation Co., Ltd., Kingston-on-Thames.

This advertisement appearing in *The Aeroplane* during December 1915 was indicative of the great changes to come in the organisation of Sopwith Aviation. Three experimental aircraft were on the stocks; the drawings for the Sopwith Two-Seat Biplane (later to become known as the 1½ Strutter) were passed on December 12th 1915 and in early 1916 the drawings for a Single-seat Scout were passed, followed in February by plans for a Triplane.

Production orders mounted early in 1916 and the works were expanded. Property bordering the premises was bought up and several cottages were demolished to make way for the new buildings. To meet the pressing needs of the Admiralty, benches were actually fitted and manned before the contractors had completed the building. An innovation was the employment of *women* on acetylene welding!

An actual photograph of a Triplane looping. Hawker showed his confidence in the type by looping the prototype three times on the first test flight.

As if anticipating the success that was to follow, the firm of T. W. K. Clarke & Company, propeller makers, opened new works in Kingston, the first of a dozen aircraft components manufacturers to site their premises within a mile or two of the Sopwith works. Formerly, Clarke's had made model aircraft in Kingston.

At this time the Society of British Aircraft Constructors (now the Society of British Aerospace Companies) was formed and Sopwith Aviation became a founder member. The management of the Society was vested in a council of seven members of which one was R. O. Cary of the Sopwith Company. The first meeting was held on April 13th. Five days later Tom Sopwith was attending a special committee meeting of the Royal Aero Club. Among the seven members was Commander C. R. Samson, D.S.O. who, as related, had first-hand knowledge of Sopwith products and was yet to have much more. At this meeting the decision was taken to extend the club premises to include a room where members could obtain lunch and dinner—a facility that members enjoy to this day.

In February the first production 1½ Strutters were being tested at Brooklands by Harry Hawker and in the Spring he occasionally took a Sopwith aircraft at the week-end to demonstrate at Hendon. On April 12th he took one of the new single-seat biplanes up to 7,200 metres (24,408 ft.) constituting a new British height record.

On May 28th, 1916 the first triplane passed inspection and its first trial flight was made two days later. Sopwith anxiously watching the latest and most unusual of products from his factory was re-assured within minutes. Hawker, liking the feel of the machine, looped three times in succession. While the Triplane showed great promise, repeat orders followed from the Admiralty for the two new biplanes in which the War Office now shared an interest.

Inter-Service rivalry in obtaining aeronautical equipment had led to the creation of an Advisory Air Board, inaugurated that very month. The Army, casting covetous eyes at the Navy's new Sopwith aeroplanes, and the hard-pressed Royal Flying Corps, facing the scourge of the Fokker with its interrupter gear permitting forward fire through the propeller arc, had their need represented by their in-

Halfway to a Pup, the Sopwith SL.T.B.P. built for Harry Hawker. It was said that the plans of this aircraft were chalked out on the floor of the Sopwith experimental shop.

creasing casualty figures. The Sopwith 1½ Strutter's great asset was a fixed Vickers gun firing-forward through the propeller arc by means of an interrupter gun gear, originally a Vickers Challenger system, but later the Scarff-Diborsky gear. Warrant Officer F. W. Scarff had also designed a universal gun ring which was used in the observer's position to mount a Lewis gun in the 1½ Strutter. Thus armed, and with the facilities for bringing these arms to bear, the Sopwith 1½ Strutter was a formidable and important fighting aircraft. Its production was pressed by Navy and Army alike.

The Sopwith factory, then expanding and already flooded with naval orders for 1½ Strutters, could not take on orders for the Army who, making vast preparations for a giant offensive—the Battle of the Somme—required urgent deliveries. Representation from the Admiralty's Air Department and the War Office's Directorate of Military Aeronautics mutually agreed with Tom Sopwith the large-scale sub-contracting of the 1½ Strutter, with a design royalty that would be paid to the parent firm for every Sopwith built in other factories.

First of the 1½ Strutter contractors, and one that was to remain on Sopwith designs for the rest of the war, was Ruston Proctor & Co. Ltd. of Lincoln, famous pre-war as manufacturers of agricultural machinery. While tailing off production of B.E.2es, they commenced constructing 1½ Strutters and delivered their first on July 11th; with successive orders they remained in production for over a year.

Orders for 100 were placed with Fairey Aviation in July but these did not materialise very quickly as it was their firm's first production order and they lacked experience in construction. Hooper & Company, motor coachbuilders of Chelsea were similarly co-opted and given an order for fifty, but they too lacked experience in airframe construction. Vickers, awarded an initial order for a hundred, allotted part of their Crayford works, where employees already had airframe building experience and they became the next contractor, to produce the new Sopwith.

The Sopwith Company were responsible for providing drawings and helping with production experience. An airframe was delivered to each contractor as a pattern. Sopwith drawing office staff was increased but this time it was not a case of advertising for personnel—which under the Defence of the Realm Acts was then illegal. Requirements were notified to Labour Offices and personnel were directed to 'work of national importance'. Sopwith now had little difficulty in obtaining personnel as the company was accorded priority, but the 'dilution of labour' was causing some concern. Women and youths did useful work but were no substitute for the additional skilled men needed. It fell upon the old hands to work long hours, lengthen production hours and eventually shift work was started.

Exhorted by the War Office to effect early deliveries, the contractors, starting a new type from scratch, were at a disadvantage; the parent factory already in production,

A present for an Ally. One of the several Sopwith Triplanes sent to the French for evaluation and which were eventually returned. The French were interested in the Triplane by virtue of its Clerget engine and its good rate of climb, but were distrustful of a triplane form.

The 1½ Strutter A1914 of an early batch built for the Navy, transferred to the Army—and captured by the Germans who tested it at Adlershof together with a captured Sopwith Triplane.

continued to deliver to the Navy. Pressure brought to bear at Cabinet level resulted in the Royal Naval Air Service being forced to hand over some of their Sopwith aircraft to Royal Flying Corps squadrons. On receipt by the R.F.C. at Farnborough, the ex-R.N.A.S. machines were renumbered which later caused difficulty in assessing production figures!

Immersed in all the planning for large-scale production of 1½ Strutters, the firm were given the added task of producing their single-seat scout in quantity for the Navy and to help further sub-contractors to meet Army orders. The Standard Motor Company already in the swing of aircraft production with B.E.12s and the Whitehead Aircraft Company, producing Maurice Farman trainers round the corner at Richmond, were both brought into the production of Sopwiths. On top of all this, Triplane production was mooted and the French wanted to put the 1½ Strutter into full-scale production—its engine was, after all, a French product.

Field Marshal Haig, Commander-in-Chief of the British Expeditionary Force, made successive estimates for the requirements of fighter squadrons as the R.F.C. battled to obtain air superiority. To meet his estimates, and keep up the supply of aircraft in the face of the fearful rate of attrition, even more contractors were sought for 1½ Strutters. The Air Board, casting their net wider, trawled Morgan & Company of Leighton Buzzard and the Wells Aviation Company of Chelsea into the Sopwith-building network.

The Sopwith 'Zoo' had not yet opened, aircraft were known by their official designations and fighters were then known as scouts. To avoid confusion it was decreed by the military in July that the Sopwith Two-seater (1½ Strutter) would be known only by that name to avoid confusion with the 80 h.p. Scout (Pup). The position on October 1st 1916:

Official Designation	Engine	Remarks
Sopwith Type 9400S	110 h.p. Clerget	33 in use, 79 on order
Sopwith Type 9400L	110 h.p. Clerget	8 in use, 6 on order
Sopwith 1½ Strutter	110 h.p. Clerget	Bomber version; 19 in use, 153 on order
Sopwith 1½ Strutter	130 h.p. Clerget	Project similar to 9400S
Sopwith 1½ Strutter	110 h.p. Clerget	Large-scale orders placed for R.F.C.
Sopwith 1½ Strutter	140 h.p. Smith	Project (later abandoned)
Sopwith Biplane	80 h.p. Le Rhone	Experimental models and production orders for R.F.C.
Sopwith Biplane	80 h.p. Clerget	
Sopwith Triplane	110 h.p. Clerget	1 flying, 115 on order
Sopwith Triplane	130 h.p. Clerget	Experimental
Sopwith Triplane	150 h.p. Hispano	Project fighter
Sopwith Triplane	200 h.p. Hispano	Project fighter
Sopwith Triplane	250 h.p. Rolls Royce	Project; 3-seat reconnaissance.

(N.B. The seaplanes are dealt with in the following chapter).

Orders for spares for the earlier types still in commission, arranging for sub-contract work and progressing the experimental models meant a vast amount of work in the drawing office and in the administration of the Company. From managing director of a small local company, T. O. M. Sopwith in the year 1916 became controller of a very large industrial complex spanning the countryside.

Another present for an Ally. This Triplane was despatched to the White City naval store on May 4th 1917 for packing for shipment to Russia. It is shown after arrival in Russia and before the Winter of 1917/18 when it was fitted with a ski type undercarriage.

CHAPTER FIFTEEN

Sopwiths at Sea—1916

Baby experimentation. A Sopwith Baby fitted with Linton Hope floats is seen at Turk's boatyard. Major Linton Hope designed these floats to reduce the Baby's take-off run, but they were not adopted in service.

While orders for Sopwith aeroplanes reached prodigious numbers during 1916, Sopwith single-seat seaplanes remained in production in the works, and there was also a constant demand for spares to maintain those in service.

In the Spring of 1916 the new improved version of the Schneider, the Sopwith Baby floatplane, made its operational debut. The carrier H.M.S. *Vindex*, with three Short and two Sopwith Baby floatplanes, escorted by units of the Harwich Force, set out to bomb airship sheds reported on the Island of Sylt. On March 24th the floatplanes were hoisted out near the Horns Reef. Unfortunately the weather was bad with frequent snow squalls and only two machines returned, one Short and one of the two Sopwiths.

Flt. Lt. Openshaw, in the returning Baby, reported that he had failed to find any Zeppelin sheds at Hoyer and so proceeded inland to Tondern where he had sighted three large sheds; diving from 3,000 feet to attack he passed over these at 1,000 feet. However, his bombs failed to release, due probably to the oil of the release gear freezing.

Flt. Lt. J. F. Hay in the missing Baby came down with engine trouble just off the enemy coast, near Hoyer. For fifteen minutes he had struggled vainly to re-start the engine, when he was seen by the crew of one of the Shorts, which landed alongside. So close to the shore was he, that Chief Petty Officer R. Mullins, observer in the Short, climbed down, jumped in the water and actually waded out to Hay, inviting him to ride back in the Short. By this time, a group of German civilians watching from the beach, had been joined by soldiers; Hay quickly followed Mullins into the water and clambering onto the wings of the Short held on to an inboard strut as Flt. Lt. G. H. Reid took-off.

The Germans, completely unaware of the attack, took the airmen to be compatriots and waved to the departing Short; politely, the others waved back. But they were out of luck. Off Sylt the Short's engine cut and a landing was made on the water. Mullins tried repairs on the spot, but a broken magneto was beyond his resources. The engine did fire spasmodically and Reid headed the seaplane out to sea, taxying on the water at about four knots. Then the floats grounded on a sandbank and they spent some time in wading and pushing the seaplane clear. When a small German sailing boat was seen, the remaining fuel was used in an attempt to chase the vessel and commandeer it at the point of their machine-gun. However, the sea grew rougher and a motor-boat with German soldiers appeared ahead, while two enemy seaplanes landed behind—and so they were compelled to surrender.

Hay's Baby, left off Hoyer, was salved by the Germans and went on tour with an exhibition of war trophies. Like all the Sopwith aircraft on delivery, the fin was marked with the Company name and address, and no doubt Kingston was ringed on the target maps of the German naval Zeppelin Service.

That March the paddle steamers *Killingholme* and *Brocklesby*, equipped to carry two or three Schneiders, were put into commission to supplement *Kingfisher* in waiting out to sea for Zeppelins. They met with little success and *Killingholme*, only a few weeks after commissioning, was torpedoed and badly damaged.

The Harwich Force set out again on May 30th. Their objective was the destruction of the Zeppelin sheds at Tondern which, as related, had been located by Openshaw. In the early hours of the 4th, off Sylt, eleven floatplanes were hoisted out from *Vindex* and *Engadine*. Openshaw again successfully reached Tondern in a Baby and this time his bombs, 2 x 65 lb, were dropped, but the morning ground mist obscured his view of the result. Meanwhile, the other floatplanes met with disaster. Wallowing in the wake of a destroyer that steamed past, one was overturned and the propellers of four others, smacking into rising water, broke up. Three others drenched with spray had their engines fail before take-off. One other Baby that did get off the water, struck the mast of H.M.S. *Goshawk* and both machine and the pilot, Flt. Lt. Walmesley, were lost.

The effect of this further failure was tempered by the success of experiments before the month was out. On May 29th five Baby floatplanes, using wheeled dollies, took off from the newly modified H.M.S. *Campania* steaming into wind at some twenty knots. Returning to her anchorage at Scapa Flow in the evening of the 30th, the eve of the Battle of Jutland, four Sopwith seaplanes that had flown out from the shore base were taken aboard, making her complement

3 Short Seaplanes, 4 Schneiders and 3 Babies. But when the Grand Fleet sailed, *Campania*, already alerted with steam raised, did not receive the vital signal to sail. In the early hours of the morning, over two hours behind Sir John Jellicoe's flagship, the *Campania* trailed unable to let her aircraft influence the battle by scouting ahead. The *Engadine*, with Beatty's Battle Cruisers, flew one flight by a Short.

During that Summer, Schneider and Baby floatplanes were withdrawn from carriers with only hoisting-out gear and maintained only on carriers with a flying-off deck. *Vindex* gave up its four Schneiders late in June to R.N.A.S. Station Dunkirk, to maintain an anti-Zepp standby patrol, supplementing the ring of defences formed by the coastal stations of Killingholme, Yarmouth, Felixstowe and Eastchurch which each had a quota of Sopwith single-seat floatplanes for standby during raids.

One of the Schneiders saw action in July. This was flown by Flight Sub-Lieutenant Graham from the Dunkirk Seaplane Station on the 21st, when a German two-seat biplane was encountered below at 11,000 feet. Diving towards the enemy machine, the aircraft approached head-on both firing and passing within twenty feet of each other. The German machine then made for the coast with the Baby following, but lack of petrol forced Graham to descend.

A Baby struck blows in Defence of the Realm on August 19th, when a Zeppelin was reported to be 35 miles east of Spurn Head at the mouth of the Humber. From Killingholme Naval Air Station Flight Sub-Lieutenant Fox was sent out in chase and he was lucky enough to sight the intruder. Climbing to 11,000 feet he passed over the airship dropping four 16 lb. bombs and two boxes of Rankin darts. The Zeppelin turned in a complete circle and the nose dropped as it turned away. Fox then had engine trouble and was forced to land, but a valve rocker arm had evidently been flung from the rotary engine into a float, puncturing it so that when he alighted on the water the port float sank and Fox was compelled to sit out on the starboard wing to keep the aircraft upright until help arrived.

Again on the night of September 23rd, when several Zeppelins roamed over England and the L.33 was brought down at Little Wigborough, an old Yarmouth-based Schneider made contact with a Zeppelin 30 miles out to sea off Lowestoft. The pilot managed to fire a complete tray of Lewis gun ammunition before losing contact. The previous day when an Albatros biplane had been sighted off Dover, two Babies, 8146 and 8160, had risen to intercept.

Next month another Baby attempted dog-fighting. Flight Sub-Lieutenant Fisher of Dunkirk Naval Air Station was patrolling in 8145 on October 23rd. When at 11,500 feet and some four miles off Ostend an L.V.G. 2-seater was sighted near the coast flying a mere 500 feet lower. As Fisher dived to attack the rear gunner opened fire, but he closed range and opened fire with his Lewis which jammed after eight rounds due to faulty ammunition, and the frustrated pilot was forced to abandon the attack.

The position of Sopwith Seaplanes in October 1916, at home and overseas was as follows:

Type	Engine	Numbers
Schneider	100 h.p. Monosoupape	33 in commission orders complete
Baby	100 h.p. Monosoupape	
Baby	110 h.p. Clerget	39 in commission, 31 still on order
Baby	130 h.p. Clerget	Projected version

Over a hundred of these floatplanes had been lost in service in various ways, some in transit through merchant ships being torpedoed. Their use for coastal anti-submarine patrol work was, however, beginning to be realised.

When the Admiralty approached the Sopwith Company in late 1916 about increasing Baby Seaplane production to meet new requirements at coastal stations at home and abroad, T. O. M. Sopwith, already fully occupied in arranging Pup, 1½ Strutter and Triplane construction within his own factory, decided to abandon all seaplane production in his own works. Both the construction and development with the projected 130 h.p. Clerget version of the Sopwith Baby was passed to the Blackburn Aeroplane and Motor Company. (This Company, in later years the Blackburn and General Aircraft Ltd, became part of Hawker Siddeley Aviation in 1963 of which Sir Thomas Sopwith is President). Baby seaplane production thereby passed from a disused skating rink at Kingston to a disused skating rink at Leeds and Sopwith Aviation, hitherto constructors of aircraft, including floatplanes, flying boats and aeroplanes, became from then onwards, constructors only of aeroplanes. From this time Sopwith became the specialists in fighting aircraft.

Baby floatplanes under construction at the Blackburn Olympia Works at Leeds, also an ex-skating rink. At the rear B.E.2Cs are being built for the R.N.A.S. and at the far right an earlier Baby appears to be in the works for overhaul or repairs.

CHAPTER SIXTEEN

Schneider Triumph—1916

The Airship Shed, Mudros, with the SS airship to which a Schneider floatplane provided escort. The aircraft in this picture is a B.E.2C. The girder work shown supported canvas to act as a windscreen when berthing the airship.

While the Western Front received its injection of new Sopwiths in 1916, the seaplane carriers in the Eastern Mediterranean, engaged in harrying the Turks, had to make do with aircraft of 1915 vintage already being discarded elsewhere as outclassed or unsuitable. But the amazing thing is that, together with Short Seaplanes, the ageing Sopwith Schneiders, in spite of the corrosive effects of salt water to which they had been exposed for over a year, in spite of the exposure of their wooden structures sweltering in the warping effects of a sub-tropical heat and with fabric constantly bathed in the harmful actinic rays of the sun, they continued to fly operationally in a variety of roles. Not least was their effect on the morale of dissident tribesmen whom the Allies were trying to coerce into rising against the Turks. Maligned at home for their abortive operations in the North Sea, in the Mediterranean in 1916 they were now a triumph for the Royal Naval Air Service and a tribute to their manufacturers.

In the Red Sea and Mediterranean, Schneiders were carried by the *Anne* (1 Short 184 + 1 Sopwith Schneider), *Ben-my-Chree* (3 Short 184s + 2 Sopwith Schneiders), *Empress* (1 Short 184 + 1 Sopwith Schneider) and *Raven II* (2 Short 184s + 2 Sopwith Schneiders). They were maids of all work, escorts for the two-seat Short floatplanes on reconnaissance or spotting for naval guns, reconnoitring in their own right and even going on bombing attacks such as on May 10th when two, accompanied by a Short, bombed targets in the Gulf of Mendalya. Pilots claimed hits on Budrum Fort, a house reported to be used by German officers, and on sheds at Dalian and Kuluk. An unsuccessful attack on the Kuteli Bridge resulted in a second visit by the aircraft three days later.

On May 25th a similar force, a Short and two Schneiders, flew from south-east of Samos up the valley of the Bijuk Mendere river to Sokia where the railway line and station were bombed. There followed a series of attacks in the course of which the station at Smyrna was damaged.

During June, the Schneiders of the *Ben-my-Chree* experienced a period of intensive flying operations following a reconnaissance that revealed enemy encampments in the Lahej Delta on the 7th. Next day, while Commander C. R. Samson (whom Tom Sopwith had met at a special meeting of the Royal Aero Club that April) and Lieutenant Wedgwood Benn (father of the Rt. Hon. Anthony Wedgwood Benn, M.P. (formerly Minister of Technology) flew in a Short, Flt. Lts. England and Bankes-Price in Schneiders bombed Waht with a 65lb bomb and 500 flechettes, and four 16lb and four incendiary bombs respectively. The following day Bankes-Price had engine trouble and only by jettisoning his bombs did he manage to nurse the machine back over the sea to a landing for a pick-up. On successive days the attacks continued until the 16th when Jeddah surrendered. It is difficult to assess precisely what the seaplanes achieved, with their small bombs, but undoubtedly they had considerable effect on the morale of the tribesmen.

Schneiders, which originally had never been intended for anything more than scouting were now bearing the brunt of aerial operations in the Eastern Mediterranean. In July their operations were intensified. With the Sopwith factory having concentrated mainly on the production of Baby floatplanes before the new aeroplanes had reached production status, there were sufficient numbers in service and store in Britain, for numbers of Schneiders, thereby replaced, to be made available for use by the East Indies and Mediterranean Stations.

Flt. Lt. Brook flying a Schneider from H.M.S. *Anne* on June 1st, carried out a reconnaissance of El Arish and dropped two bombs—one of which fell on the outhouse of an ammunition depot—from 3,000 feet in spite of heavy and fairly accurate anti-aircraft fire. Next day, a Schneider

Royal Marine gunners on the fringe of Mitylene Aerodrome, Thermi, from which 1½ Strutter bombers and fighters operated.

In European waters, the main shipborne aircraft continued to be Sopwith Schneider carried aboard light cruisers of the Harwich Force as shown. However, the wake of the ships proved a hazard and often capsized the floatplanes before they could take off.

accompanying a Short Seaplane failed to return. A search by the Short revealed it a half-mile off Acre, drifting inshore in a sinking condition. The Short landed and took off the pilot, Flt. Sub-Lt. Mann; as it was impossible to salvage the Sopwith it was fired at and sunk. From July 7th, the *Ben-my-Chree*'s Sopwiths were in action off Beirut and during the month, a small military encampment at Tel Kale and lighters in Tarsus Chai were among many targets attacked by these ageing single-seat floatplanes.

The Schneiders were unfortunately easy prey for the new German aeroplanes being drafted to the East. Two Schneiders escorting a Short from the *Ben-my-Chree* were shot down on September 17th. An enemy machine, superior on speed and climb, overtook the Short and flew between it and the escorting Sopwiths. Flt. Lt. Bankes-Price on one of the Sopwiths was first to go down and his Schneider hit the water in a sheet of flame; Flt. Sub-Lt. Nightingale in the other, escaped with his life. His Schneider, although repeatedly hit in the tail by fire from the enemy, did not go out of control. Forced down when his fuel tank was hit, he made a safe landing and was rescued by a monitor, while a trawler took the Schneider in and and brought it back into commission.

A plea was made for the later model Baby floatplanes of which the first had been received but others were not forthcoming for several months. Meanwhile, the Schneiders

Typical of the landplanes operating from Mudros is this Pup with tripod mounting for a Lewis machine gun.

continued to act as fighter escorts and in general they could hold their own against enemy seaplanes, but not against the latest enemy aeroplanes. A new escort task was actually allotted to them in October—that of airship protection.

The blimp S.S.8 (Submarine Scout No. 8) did useful long-range patrol work from Mudros on anti-submarine and mine-spotting duties. While carrying out a patrol at 500 feet of the recently swept channel to Mudros harbour an enemy seaplane attempted to drop bombs on to its envelope from several thousand feet above. A Schneider set out from Mudros in pursuit, but the enemy disappeared into a thick haze. From then onwards it was arranged that a Schneider would escort the airship on patrol, but to reduce wear on the seaplanes, the airship was only to set out when there was good visibility or for an important objective such as searching on receipt of a submarine sighting report. A high value was evidently still placed on these Sopwith seaplanes.

This escort apparently proved successful for the airship was not attacked again in the air, but on November 21st the wily enemy attacked it in its shed. While an L.V.G. biplane approached from the south-west, a Friedrichshafen floatplane came in from the east. One bomb from the aeroplane fell within 50 yards of the southern end of the shed and splinters pierced the blimp's envelope. Flt. Sub-Lt. Brandon raced for a Schneider which mechanics had already got started and set off in pursuit of the seaplane which, having been met by heavy anti-aircraft fire deposited only a single bomb, and that a mile from the shed. Climbing to the Friedrichshafen's height at 8,000 feet, Brandon fought the enemy down to 3,000 feet by which time he had expended all his ammunition. Although he had seen tracer ammunition enter the fuselage of the enemy, it flew serenely on. Short of ramming, an idea which he contemplated and fortunately rejected, there was nothing he could do and he returned to base. However, the seaplane had evidently been hit in a vital spot, for a naval vessel discovered it abandoned in a bay, east of Tenedos. The crew had escaped ashore and as the water was too shallow to attempt salvage it was destroyed by gunfire to forestall any salvage attempts by the enemy.

Schneiders continued in their multi-roles. On December 10th one from 'A' Flight No. 2 Wing at Thasos set off

A typical late production Schneider with increased fin area and ailerons replacing warping wing control to bring it up to Baby standard. Four, Nos. 3707, 3709, 3765 and 3806, were sent to Canada (less engines) to form the nucleus of a Royal Canadian Naval Air Service.

after a hostile seaplane, and chased it to the enemy seaplane base at Gereviz. Next morning, a Schneider with light bombs, together with a Farman carrying a 100lb bomb, bombed this base, being escorted by Nieuport Scouts.

On December 27th the Schneiders had one of their greatest successes. The Chikaldere bridge carrying the Constantinople-Baghdad Railway was the target for the seaplanes of *Ben-my-Chree* and *Raven II*. Commander Samson leading on the veteran Schneider No. 3770, accompanied by Flt. Lt. A. W. Clemson on No. 3778 and Flt. Lt. Brooke on a newer machine, Baby No. 8188, with Flt. Sub-Lt. Smith and Wedgwood Benn on Short 8080 and two Shorts from *Raven II*, the bridge was bombed from four hundred feet and the bridge guard was scattered with machine-gun fire. In further attacks that day Clemson, using the same Schneider, scored direct hits on the permanent way with both his bombs. Most of the aircraft received hits from ground fire, but all returned without injury. As a result, traffic on this important railway was held up.

While these events have highlighted some of the particular operations carried out by Schneiders, much routine flying of reconnaissance, spotting for guns, anti-submarine patrols and other duties continued. The *Ark Royal* at this time was acting as a depot ship for receiving new machines from England, carrying out repairs of Shorts and Sopwiths from the carriers and conducting tests and training. At the end of the year *Empress* steamed to Mudros on orders received by wireless to exchange her old aircraft for new and took over aircraft from the *Ark Royal*.

By this time Baby floatplanes were coming through to replace the Schneiders but more important, at last a trickle of the finest of the Allies fighting machines on the Western Front were reaching the Aegean Islands. A ship with a cargo of crated 1½ Strutters, of both types, bombers and fighters, had arrived at Mudros. These new Sopwiths were quickly assembled, and put into service. The first offensive use of the new aircraft was on the last day of 1916. A single 1½ Strutter dropped four bombs from 10,200 feet on to the Buk bridge. Its escort was a single Schneider seaplane; epitomising the old and new in Sopwith aircraft types. But even at the end of 1916, the Schneider seaplanes, built early in 1915, still had useful work to perform. Their exploits went unsung, for they served in remote corners of Asia Minor in an area of activity known to the Imperial General Staff as a Minor Theatre. The Sopwith design that had been acclaimed for winning the Schneider Trophy contest at Monaco in 1914, proved its worth three years later and triumphed at the other end of the Mediterranean in conditions obscured by the fog of war.

Sopwith 1½ Strutters operating from one of the Aegean island bases that came under No. 2 Wing of the Royal Naval Air Service.

CHAPTER SEVENTEEN

Naval Sopwiths in France, 1916-1917

'It was not until the spring of 1916 that the first real fighting machine in the British Services was involved. This was the Sopwith one-and-a-half Strutter'—the foregoing is not the biased view of an author enthusing over his subject, but the considered opinion of the writer of a naval technical history. He considered that it was probably the finest fighting machine in the world of its day. However, he qualified these opinions by remarking that until taken in hand by the R.N.A.S. Armament Section it was not suited for the Front.

This may well be true. Any military aircraft is merely the vehicle to enable a weapon to be brought to bear. Sopwiths were the airframe specialists, having to accept such engines as were available in sufficient quantity and the standard machine-guns of the period. That Sopwiths were aware of the essentials in bringing the weapons to bear is evident from the endeavours of Harry Kauper who experi-

Following the 1½ Strutter into active service was a prototype Pup (then known generally as the Sopwith Scout) which went for service trials at R.N.A.S. Furnes in France, in May 1916. Production to a standard agreed between Sopwiths and the Admiralty brought a new official designation—the Type 9901; the type number was conditioned by the serial number of the first production aircraft which, in view of the pressure of work at the Sopwith factory, was placed with Wm. Beardmore & Sons at Dalmuir, Dunbartonshire, Scotland.

Next to be in action was the Triplane—and in a remarkably short time. Passed by the Sopwith experimental department at the Brooklands sheds on May 28th, it was taken into the air by Harry Hawker two days later and it is reported that he looped it three times within three minutes of take-off. In mid-June, delivered and armed, this prototype (N500) arrived at 'A' Squadron R.N.A.S. at

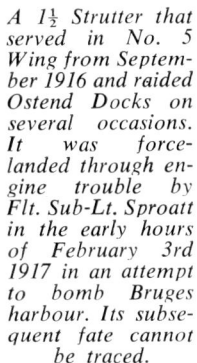

A 1½ Strutter that served in No. 5 Wing from September 1916 and raided Ostend Docks on several occasions. It was force-landed through engine trouble by Flt. Sub-Lt. Sproatt in the early hours of February 3rd 1917 in an attempt to bomb Bruges harbour. Its subsequent fate cannot be traced.

mented and produced a successful gun interrupter gear, permitting fire through the propeller arc—the Sopwith-Kauper gear. One possible shortcoming was in their lack of knowledge of two-seater aircraft fighting tactics but these were probably not then appreciated by the flying Services themselves. Not until the Bristol Fighter came along with its gunner's cockpit within shoulder tapping distance of the pilot was it realised what a closely knit team, exchanging touch signals, could achieve in aerial fighting. Had the Sopwith 1½ Strutter not had the fuel tank spacing out the two crew members, it might have been even more effective. Its testing time was at hand.

The first Sopwith 1½ Strutters reached No. 5 Wing, R.N.A.S. at Coudekerque on the French coast on April 24th and next day an offensive patrol was conducted by Flt. Sub-Lts. R. R. Soar (pilot) and F. Potts (observer) on 9878. The first action was on May 21st when Flt. Lt. B. L. Huskisson and Lt. St. John engaged and drove off three enemy machines; one of these they reported as a new enemy two-seater with the upper surface of the wings painted a chocolate colour.

Furnes and was sent up on an interception within fifteen minutes of landing. When flown by Flt. Sub-Lt. R. S. Dallas on July 1st two large enemy biplanes were encountered at 12,000 feet north of La Panne. Overtaking one of the machines Dallas fired 40 rounds and observed hits, resulting in the enemy machine nose-diving towards the sea. The second machine then attacked the triplane which had a gun stoppage and Dallas returned having participated in the first known action by a British triplane.

Two things plagued the operation of the Sopwiths at this time, engine failure and gun stoppages. It was the latter that thwarted further success in the first dog-fight involving Sopwith aircraft. Two 1½ Strutters on July 9th, on patrol between Westende and Ostend, spiralled down from 14,000 to 10,000 feet to attack a small formation of enemy machines, but after getting off some rounds the pilot's Vickers gun on one Sopwith, and the observer's Lewis gun on the other, jammed.

The Sopwiths tended to get blamed more than the weapons, from the simple fact that stoppages never occurred with such frequency on other aircraft with similar weapons.

The fact was that the Sopwiths were flying at higher altitudes than other aircraft. Height in aerial fighting was of paramount importance, apart from making it more difficult for enemy anti-aircraft gunners, speed in joining combat could be gained by diving and each side sought the advantages of height.

At the increased altitudes, the viscosity of lubricating oils increased with the cold and oil was even known to freeze causing malfunctioning of the guns. Also, in the rarefied atmosphere, sufficient energy was not always imparted to the recoiling portions of the Lewis gun and it did not then complete the reloading cycle. The Lewis ammunition drums at the time held only 47 rounds, necessitating frequent changing. Later, drums of 97 rounds were supplied.

The canvas belts for feeding the forward-firing Vickers on the 1½ Strutter and Pup, and being fitted to the Triplane, absorbed moisture and when this froze it led to stoppages. Ammunition for the Vickers had only to be slightly defective, with a hangfire of a mere 1/250th of a second, for a round to hit the propeller blade—and on average about one in every 20,000 rounds did so!

It seemed as if the fates were conspiring to negate the efforts of the men who at last had efficient fighting machines in their hands. Flt. Lt. T. F. N. Gerrard, on July 28th, was next to experience frustration. Sighting an enemy machine returning to the German Lines over Nieuport, he dived from 14,000 to 10,000 feet. While manoeuvring behind the enemy, trying to bring his guns to bear, a terrific blast, presumed to be anti-aircraft fire, sent his machine spinning out of control. Only with the greatest difficulty did Gerrard get it down in one piece, to find fabric stripped from the wing. What was significant, was that the structure, in the face of such concussion, had remained firm.

That same day the attention of Flight Sub-Lieutenant Hervey, in a 1½ Strutter, was drawn to heavy Allied anti-aircraft fire, some six miles to the east of Calais at about 10,000 feet, bursting around an enemy machine. Approaching the area, the observer, Lt. St. John, fired a recognition signal to advise the gunners that they were attacking and the ground fire ceased. The enemy made for his lines and a running fight ensued. For once the naval pilots found they had a two-seater of superior speed to their opponent and for the pilot to keep his forward-firing Vickers gun in action while remaining behind the enemy, he had to stall the Sopwith. As it dipped, St. John was able to bring his Lewis gun to bear firing over the top plane. The enemy observer was probably hit as he ceased firing back, but the machine escaped as the chase had to be abandoned through lack of ammunition.

The first fifty 1½ Strutters had left the Sopwith works as fighting machines; the next order, for 100, were a mixed batch, with thirty-one of them ordered to be completed as bombers. This meant that the rear cockpit was eliminated and provision was made for the carriage of 4 × 65 lb bombs stored internally. Bomb doors were provided, opened by the weight of the falling bombs and then pulled shut by rubber cords. Eleven machines were also specially fitted out for the carriage of annalite bombs, that needed special handling.

Both types of 1½ Strutters were delivered to No. 5 Wing at Dunkirk where plans were being formulated for a bombing offensive along the German-held Belgian coast. The first determined attack in the new offensive was on August 2nd, before sufficient bomber versions of the 1½ Strutters were available, but five of the 2-seat 1½ Strutters

An early production 1½ Strutter. First of the Sopwith aircraft to employ a gun-gear to permit firing through the propeller arc, the 1½ Strutter had a Scarff - Dibovsky gear in general, but a few had the Vickers - Challenger gear.

were provided to escort ten Caudron G4s and a single Farman attacking an airfield and an ammunition dump.

New techniques were evolved for the 1½ Strutters in formation flying and attack. Their first target, the airfield at St. Denis Westrem, was approached in 'V' formation, with the aircraft reforming into line astern to drop their bombs over the sheds and hangars in succession and then reforming into 'V'. These manoeuvres were controlled by a pilot in another Sopwith, acting independently of the formation, who gave the necessary signals. Presumably this Sopwith pilot was the first Master Bomber in air history. Another target, an ammunition dump near Ghent, was attacked by Sopwiths dropping 31 bombs.

As in 1914, so in 1916, the Zeppelin was still something of a bogey to the Navy. True, the Admiralty had been relieved of its responsibility for the aerial defence of the United Kingdom, which had passed to the War Office on February 16th, 1916, but in the Zeppelin the German Navy possessed a fleet reconnaissance medium far superior to the Royal Navy. The German naval airship bases were well out of range in Germany, but the German Army's airship sheds in Belgium were thought to be used occasionally by raiding naval Zeppelins as an advanced base. They

An early production Pup, built in the Sopwith Works at Kingston for the Royal Naval Air Service, shown at Brooklands for testing before delivery to the Service during the winter of 1916-1917.

were placed on the target lists of No. 5 Wing, R.N.A.S.

On the morning of August 9th Flt. Sub-Lts. R. H. Collet and D. E. Harkness each took a Sopwith bomber to attack the airship sheds at Berchem St. Agathe and Evere, near Brussels. Collet, reaching the first target, saw the eastern door of the shed open, revealing nothing inside, so he flew on to Evere where he left the shed with dense smoke issuing. Harkness, seeing the Evere shed smoking, went back to Berchem St. Agathe to deposit the remainder of his bombs on the empty shed.

After the Dunkirk-based Sopwiths had set an ammunition dump at Lichtervelde afire on the 18th, another attack was made on airship sheds, this time at Cognolee, near Namur on the 25th, over a hundred miles distant. This time three Sopwith single-seat bombers set out and Collet, participating again, was the only one to reach the target and return. Flt. Lt. Wood, losing touch with the other two in a heavy bank of cloud was unable to pick up any landmarks and after searching vainly for eighty minutes, wisely made for the coast and followed it back to Dunkirk.

Both Flt. Sub-Lt. C. W. Jamieson and Collet came under heavy anti-aircraft fire over their target and were forced to evade in cloud immediately after dropping their bombs, thereby being denied the opportunity of assessing their efforts. As it was, no direct hits were made and Jamieson in 9396, a two-seater of No. 5 (Naval) Squadron, used as a single-seater, came down in Holland.

The Dutch, bounded landwards by Germany and German-held Belgian territory, maintained a strict armed neutrality. Unable to buy sufficient arms and aircraft from the belligerents who needed all they could produce for themselves, their own air arm was made up largely of aircraft which they had interned. Already Sopwith Baby Seaplane 8140, down in Dutch territorial waters, had been taken into service by the Dutch Navy; now the Dutch had the first of several Sopwith aircraft they were to acquire.

Shipbuilding yards at Hoboken, Antwerp, were next to be attacked by No. 5 Wing's Sopwiths. This was on September 2nd when four set out at 3.10 p.m. under cloudy skies. One aircraft was forced to return early, while the other three continued, steering a compass course above the clouds. They descended to bomb through a hailstorm, catching the yard completely unawares. Next day, the airfield at Ghistelles was their target and on the 9th three Sopwiths revisited the Lichtervelde dump.

Gradually the Caudrons used by the Navy were being replaced by Sopwiths, the two types, bombers and fighters, operating together. On September 22nd three 1½ Strutter

Since the number of this aircraft, N5180, has been painted on surviving and replica Sopwith aircraft, it is confirmed that this is the original, first production naval Pup, Admiralty Type 9901.

In November 1916 the Sopwith Company patented a padded windscreen surround to prevent injury " to the occupants of an aeroplane or the like, by violent contact with the windscreen". This did much to minimise injury in such mishaps as this which, due to engine failings, were not infrequent. The aircraft is a Pup.

bombers, detailed to bomb St. Denis Westrem airfield, were escorted by two 1½ Strutter fighters. An enemy machine that tried to interfere over Ostende was engaged by one of the Sopwith fighters and driven off; in this case the pilot got about a hundred rounds away before his Vickers gun jammed. That same day Flt. Sub-Lts. N. Keeble and E. B. C. Betts carried out a photo reconnaissance of Ostend and district, including the Tirpitz battery. Although flying at 13,000 feet, their aircraft was twice hit by anti-aircraft fire but returned safely.

Further raids on enemy airfields continued, 36 × 65 lb bombs and 40 × 16 lb bombs were dropped by six 1½ Strutters on Ghistelles and Hanelzaeme aerodromes on the 23rd. During that day Flt. Sub-Lt. Thom, while escorting a photo reconnaissance aircraft, attacked an enemy seaplane with his Vickers gun, getting off some thirty rounds, and while banking away Gunlayer Symonds in the rear cockpit put in a further sixty rounds from his Lewis gun. The seaplane dived away and was presumed shot down.

While the tempo of operations by 1½ Strutters increased, the Pup came into the arena. Flt. Sub-Lt. S. J. Goble on September 24th encountered two L.V.Gs at 12,500 feet in the vicinity of Ghistelles. Approaching unobserved on a Pup to within fifty yards of one of them, he opened fire with a short burst of about ten rounds and saw the enemy observer jump up to man his gun. Side-slipping under and across the enemy machine, he fired some thirty rounds. The L.V.G. side-slipped, caught fire and spiralled down.

Next day when Flt. Sub-Lt. S. V. H. Trapp engaged a German two-seater in his Pup, his Vickers jammed. Turning away and trying to clear the gun, he was thrown into a dive by a close anti-aircraft first burst from which shrapnel hit the engine causing it to vibrate badly, forcing him down at Malo.

Flt. Lt. R. S. Dallas, D.S.C., still flying the prototype Sopwith Triplane and giving the enemy the impression that at least a squadron of them was in service, observed on September 30th a German two-seater with its crew apparently bent on artillery spotting. Diving 1,500 feet he fired 60 rounds sending the enemy machine down out of control. Continuing his patrol, he saw an enemy seaplane stalking a Belgian machine and drove this away.

The R.N.A.S. at this stage had the task of forming squadrons from personnel at Dunkirk into fighter squadrons to operate under R.F.C. command; thus was born the famous 'Naval Eight'—No. 8 (Naval) Squadron which formed on October 25th and was to fly five different types of Sopwith fighters.

Meanwhile, Dallas continued to hunt the enemy in the lone prototype Triplane N500. On November 6th, when at 14,000 feet over Westende, Dallas encountered three enemy two-seaters circling together as if practising tactics. Approaching unobserved by banking round with them, Dallas closed behind one and fired. Then, as so often happened, his Vickers gun jammed after 40 rounds and he was beset upon by all three. Using the superior climb of the Triplane to escape, he sought the safety of cloud, but on emerging

The Sopwith Baby with Linton Hope floats shown on flotation and taxying tests on the Thames in 1916. In the normal form, to drawings approved by Herbert Smith on July 13th 1915, the Baby's contract price was £1980 complete.

he nearly collided with another enemy machine. Not wishing for further involvement with a jammed gun, he dived and made back to Dunkirk.

After initial delays, production Triplanes started arriving at Dunkirk in early 1917 and No. 8 (Naval) Squadron, whose personnel were withdrawn to re-equip with Triplanes, together with No. 1 (Naval) Squadron. Dallas, now using a production Triplane, N5436, shot down an Aviatik 2-seater with it on February 1st.

While the Pup had been showing its paces with the R.F.C., and the Triplane showed equal promise with the R.N.A.S., the 1½ Strutter continued to be the workhorse of the R.N.A.S. along the French and enemy-held Belgian coast. On February 1st Flt. Lt. H. G. Holden with Flt. Sub-Lt. Betts, R.N.V.R. in 9417 took important photographs revealing ammunition stores along the Bruges canal, and at Bruges itself, the harbour was shown to be congested with shipping—eight destroyers, over ten torpedo boats and three U-boats. Prints also showed the harbour to be covered with floating blocks of ice.

The weather next day was unsuitable for flying and the 3rd was not much better, but the R.N.A.S. decided to strike at the shipping. Twelve aircraft of Nos. 4 and 5 Wings were ordered to strike but only two of the seven 1½ Strutters participating reached their target. Understandably, as there was 22½° degrees of frost which adversely affected their engines. The attack was repeated on the afternoon of the 7th using Sopwith 1½ Strutters only, five bombers (9394, 9395, 9397, N5081 and N5114) escorted by two fighters (N5093 and N5102). At the same time a photographic 1½ Strutter (N5172) took shots revealing for the first time that the Darse at Bruges was being used by German warships. Attacks continued at intervals throughout the month.

Sopwith 1½ Strutters were also used as artillery observation aircraft at times, and one such occasion was on February 16th when Flt. Sub-Lt. Wyatt with Lt. Greenwood, using No. 9897 fitted up with wireless telegraphy apparatus, attempted spotting for naval siege guns.

Holden and Betts continued to use the 1½ Strutter for photo reconnaissance with excellent results. Setting out on 9419 at 11.30 a.m. on February 26th they exposed 25 plates along the Zeebrugge-Bruges Canal, showing new defence emplacements, and the same flight revealed a new battery between Breedene and Clemskerke and new seaplane sheds at Ostend. Off Wenduyne their aircraft was attacked by three machines and Betts got through 1½ trays of Lewis gun ammunition in warding them off. Then heavy and accurate anti-aircraft fire was experienced making several hits, and on landing they found a longeron completely broken through by shrapnel.

A Standard production Sopwith Triplane (believed N5364) at Farnborough. By an arrangement between the Services, the R.N.A.S. took over Army contracts for Triplanes, while the R.F.C. had Spads originally ordered for the Navy. In all, under 150 production Triplanes were built and they equipped Nos. 1, 8, 9 and 10 (Naval) Squadrons. Only one, N5430, was delivered to the R.F.C.

Sopwith seaplanes were still playing their part along the coast. The R.N.A.S. seaplane station at Dunkirk was now operating some of the new Blackburn-built machines, for example when on February 28th a German U-boat was reported on Breedt Bank and a Short Seaplane was sent to investigate, Sopwith Baby N1017 provided the escort.

Apart from the actions highlighted a great amount of unspectacular routine patrolling was now being carried out by the Sopwiths. Whereas a year previous not a Sopwith aircraft had been with the R.N.A.S. in France or on the Western Front, now three different types of Sopwith aeroplane and one type of Sopwith seaplane were in active service, performing a variety of duties more efficiently than any other type then available. The Army, casting covetous eyes at the Navy operating the cream of the equipment alongside their own, made further pleas for reinforcement and the R.N.A.S. responded by detaching squadrons for temporary operations with the R.F.C.; squadrons equipped with the cream—Sopwiths.

CHAPTER EIGHTEEN

Army & Navy Sopwiths on the Western Front

Navy 1½ Strutters handed over to the Army for operations on the Western Front. The second aircraft bears the Royal Naval Air Service number 9681 which was re-numbered A891 after transfer to the Army.

Major-General Hugh Trenchard, General Officer Commanding the R.F.C. in the Field, had been pressing for the new Sopwith 1½ Strutters to arrive by July 1st, the opening day of the Battle of the Somme. As it was, in spite of all the energy that had been expended at the works, not one complete squadron was deployed in time. In fact, on that fateful day, only four Sopwith aircraft were actually available for operations.

It was normal practice for R.F.C. squadrons to form up in England and go to France as a complete unit. But there were insufficient Sopwith 1½ Strutters available from War Office contracts to equip a squadron, and Sopwith factory-built machines had to be switched from R.N.A.S. Depots. Regarding the new Sopwiths of much importance, the Royal Flying Corps considered that a few were better than none and No. 70 Squadron, instead of waiting until it was up to establishment, worked up a flight at a time as aircraft became available. 'A' Flight, under the squadron commander, Major G. A. K. Lawrence, D.S.O., reached France on May 24th and 'B' Flight on June 29th, but at their allotted war station at Fienvillers, behind the Western Front, only four aircraft were available to participate on the opening day of the battle; three days later more arrived.

In such small numbers, the new Sopwiths could not exert much influence. Like the tanks which made their operational début at a later stage of this battle, they were used in 'penny packets' instead of achieving surprise and success by weight of numbers. No. 70 Squadron was part of the Ninth (Headquarters) Wing under Lt.-Col. H. C. T. Dowding (later Lord Dowding, who was C.-in-C. Fighter Command during the Battle of Britain in 1940). Dowding's terms of reference as commander of G.H.Q's own Wing was 'Strategical reconnaissance for G.H.Q., offensive action against the enemy air service and the long-distance bombing of enemy communications'. This was a formidable task for his existing equipment, No. 60 Squadron at Vert Galand with Moranes, which could not give adequate protection to the two other squadrons at Fienvillers—No. 21 with 14 antiquated R.E. 7s and 5 B.E. 2c/es and No. 27 with 17 Martinsyde Elephants, which had failed as scouts but were proving useful as light bombers.

No. 70 with its 1½ Strutters offered Dowding new possibilities. They would be ideally suited for the long-distance reconnaissance and bombing; moreover, with their armament, they should be capable of looking after themselves. On the opening day of the Battle the four aircraft of 'A' Flight were ordered to reconnoitre the Cambrai area at 06.00 hrs. Eighty minutes later they were safely back, their crews reporting no opposition and no unusual movement on the ground. At least some Sopwiths had played some part in this momentous battle.

The ability of the Sopwiths at this stage to look after themselves was such that Dowding detailed them frequently for escort duties to look after other types of aircraft. They were holding their own well and although A384 was soon lost it was to ground fire, not opposition in the air. They worked closely with No. 27 Squadron; first reconnoitring and reporting enemy activity and then escorting No. 27's Martinsydes to bomb the targets.

The Sopwiths so far used, diverted from Admiralty contracts on production, did not have the new Scarff No. 2 gun ring. When the third flight ('C') of No. 70 Squadron arrived at the Front on July 30th, they were ex-R.N.A.S.

A minor mishap to A6150, the first of the Sopwith Pups built by Whitehead Aircraft.

service machines with the new Scarff ring incorporated as a Service modification. Trenchard, so impressed with the squadron's glowing reports of the new ring, ordered it for all future R.F.C. two-seaters.

On August 6th a patrol of No. 70 Squadron led by Captain W. D. S. Sanday completely thwarted an enemy raid. Meeting a formation of ten enemy bombers they fought them back to their own airfield where they landed back with their bombs.

Such was the high value placed on the new Sopwiths that next day one is believed to have been crashed deliberately! This paradox can be explained by the following circumstances: forced down at Walincourt, south of Cambrai, 2nd Lieutenant C. W. Blain in A380, realising that the precious new Sopwith was about to fall into enemy hands if he landed safely, veered towards some trees—so other members of 'B' Flight circling anxiously above reported—and deliberately smashed his machine.

The Sopwith 1½ Strutter was not one of the aircraft types on which the Victoria Cross was won, but many deeds worthy of the award were no doubt performed. On July 24th 2nd Lts. A. M. Vaucour and A. J. Bott in a No. 70 Squadron 1½ Strutter had their fuselage canvas set alight from anti-aircraft fire. Bott fought the fire by beating the flames with his hands, and by tearing strips of burning canvas off, he stopped it from spreading. Then the aircraft was attacked by enemy aeroplanes. During the combat an enemy bullet severed a petrol pipe and Vaucour was forced to land. Gliding over No Man's Land he alighted in the British lines south of Carnoy from where the squadron was 'phoned and Bott was taken to hospital. 1st Class Air Mechanic H. P. Warminger was sent out from Fienvillers by truck with tool-kit and petrol cans and had the aircraft serviceable by early morning. By then the truck had departed, so Warminger climbed in the rear cockpit for a lift back to their airfield. During the flight the aircraft was set upon by three enemy scouts; the mechanic was wounded and again the aircraft was put out of action. Again Vaucour landed his aircraft safely, but Warminger died of his wounds.

The ability to make such forced landings successfully was no doubt aided by the machine's airbrakes, an innovation not exclusive to, but certainly unusual, in aircraft of the period. Situated by the port and starboard roots of the lower wings were panels, actuated by a lever in the pilot's cockpit, that could be pivoted upwards. Thus, although flaps in the general sense of the word, they were not flaps by aerodynamic inference for their function was to provide resistance, not lift.

There were times when the Sopwiths did not have things all their own way. An offensive patrol of seven on September 15th, led by Captain G. L. Cruikshank, was split up by enemy fighters. Cruikshank, after a brilliant duel, fell into Havrincourt Woods a victim of no less an ace than Oswald Boelcke. One Sopwith, with the engine out of action, landed in enemy territory with a mortally wounded observer; two others that returned had dying observers. The German single-seaters, with their slightly superior speed, were getting the measure of the Sopwiths by attacks on their blind spot under the tail. Diving down out of range, they came up behind and below the Sopwith's tail, where the latter's gunner, even if he could see the enemy machine, had difficulty in bringing his gun to bear. What was needed now was a fast single-seat scout to match the enemy's new Albatros and Halberstadt fighters, and the British had this with the new Sopwith scout—the Pup.

On October 15th No. 45 Squadron under Major W. R. Read, the second to be equipped with 1½ Strutters, joined No. 70 Squadron at Fienvillers. Eleven days later more joined the battle from another direction. Not only had the Navy been forced to hand over many of its 1½ Strutters to the Army—in fact 70 altogether, but they were now detaching complete units from Dunkirk to serve under R.F.C. Wings. The first was No. 8 (Naval) Squadron under Squadron Commander G. R. Bromet, whose earlier experience with Sopwiths had been with seaplanes. This squadron had three flights of six aircraft, 'A' Flight with Nieuport 17Bs, 'B' Flight with Sopwith Pups and 'C' Flight with Sopwith 1½ Strutters.

The naval pilots, soon involved in the dog-fighting

Another view of the first Whitehead Pup. On early Pups the Sopwith-Kauper synchronised gun gear, patented retrospectively in 1917, was fitted. Kauper, the Sopwith engineer who devised the gear, received a handsome royalty on the gears of which 3950 were made and 2750 installed.

A Sopwith Pup being groomed for flight by Royal Flying Corps personnel at a station in England.

over the Western Front, preferred the Pup to their other aircraft; on November 16th the 1½ Strutters, and later in December the Nieuports, were replaced by Pups, to form the first complete Pup squadron in action.

There was little wrong with the Pup airframe and the aircraft soon proved its superiority in manœuvre, but engine troubles and gun jamming continued to plague the pilots of 'Naval Eight'. Flt. Sub-Lt. Trapp's report is typical. While on offensive patrol at 15,500 feet he saw three enemy machines in formation. Coming under the tail of one, he opened fire at an estimated fifteen yards. After a burst his gun jammed, and while reaching up to clear it, he came so close to the enemy machine that he had to jerk the controls into a right-hand vertical bank. Continuing round and clearing the gun, he found himself directly behind another of the machines and opened fire. He saw tracers enter the fuselage of the enemy which went into a nose-dive and then righted, only to flutter down completely out of control.

A man who was to become a Triplane ace the following year, R. A. Little then a Flight Sub-Lieutenant, started making his name on Pups in 'Naval Eight'. On December 4th he was escorting F.Es on a bombing raid when he saw an enemy scout approaching. Intercepting the enemy a half-mile from the bombers, he fired a burst at close range, only to have his gun jam. Breaking off the fight to clear his weapon, he was assailed by the German scout. As his engine was running badly he could not make use of the Pup's climbing ability and dived down, crossing the lines at 5,000 feet and then landing on the first available field. There, he cleared his gun, tended his engine and took off again and at 10,000 feet met other F.Es being harassed by enemy fighters. Flt. Lt. Goble, who had set out on escort duty with Little, fired on three Halberstadts trying to attack some F.E.s from the rear at 12,000 feet, and then climbed up to 13,300 feet to attack three more.

Nevertheless, in spite of the shortcomings of guns and engines and the limiting factor of the weather which permitted flying on only eight days of December 1916, the squadron had an unrivalled record. Their claims for December are summarised in the table at top right:

* For an account of this flight see *'von Richthofen and the Flying Circus'*, Harleyford, 70s. or $11·95.

Dec.	Pilot	Aircraft	Enemy aircraft
4	R. A. Little	Pup N5182	Halberstadt biplane
4	S. J. Goble	Pup N5194	Halberstadt biplane
4	C. R. Mackensie	Nieuport 8750	Type M biplane
4	G. G. Simpson	Nieuport 3956	Type M biplane
11	S. J. Goble	Pup N5196	Type L biplane
20	A. S. Todd	Pup N5193	Type K biplane
20	R. A. Little	Pup N5182	Type K biplane
20	R. R. Soar	Pup N5180	Halberstadt biplane
20	R. R. Soar	Pup N5181	2-seat scout
20	C. E. Hervey	Pup N5184	A white biplane
26	B. L. Huskisson	Pup N5186	Type K biplane
26	J. C. Croft	Pup A626	1-seat scout
26	R.J.O.Compston	Nieuport 8750	Albatros scout
26	N. E. Woods	Pup 3691	Halberstadt biplane
26	C. E. Hervey	Pup N5197	Halberstadt biplane
27	C. R. Mackensie	Pup N5198	Type L biplane

Types 'K', 'L' and 'M' were aircraft unknown by name but recognised from sketches held by the intelligence officers.

The A626 Pup listed is significant. It was the first contract-built Pup for the Royal Flying Corps, having been built by the Standard Motor Company. Since No. 8 (Naval) Squadron were operating under the R.F.C., it was entitled to re-supply from R.F.C. sources, and so commenced a form of air services integration that culminated in the formation of the R.A.F. The Pup A626 was significant in other ways. Flown by Flt. Lt. J. C. Croft on January 4th 1917, it was forced down by Lieutenant Mallincrodt of Jasta 10 and before Croft could destroy the machine, he was taken prisoner. So the Germans gained an early flying example of Britain's latest and most effective fighting machine.

That day Flt. Lt. A. S. Todd was also lost, in Manfred von Richthofen's first 'tangle' with a Pup and, in fact, with the Navy. He did it the justice of reporting that only because it was a case of three against one, was it possible to gain an advantage.* Furthermore, it was nine months before he 'tangled' with another!

The naval pilots were delighted with their Pups. They reported that hostile aircraft were becoming more numerous and more aggressive, but that with the Pups they had the edge over their opponents. Flt. Lt. D. M. B. Galbraith was awarded a bar to his D.S.C. for exceptional gallantry in attacking six enemy machines single-handed in his Pup, driving down two and putting the rest to flight.

The pilots had great confidence in the flying qualities of their machines. During January Flt. Lt. E. R. Grange, hit in the right shoulder, lost the use of his arm and a great

Another view of A1914, shown in Chapter 14, the Sopwith 1½ Strutter captured intact by the Germans.

deal of blood, yet he was able to land his aircraft safely. Flt. Sub-Lt. Lawson, hit in the right buttock and with his Pup badly knocked about with bracing wires on the port side shot away, brought his machine back to his airfield without further damage.

On February 1st the Squadron handed over its Pups to No. 3 (Naval) Squadron who took over their duty with the R.F.C., while the personnel returned to Dunkirk to reform with Triplanes. The R.F.C. had ordered Triplanes but it was agreed in February 1917 that the Spads ordered for the R.N.A.S. would go to the R.F.C. and the Triplanes for the R.F.C. to the R.N.A.S

The R.F.C. had the first Pup squadron of their own— No. 54—arrive at the Front on Christmas Eve 1916 and go into action early in 1917. On February 13th 2/Lt. F. N. Hudson, engaging an enemy machine on his Pup near Le Transloy, sent it down in a nose-dive. The 1½ Strutter was still having its day and on March 1st Captain A. G. Saxty, with 2nd Class Air Mechanic McMullen manning the Lewis gun, drove an Albatros Scout down. Four days later one of the Pups of No. 3 (Naval) Squadron was heavily engaged and thereafter the tempo of operations increased with the Spring weather and preparations for an offensive.

It was not only in the air that the enemy came to fear the Sopwiths, but also on the ground. Sopwith Pups of No. 54 Squadron attacked an enemy balloon on April 5th, bringing it down in flames. On returning from the venture Captain Pixley attacked a lone rider on horseback; both fell. Captain Hudson, opening fire from 200 feet at a group of about a hundred men unloading boxes from open trucks at Gouy railway station, spattered them with fire, observing the majority scatter leaving some lying on the ground. Thus started ground strafing tactics by Sopwiths.

Next day Hudson drove an enemy machine down and so did Lt. Stewart of his squadron, while the naval Pups of No. 3, escorting B.Es on a bombing raid, engaged three enemy machines that tried to intercept. Flt. Lt. L. S. Breadner sent one down to crash near Bois de Bourlon and Flt. Sub-Lt. J. S. T. Fall sent down another in the same locality. 2/Lts. Cook and Murison, out on long-distance reconnaissance on a 1½ Strutter from No. 45 Squadron, destroyed an enemy machine near Lille and drove another down near Templeuve.

That day another Pup Squadron came into action. As many Sopwith-equipped squadrons as could be mustered were required at the Front for the forthcoming Spring

The captured Sopwith 1½ Strutter which had a Nieuport Type gun ring fitted in the rear cockpit.

Sopwith Triplane of No. 1 (Naval) Squadron which used Triplanes throughout the whole of 1917.

offensive—the Battle of Arras. Production was still a limiting factor. Initial production from the Standard Motor Company had gone to equip No. 54 Squadron and to supply their backing at Aircraft Parks in France. Standard Motors 38th and 40th production machines, A663 and A665, were issued as the initial equipment of No. 66 Squadron on February 5th. Since Whitehead Aircraft at Richmond had just commenced their output of Pups, their initial production was allotted to bring No. 66 up to establishment. However, first there were formalities to observe.

The first Whitehead-built Pup (A6150) was required for service evaluation to approve their manufacture. The second (A6151) had been sent to Brooklands on February 2nd for Sopwith engineers to run their eyes over, but it evidently met with approval as Sgt. W. H. Dunn flew it off next day to No. 66 Squadron, forming up at Filton. The Royal Aircraft Factory gave assistance by erecting the next few aircraft from output.

On the 8th Lt. Johnstone collected A6152 from Farnborough while another pilot collected A7302 from Standard Motors at Coventry. Other deliveries from both factories continued during the month until No. 66 had a full establishment of 18. One of these, A7309, was the subject of a book—Sir Gordon Taylor's *Sopwith Scout 7309*— which correctly should have been recorded as 'A7309'.

Reaching France on March 6th, No. 66 Squadron went into action exactly a month later and drove off three hostile machines interfering with an artillery spotting aircraft and shot the observer from one enemy machine out of his cockpit.

A No. 54 Squadron pilot (now Air Chief Marshal Sir Robert Foster, K.C.B., C.B.E., D.F.C., R.A.F. (Retd.)— wrote of his experiences with the Pup at this time:

'From the flying point of view, the Sopwith Pup was certainly the finest aircraft of its day—no vices, beautiful to handle, and with only 80 horse power in its Le Rhône engine of outstanding performance. We attained 18,000 ft. with regularity, and could get even higher. In operations our best chances came from climbing above the maximum height obtainable by the German fighters and then hoping to make a surprise attack. The Germans were always superior in level speed and in the dive, but the Pup was much more manœuvrable and we could turn inside any German fighter of the day. The winter of 1916/17 was bitter in Northern France; and at 18,000 feet everything froze—the engine throttle, the gun, and the pilot. With open cockpits, no oxygen, indifferent flying clothes, a number of us were frost-bitten, an experience which one found extremely painful.

'The aircraft itself always behaved in a most gentlemanly way, but it needed careful handling—a dive of 160 m.p.h. was fast enough, and at 180 m.p.h. the wings were definitely flapping, and a gentle recovery was essential, since to lose a wing when one had no parachute offered no future. The Pup's one disadvantage was its extreme lightness. When operating in strong winds, our Squadron's practice was to call out all available personnel when a patrol was landing back: the men would be spread out in two lines on the airfield between which the aircraft would land, and have their wing-tips seized before a gust could blow them over. In this connection, this entry is in my log book for the 24th March 1917 when we were stationed at Chipilly on the Somme.

'24.3.17 Sop Scout 646 (A646) 2 hrs. 20 m. Escort to F.E.s No scrapping, V-cold and high wind. Crashed landing.

'This was a shame-making incident. I was the last to land; all the squadron was out in its two lines; and the wind was gusting very strongly. After a number of unsuccessful efforts, I eventually got on the ground, but, while one mechanic managed to catch a wingtip, the man on the other side failed to do so. So I gracefully turned over on to my back in front of the whole squadron. All my Squadron Commander (to whom we were very much attached) said to me was "I knew you would make a balls of it. But why waste the squadron's time making thirteen shots to land? Why not crash the first time?".'

With the supply of 1½ Strutters assured from the outpourings of the various constructors towards the end of 1916, so the R.F.C. in the Field turned to measures calculated to increase its efficiency as a fighting machine. As 130 h.p. Clergets became available, they were substituted for the 110 h.p. Clergets initially installed; the first tested at No. 2 Aircraft Depot on December 20th climbed to 10,000 feet in under 17 minutes. Less successful was an attempt, like No. 3 Wing R.N.A.S., to fit a smaller tail; a 12-foot span tail caused the controls to be sluggish and gave difficulties in recovering from dives. Equally unsuccessful was an evaluation conducted by No. 2 Brigade on a special 1½ Strutter, delivered to No. 46 Squadron on February 11th 1917, fitted out as an artillery observation aircraft. That same month, a Ross interrupter gear (designed by Captain Ross of No. 70 Squadron) was fitted to A3431 and next month a special experimental compass was tried out in A1099 of No. 45 Squadron.

Another 1½ Strutter squadron, No. 43, had arrived at the Front on January 17th and 'Naval Eight', arriving back to the R.F.C. on March 28th, started repeating with their new Triplanes, the successes they had achieved with the Pup. Another naval squadron arrived and when the Battle

The King of Montenegro, during a visit to the Western Front, inspects Sopwith 1½ Strutters of No. 70 Squadron which was the first unit to receive these aircraft. In this case the 1½ Strutter has a Scarff gun ring fitted in the rear cockpit which became standard. The censor has scratched out the serial number from the negative.

of Arras opened on April 9th 1917, 32% of the fighter squadrons were equipped with Sopwiths. These were:

Squadron	Location	Commander	Equipment
No. 1 (Naval)	Chipilly	Sqn. Cdr. F. K. Haskins, D.S.C.	18 Sopwith Triplanes
No. 3 (Naval)	Marieux	Sqn. Cdr. R. H. Mulock, D.S.O.	18 Sopwith Pups
No. 8 (Naval)	Lozinghem	Sqn. Cdr. G. R. Bromet, D.S.O.	18 Sopwith Triplanes
No. 43 R.F.C.	Treizennes	Major W. S. Douglas, M.C.	18 Sopwith 1½ Strutters
No. 45 R.F.C.	St. Marie Cappel	Major W. R. Read, M.C.	16 Sopwith 1½ Strutters
No. 54 R.F.C.	Chipilly	Major K. K. Horn	18 Sopwith Pups
No. 66 R.F.C.	Vert Galand	Major O. T. Boyd, M.C.	18 Sopwith Pups
No. 70 R.F.C.	Fienvillers	Major A. W. Tedder*	18 Sopwith 1½ Strutters

* Later Lord Tedder, 1st Baron of Glenguin, Marshal of the R.A.F.

Sopwith aircraft were now positioned to play a major part in attaining air superiority over the Western Front. Three different types were in service and already there was promise of an even better machine to follow. The first of the Camels had already flown.

CHAPTER NINETEEN

The Sopwith Sailors (No. 3 Wing R.N.A.S.)

The Air Department of the Admiralty had advanced ideas on strategic bombing. The way had been pointed in 1914 by the early raids with Sopwith Tabloids, but as the Germans advanced into France and Belgium overrunning Antwerp and Ostend, targets in the German homeland could no longer be reached by bombers from behind the Allied lines with the aircraft then in service. With the 1½ Strutter the picture changed; the Admiralty arranged with the French for a Wing to be stationed in the area of Nancy from where important German industrial and military centres could be reached by their new aircraft.

No. 3 Wing, disbanded in the Dardanelles late in 1915 as the Dardanelles Operations came to a close, reformed at Manston (then Manstone) under Captain W. L. Elder, R.N., who went to France in May 1916 to make arrangements for the Wing's reception. The initial establishment was for a strength of twenty Sopwith 1½ Strutter bombers and fifteen Short Bombers (landplane adaptation of the successful Short Seaplane) and twenty Sopwith 1½ Strutter Strutters fighters and twenty-five French fighters, set out to raid the Mauser arms factory at Oberndorf. During the round trip of 223 miles, Flt. Cdr. Jones with Sub-Lt. Downs brought down one enemy fighter and another Sopwith damaged an enemy machine. Losses were one 1½ Strutter fighter and two R.N.A.S. Breguet bombers whose crews were taken prisoner. Most of the bombs fell on new buildings not yet brought into use.

Eleven days later the R.N.A.S. contribution to another Allied raid was wholly by Sopwiths; two flights of bombers and one of fighters raided iron works and factories at Hagendingen. Returning, Flt. Sub-Lt. Smith and Gunlayer Clegg, who had damaged the enemy fighter in the earlier raid and had been denied the opportunity on this occasion of meeting enemy aeroplanes, attacked a German kite balloon. Three out of five blast furnaces were left damaged. By this time it was realised that only the Sopwiths were capable of this type of operation and the Breguets were withdrawn from operations and used only for training.

An Admiralty Type 9400 two-seat fighter built by Mann, Egerton and Company of Norwich. It was the 'W'-struts supporting the top wing centre-section above the fuselage that gave rise to the nickname 1½ Strutter which became officially adopted.

fighters, to be effected by July 1st. Further expansion was envisaged to a strength of a hundred. Thus was planned the world's first strategic bombing force, and its weapons were predominantly Sopwiths.

The French allotted airfields at Luxeuil and Ochey for the Wing which was far below establishment, but made its first operation on July 30th when three aircraft joined six French Air Service bombers in attacking benzine stores and barracks at Mulheim. In spite of heavy opposition from anti-aircraft fire, the bombers returned safely.

Following this first rather limited operation, further build-up of the force was thwarted by the very excellence of the Sopwith. The Army having staked their claim for a share of the cream of the aircraft, the Navy had to forgo some deliveries of 1½ Strutters. By the end of August only 22 aircraft were on No. 3 Wing strength and since the Shorts had been found unsuitable, some Breguets were acquired as an interim measure.

Their first major operation was on October 12th when fifteen naval bombers, Sopwiths and Breguets, accompanied by nine French bombers, and escorted by six Sopwith 1½

On November 10th a raid by nine single-seat Sopwith bombers on the Volklingen Iron Foundry works and blast furnaces, followed an earlier French raid. In all 35 × 65 lb bombs were dropped but visibility was poor. Six enemy machines were encountered and three were claimed shot down by the escorting two-seat Sopwiths, in which one observer sustained minor wounds. One bomber was missing but the pilot turned up later after having flown his aircraft to extreme range and unwittingly committed an 'act of war' further into France than any *German* had ever penetrated!

The officer concerned, Flt. Sub-Lt. G. S. Harrower, finding his compass to be faulty, followed the other aircraft to Volklingen. Losing sight of his comrades in cloud on the return journey, he had taken directions by the sun and flown low to try and recognise landmarks. Seeing only unfamiliar country, he had decided to carry on flying south-westerly by the sun as far as possible to ensure a good penetration of the French lines. He eventually came down in Central France, three times the distance of his airfield to the target, in a field near Moulins! It took ten days before,

An Admiralty Type 9700 single-seat bomber built by Mann Egerton. While the name 1½ Strutter came into general use, the type numbers 9400 and 9700 were used officially to differentiate between two-seat and single-seat versions respectively.

in answer to his telegram, a truck from Luxeuil arrived with a mechanic and fuel, together with a sarcastic note from his squadron commander asking if he had been making for Cannes.

But the machine never did return in one piece. Some thirty minutes after take-off, flying over the Côte d'Or, the engine failed and Harrower was again forced to descend. Picking a suitable field, he became aware of horses well away from the spot on which he had chosen to land, but as he was about to touch-down he perceived one galloping towards him. As far as possible he tried to alter course, but the horse, mad with fright, raced towards the aircraft. With insufficient speed to gain further lift and aware of the dangers of cartwheeling if he turned sharply, Harrower kept straight on, hoping the horse would turn. Alas, it did not; there was a collision and the horse was killed and the aircraft wrecked. Harrower was unhurt.

For months later the British Claims Commission at Boulogne were still studying with incredulity and their usual parsimony the claim of Monsieur Roger de Salverte, farmer of Rouvres near Fauverney. A sailor in a Sopwith, descending upon a horse in Central France, far from the battlefronts, was beyond their experience. On ascertaining the full circumstances they pointed out to the unfortunate farmer that, 'Damage done by aeroplanes engaged in active operation against the enemy, and when proceeding on or from such operations (including raids, reconnaissances, artillery observation, etc.) are classified as *Fait de Guerre*'.

Thus a Sopwith, by the excellence of its range, had carried the 'fates of war' to the heart of France.

The day following the Volkingen raid, similar targets were attacked by nine Sopwith bombers each dropping 4 × 65 lb bombs, escorted by seven Sopwith fighters. One bomber failed to return, but the pilot, landing in friendly territory, turned up later. Two attacks on Dillingen in late November/December closed operations for 1916.

The 1917 bombing season started on January 25th with an attack on Brebach, about three miles south-east of Saarbrucken, where a group of ironworks including blast furnaces was their objective. Although two aircraft returned early a sizeable force of 13 bombers reached the target area and several bombs were seen to score hits. Five fighter 1½ Strutters accompanied the bomber versions—and were needed. A Roland was driven off by Flt. Sub-Lt. Pattison whose gunner was Petty Officer Herbert J. L. Hinkler who had previously worked as a mechanic in the old skating rink at Kingston on early Sopwith aircraft and later to earn world-wide fame as an aviator. Flt. Lt. Potter with Gunlayer Air Mechanic Dell, engaged by four enemy machines, shot down one but the remaining three sent their 1½ Strutter down with a damaged engine; fortunately they managed to glide to the Allied side of the lines. One of the bombers piloted by Flt. Sub-Lt. J. M. Sharman, engaged an enemy monoplane with his single forward-firing Vickers gun and shot it down out of control, a fact for which three other officers of No. 3 Wing vouched.

A Sopwith 1½ Strutter in England before handover to the Services. Next to the Camel, more 1½ Strutters were built than any other Sopwith type; the total being over 5400.

N5116 of No. 3 Wing shown with bomb doors open. On the left is the bomber prototype, aircraft 9700

Operations were conditioned by the numbers of men and machines. Captain Elder was beset by limiting factors. His build-up of aircraft was slowed down by delays in deliveries, due to the deflection of 1½ Strutters to the R.F.C. Then he was detailed to hand over a number of his own 1½ Strutters to enable the French, already receiving some direct, to increase their bombing effort. But this was not his main worry.

To the Admiralty he wrote on January 15th 1917— 'The shortage of engines, or more than anything else the shortage of engine spare parts to put the engines I already have in running order is chronic. Cannot something be done to supply the few necessary spare parts so that the material that is already available at least can be used . . . I am continually having it rubbed into me by the French the benefit of this wing and the necessity of very largely increasing it'.

Later that month Elder had instructions to hand over nine of his best pilots to Wing Commander C. L. Lambe at Dunkirk who was given the task of forming naval squadrons of the new Pups and Triplanes to aid the R.F.C. Writing to Lamb he told him 'I have given you the pick of the Sub-Lieutenants from this Wing, and in so doing I am sorry to say, have absolutely upset my organisation and squadrons'. He wrote of one pilot being sent, J. J. Malone, as of 'exceptional ability'.

Yet another thorn in Elder's side was the variation in ranges of his 1½ Strutters which were of three types with different sized tanks giving ranges as follows:

Bomber Type 9700	(54 galls)	372 miles
Fighting Type 9400L	(57½ galls)	362 miles
Fighter Type 9400S	(37½ galls)	275 miles

(The range figures were official figures less 30% of absolute range to allow for wind, weather or any other operational diversion).

While Elder pressed for the long-range types he was saddled with a few short-ranged types and modifications were carried out in the unit's own workshops. Apart from the base at Luxeuil there was a forward airfield at Ochey and most of the fighters with their more limited range were stationed there.

On some matters the unit corresponded direct with Sopwith Aviation. There were some control difficulties and Squadron Commander C. D. Draper (known in more recent years as 'The Mad Major' for flying under several Thames bridges) suggested reduced tail area. A single Pup, No. 9496, had been received by the unit at this time and was under repair; it was suggested that the Pup's tail be tried on a 1½ Strutter and Lieutenant E. R. Peal, D.S.C., the unit's engineer officer, wired the Admiralty about taking this measure. He received the rather ambiguous reply to the effect that they approved the Pup's tail being fitted— and that this had been tried before and found unsatisfactory! Peal then wrote direct to Sigrist at Kingston to ask what experiments had been made with reduced tail surface on 1½ Strutters. This matter apparently lapsed when the unit disbanded shortly afterwards. In any case, the Wing had been asked to deal with Sopwith's Paris office on such matters. To Elder the position was exceedingly frustrating. He had in the 1½ Strutters at his disposal the power to strike at German industry with a large formation, but many

French - built 1½ Strutter. This aircraft is shown after transfer to American Forces for training purposes.

French Sopwith 1B1 of Escadrille SOP111 *which forced-landed in Dutch territory on July 7th 1917.*

of his aircraft were under repair or awaiting engine spares. In February 1917 Curtiss R2s were in transit to the Wing; not because they were better machines, but because they did not use Clerget engines.

By February No. 3 Wing had a hundred aircraft, tabled as shown in an Appendix at the end of this book, from which it can be seen that the world's first strategic air force was predominantly a Sopwith one—in fact 94% of the aircraft were Sopwith.

But only 37% of the total force was operational and of those the 9400S machines had insufficient range. The intention was to organise the Wing into five squadrons, three of Bombers (1½ Strutters) with one of Curtiss R2s to be formed and one of Fighters (1½ Strutters), plus a flight of Handley Page 0/100 twin-engined bombers. A school was also established and to make up his pilot deficiencies Elder suggested to the Admiralty that he be sent a dozen cadets as 'makee learnee pilots'—to use his own words—to make up his numbers as a few of the Sopwiths and the Curtisses had dual controls and were organised into a Depot School.

A large workshop completed Elder's 'private air force', a formation run essentially on naval lines, although miles inland, with a White Ensign hoisted at each airfield. This is typified by a note Lt. Peal wrote on February 4th to Warrant Officer Badley who was in charge of the Engineer Workshops:

'It is reported that ratings have been seen adrift during working hours, at the coffee barrow on the road, outside the Depot workshops. Please see that none of the ratings under you leave, except during the Stand Easy period, which is sounded on the Ship's bell at the Aircraft Depot Office'. Such was the way of life with the 'Sopwith Sailors' as No. 3 Wing personnel were colloquially known.

On March 4th in another attack on Brebach, Pattison and Hinkler in Fighter No. 9410 and Flt. Sub-Lt. H. Edwards with Gunlayer Walker in N5173 each claimed enemy aircraft shot down. Not until the 22nd did the Wing operate again when 1,560 lb of bombs were dropped on Burbach in the Saar Valley. Only the Sopwiths were being used, the Curtisses having been found unsuitable.

No. 3 Wing commenced strikes on German cities in April; Freiburg in Brisgau being the target on the 14th. Bombs dropped on the Karlsruhe railway line dislocated the permanent way and damaged stations. Twenty soldiers, sheltering under a subway, were killed or wounded when the roof was destroyed by a bomb and other Service personnel were killed in the Ophthalmic Institute. However, just as the attacks were beginning to tell, the Wing was broken up. Captain Elder was ordered to return to England on June 4th 1917 and two engineer Warrant Officers, L. J. Lester and H. Buxton, having arranged for the disposal of the aircraft, mainly to Dunkirk, closed down the naval station at Luxeuil on June 30th.

The Wing's disbandment was undoubtedly due to pressure exerted by the Army on the War Cabinet. The arguments put forward on the cost of the force in relation to what it achieved, by estimates from intelligence, were forceful and perhaps accurate, but the effect of such a force on the German war effort cannot be measured purely by the damage it created. The German deployment of fighters and guns to ward off the raiders meant a diversion of effort from air support on the Western Front, probably balancing the effect in deploying No. 3 Wing. In that case, the damage and disruption that they created was a 'war bonus'.

R.N.A.S. Probationary Flight Sub-Lieutenants under gunnery instruction, shown training with Vickers (on table left) and Lewis (centre) machine guns — the armament of 1½ Strutters. Both guns used .303 or .300 ammunition, but the Vickers was belt-fed and the Lewis drum-fed.

French Strutters

CHAPTER TWENTY

In the overall picture of the 1914-1918 War air operations, it would be true to sum up the Sopwith 1½ Strutter as an aircraft widely used by the French and that it also served operationally in a few British squadrons. The French built three times as many 1½ Strutters as the British, equipped over six times as many squadrons as the British with the type, and used it operationally in the Western Front for nearly a year longer than the British.

Early in the war the French had become convinced that, by their configuration, pusher aircraft were ideally suited for reconnaissance and bombing. Farman and Voisin types, and early Breguets, all pusher types, were ordered in quantity and, to an extent, the tractor biplane was thereby neglected. When the Sopwith 1½ Strutter arrived at the Front in early 1916, the French realised that they had been outpaced in this sphere of development by the British.

For the French it was a bitter pill. In aviation they had led the world and in 1916 both the Royal Naval Air Service and the Royal Flying Corps used Bleriots, Breguets, Caudrons, Farmans, Moranes and Voisins to make up their deficiencies and of the British aircraft types, the great majority used French engines—Clergets, Gnômes, Le Rhônes and Canton Unnés. French interest in the new Sopwith 2-seater meant that it was not only a success in British aviation, but in World aviation.

Swallowing their pride, the French asked for details of the Sopwith, arranged direct delivery of numbers from the Navy and obtained licence rights from Sopwith Aviation for home production. It was this that led T. O. M. Sopwith to open a Paris office, at 21 Rue du Mont-Thabor and place the Fenn brothers in control. The executives were registered as:

Président: T. SOPWITH, Directeur Général: R. O. CARY, Directeur à Paris: ALAN R. FENN, Sous Directeur à Paris: M. H. FENN.

While the Paris office of Sopwith Aviation arranged for the supply of drawings for French industry, the French Army were taking delivery of ex-R.N.A.S. 1½ Strutters from No. 3 Wing's base at Luxeuil and at Dunkirk. The first to go to the French were two fighter versions, followed by a single-seat bomber, for evaluation. They then requested sufficient of the bomber version to re-equip *Escadrilles* F29 and F123 of the *4me Groupe de Bombardment* (*G.B.4*) which shared Luxeuil with No. 3 Wing, R.N.A.S.

Using two of the Sopwith bombers, Capitaine L. de Beauchamp and Lieutenant Daucourt took off from Luxeuil at 11 a.m., September 24th 1916, on the first French operations with Sopwith aircraft. Climbing to 14,000 feet the two bombers made for Krupps Works at Essen where they dropped 6 × 120 mm Gros-Andrean bombs apiece and returned safely. On November 17th, de Beauchamp made another epic flight with his Sopwith. Leaving Luxeuil at 8 a.m., so timed to be over Munich at midday where he dropped his bombs, he crossed the Alps and landed at Santa Dona di Piave, twelve miles north of Venice.

Meanwhile Sopwiths of *G.B.4* were accompanying R.N.A.S. Sopwiths of No. 3 Wing to targets in the Saar.

French Air Service aircraft did not normally bear roundels on the fuselage side, but the insignia of their unit; in this case Escadrille SOP 66's motif is shown on a French-built Sopwith 1B1.

But there was a limit to the numbers of 1½ Strutters that the R.N.A.S. could spare, particularly as they were also supplying R.F.C. squadrons with this aircraft; in the main it was bombers (Type 9700) to the French and fighters (Type 9400) to the R.F.C. To equip further units the French Army had to await their home production. Contracts were placed with seven manufacturers: Amiot, Bessoneau, Darracq, Hanriot, Lioré et Oliver, R.E.P. and Sarazin Frères.

Whereas the British had two main versions, the French had three designated as follows:

SOP 1A2 (The SOP is self-explanatory, the initial 1
SOP 1B1 denoted Type 1 and the A and B 'Artillery'
SOP 1B2 and 'Bombardement' respectively, whilst the
final figure related to the number of seats.)

Sopwith's French office had a difficult task. The French were inclined to criticise the Sopwith and wanted to introduce a spate of modifications on their models. When built and tested they maintained that their Sopwiths had a better performance than the British-built models. This was probably true. They had the pick of engines, whereas the British had to use such French engines as the French would let them have, or their 110 h.p. and 130 h.p. Clergets in limited production in England. The French models had the 135 h.p. Clerget 9Bb, the 135 h.p. Le Rhône 9Jby or the 145 h.p. Clerget 9Bc.

At one stage tht French considered the 1½ Strutter as structurally unsound and in March 1917 issued a series of instructions for certain parts to be strengthened. This followed a machine breaking up in the air over Luxeuil, witnessed by British and French officers, whose respective views differed sharply. To quote the official French report by Captain Feugere.

'On February 19th, the pilot Lumiere was making a flight on a Sopwith monoplace de bombardement, English built, belonging to the Groupe de Bombardement 4.

'After having made several turns pretty sharp at 700 or

A Sopwith 1B1 of Escadrille SOP 107 forced down in German territory and camouflaged by cut grass to prevent it being detected and destroyed by other Allied aircraft.

800 metres, he then descended very steeply with full motor; the machine straightened itself out very sharply. At this moment the eye-witnesses saw one part of the machine (some say it was the tail, the others say it was a wing) take an abnormal position in relation to the fuselage. Consequently to which the machine dived vertically with the motor full on, down to the ground where it crashed and killed the pilot. From the beginning of the fall the right wing became quite separate from the rest of the machine.'

The French view was that the machine broke up in the pull-out through structural weakness; the R.N.A.S. view was that in such a sharp pull-out from a steep power dive, any aircraft would be overstressed. Nevertheless, the French insisted on modifications, and Alan Fenn left on February 28th for England to discuss this with Sigrist.

As a result of changes such as occasioned by this accident, and other modifications, French production did not get underway until the spring of 1917, by which time the 1½ Strutter was becoming outclassed. And by this time, the French had some two-seat tractor bombers of their own. One of the first French units, apart from *G.B.4* at Luxeuil, to have the 1½ Strutter was *Escadrille* M.F.7 which exchanged its Maurice Farmans for Sopwiths in April 1917—only to have them replaced by Breguet 14s the following month. However, the Breguets were not in sufficient numbers at this time and the Sopwiths came into general use as a stop-gap in many French units, until the Breguets arrived. The French were also introducing the Paul Schmitt bomber, but the Sopwith evidently proved superior as *Escadrille* P.S.128, formed May 13th, discarded their new bombers for Sopwiths at the end of July.

A Sopwith 1A2 captured by Austrian troops being prepared for use by Austro-Hungarian airmen.

Throughout 1917 bomber squadrons of the French Army continued to re-equip with French-built Sopwith 1½ Strutters. In addition, *escadrille* specialised in spotting for long-range artillery were formed, known as S.A.L. units-*Sections Artillerie Lourde;* these were SOP 204, 206, 207, 208, 212, 216, 222 and 223. Outclassed by early 1917 the Sopwith 1½ Strutter may have been, but the fact is that 60 *escadrilles* (an *escadrille* was the French equivalent of a British squadron) still had Sopwith 1½ Strutters in 1918—a time when the British had long since discarded the type for operations on the Western Front. On September 16th 1917, twenty-four Sopwiths in a mass raid had attacked Stuttgart, shooting down three defending fighters.

French use of the 1½ Strutters was not restricted to the Western Front. A detachment of SOP 219 went to Italy, presumably to demonstrate artillery co-operation. Commanded by Capitaine Fontenilliat it was known as the *Espinasse Escadrille* (a General Espinasse had been killed in the Battle of Magenta in 1859 during Napoleon III's intervention in Italy on behalf of Italian unity.) After the Italian collapse during the Battle of Caporetto, when Allied reinforcements included R.F.C. Sopwith Camels, the French sent Escadrilles SOP 36 and 206 with 1½ Strutters to support their troops. One unit, SOP 582, was prepared for the Russian Front.

Up to October 1917 the French 1½ Strutters were finished in clear dope and thereafter in camouflage. In all 4,200 were built in France. When generally replaced in squadron service in the first half of 1918, a large number were relegated to training and a purchasing commission of the American Expeditionary Force bought 514 in the spring

A French-built Sopwith 1A2 of the American Expeditionary Force at La Valdohon, January 25th 1919.

of 1918. And even the British, rebuilding 1½ Strutters for shipboard use that year, had to turn to the French to buy up a small quantity!

It is indicative of the wider and prolonged use the French made of the 1½ Strutter, that whereas only one reached the British Civil Register post-war, the French registered 48.

French Escadrilles equipped with 1½ Strutters

SOP 5	Sep 17 to Jul 18	SOP 216	1917-Apr 18
SOP 7	Apr 17 only	SOP 217	1917-Apr 18
SOP 9	July 17 to May 18	SOP 219	1917-Feb 18
SOP 13	Sep 17 to Mar 18	SOP 221	1917-May 18
SOP 17	Sep 17 to Mar 18	SOP 222	1917-May 18
SOP 24	1917-Mar 18	SOP 223	1918
SOP 28	1917-Mar 18	SOP 226	Early 17-Apr 18
SOP 29	Dec 17-Jan 18	SOP 227	Oct 17-May 18
SOP 36	Jul 17-Jan 18	SOP 231	Dec 17-Apr 18
SOP 39	Oct 17-Feb 18	SOP 232	Jan-May 18
SOP 43	Late 17-May 18	SOP 234	Jan-May 18
SOP 47	In 1918 only	SOP 235	Apr-May 18
SOP 51	In 1917 only	SOP 236	Mar-May 18
SOP 55	Late 17-Jan 18	SOP 237	Mar-May 18
SOP 60	Late 17-Jun 18	SOP 238	Mar-May 18
SOP 61	Oct 17-Jan 18	SOP 250	Nov 17-Sept 18
SOP 66	In 1917 only	SOP 251	Nov 17-Aug 18
SOP 104	Sep 17-May 18	SOP 252	Nov 17-May 18
SOP 105	Oct 17-May 18	SOP 255	Jan-Jun 18
SOP 106	1917-May 18	SOP 260	Jan-Jun 18
SOP 107	Jun 17-Early 18	SOP 263	Jan-May 18
SOP 108	Jun 17-late 17	SOP 269	Jan-Jul 18
SOP 111	Mar-Oct 17	SOP 270	Feb-Jul 18
SOP 123	Oct 16-1917	SOP 271	Feb-Jul 18
SOP 128	Jul 17-Jan 18	SOP 273	Feb-Jun 18
SOP 129	Jan 17-Mar 18	SOP 276	Feb-Jun 18
SOP 131	Jun 17-Mar 18	SOP 277	Feb-May 18
SOP 132	Jun 17-Mar 18	SOP 278	Feb-Jun 18
SOP 134	Jan-Apr 18	SOP 279	Mar-Jun 18
SOP 141	Dec 17-Aug 18	SOP 280	Mar-Jun 18
SOP 204	Oct 17-Feb 18	SOP 281	Mar Aug 18
SOP 206	Oct 17-May 18	SOP 282	Mar-Sep 18
SOP 207	Oct 17-Apr 18	SOP 283	Mar-Jun 18
SOP 208	1917-Feb 18	SOP 284	1918
SOP 212	1917	SOP 285	Mar-Jul 18
SOP 214	1917	SOP 287	Mar-Jun 18

R.N.A.S. Sopwith 1½ Strutters transferred to the French Army

Quantity	Type	Serial Numbers and Remarks
2	9400S	9895 and 9413 (Delivered with 110 h.p. Clergets)
1	9400L	9651 held at Dover July/August 1916. Transferred without engine from No. 3 Wing.
7	9700	9655, 9657, 9661, 9664, 9666, 9720, 9742 transferred without engines, all ex-3 Wing.
10	9700	9669, 9673, 9700, 9706, 9714, 9729, 9736, 9738, 9745, N5514 transferred with 110 h.p. Clergets.
16	9700	N5088, N5091, N5094-5095, N5097-5098, N5100-5101, N5103, N5113, N5115-5116, N5122-5123, N5128, N5149, transferred with 110 or 130 h.p. Clergets ex-3 Wing.
34	9700	N5092, 5104, N5118, N5125-5127, N5129-5148, N5157-5158, N5160, N5502, N5507, N5511, N5522-5523 transferred without engine.
6	9400	Un-numbered.

Total: 76 transferred to French.

The pilot of a French Sopwith 1A2 being inspected by General Franchet d'Esperey on the Macedonian Front.

CHAPTER TWENTY-ONE

Set for Success

Local liaison. A Sopwith Camel, latest of the Sopwith products, performs aerobatics over Kingston-upon-Thames during War Bonds Week held in the town. This view of the town centre is about 500 yards from the Sopwith Works. One of the Mk. 1 tanks can be seen at the bottom right of the photograph.

The year 1916 had been one of tremendous expansion for the Sopwith Aviation Company. Their aircraft were in demand by the Army and Navy and by the Allies. While some firms had experienced grave labour troubles, relations between management and workers at the Sopwith Works were consistently good. Indeed, on January 29th 1916, marking the third anniversary of the establishment of the works, the workers arranged a dinner at the Castle Hotel, Richmond, for T. O. M. Sopwith, and R. O. Cary the general manager. The Mayor of Kingston was also invited. Music was provided by the Sopwith Works Band in which Cary himself took a great interest and, after dinner, he gave a violin solo.

In the hundred-fold expansion by 1917 that had been effected in four years, Sopwith Aviation had become a community as well as a manufacturing organisation. Apart from the orchestra, there was a Sopwith Choral Society in which May Sopwith took a special interest. A Sopwith Athletic Club had premises on Fife Road, Kingston; an off-shoot of this was the Sopwith Harriers.

Both directors and employees continued to give generously to war charities. The Company gave £100 to the Kitchener Memorial Fund in 1916, and in January 1917 collections raised a further £210 for the R.N.A.S./R.F.C. Winter Comforts Fund. Next month, when Lady Drogheda organised an exhibition in aid of war charities at the Grosvenor Galleries, T. O. M. Sopwith gave £50, arranged for workmen to assist and lent a Triplane for exhibition.

The work in hand in the factory was now far more stabilised. Clear of floatplane construction, tailing off 1½ Strutter production, and concentrating on Triplanes and Pups, the work became more straightforward. The experimental department, separated from the main works by utilising the original skating rink and by using garages where the great store of Bentall's now stands, was free to concentrate on a whole series of experimental types.

There was an anxious time in early 1917 when the French report of a structural failure on a 1½ Strutter coincided with a similar report from the R.F.C. in the Field. In the latter case the pilot survived and an investigation could be made. It proved the strength of the aircraft, rather than any weakness. A 1½ Strutter, with a rear cockpit fitted with a Scarff gun-ring, mounting a Lewis gun with drums of ammunition, let alone the weight of the gunner, could hardly be expected to perform the aerobatics of a single-seat scout—but some pilots did expect this! Captain E. Lubbock of No. 45 Squadron was flying A1084 when he was attacked by enemy aircraft from above and behind. Turning away sharply, he dived and looped to shake off his adversaries and finding them still behind, looped again. On top of the second loop the stick flopped forward and the machine nose dived vertically for 1,500 feet. Lubbock noted that the indicated speed on the counter was 60 m.p.h.—on the *second* time round!

Pulling out of the dive which had finally shaken off his pursuers, Lubbock, on looking round, was aware of the frantic gesticulations of his gunner who, after attracting his attention, pointed to the tailplane. On the starboard side the main spar appeared twisted and the leading edge folded back. In fact, the main spar had sheared at the junction with the fuselage. What is even more amazing is that having broken, the remaining members held the structure together while Lubbock glided back to land. A full report, made by the squadron, was examined carefully in the works. It was found that if the main spar junction was strengthened and all members strengthened to the same safety factor, the aircraft would never get into the air, it would be that much

Used by Harry Hawker for aerobatic displays, this Sopwith Bee was powered by a 50 h.p. Gnôme engine. It was built from Pup components, and had a wing-warping control system.

overweight. It just has to be accepted that even a Rolls-Royce will overturn when cornering at a certain speed.

But in the single-seat scout squadrons, pilots found all they wanted in manoeuvrability with the Pup, and even today the Pup is spoken of by veterans as the most delightful aeroplane to fly that was ever built. By sub-contractors it remained in production until the end of the war, functioning as a trainer after being out-paced on the Western Front.

Then there was the Triplane. The most remarkable thing about the Sopwith Triplane is the fame that it achieved for the relatively few built. As a service aircraft it was unique, being the first triplane ever to become operational.

Pups and Triplanes were being built side-by-side in the works when King George V and Queen Mary visited the works on Thursday, April 19th 1917. The Royal Party, including the Maharaja of Bikanir, drove over from Windsor and were piloted round the works by an equal number of the Sopwith executives, T. O. M. Sopwith, R. O. Cary and Fred Sigrist. After spending ninety minutes in the Kingston shops, the party drove to Brooklands where Hawker threw a Camel about the sky to demonstrate the latest of the Sopwith products. Afterwards, he was presented to the King.

In service, it was not just the Pups and Triplanes that were involved in operations. The Navy's 1½ Strutters were being withdrawn on the Western Front, only to be shipped to the Aegean for further operations. For the Baby floatplanes, there was active service right up until the end of the war. During 1917 they were up-dated by the installation of a synchronised gun-gear and were also introduced into a new theatre of operations—the Adriatic. Baby floatplanes

This earlier SL.T.BP biplane was another of the Hawker 'Runabouts'. It was used as a personal transport by Hawker for visiting firms to test-fly Sopwith aircraft.

were included in the initial establishment of R.N.A.S. Station Otranto in Italy. They supported the patrolling role of Short Seaplanes in denying the use of the Adriatic to German U-Boats. By mid-1917, with additional bases, the R.N.A.S. formation had grown into No. 6 Wing and the number of Baby floatplanes and 1½ Strutters was increased.

First action for Sopwith aircraft in Italy came at 7.40 a.m. on July 11th 1917, when Austrian flying boats bombed Otranto from about 7,000 feet demolishing two houses and wounding two Italian sailors. Two 1½ Strutters ascended immediately and one, N5232 piloted by Flt. Lt. Biles, exchanged fire with one of the enemy flying boats. On September 6th, the first Baby dropped its 65 lb. bomb on a submarine spotted submerging, and six days later Flt. Cdr. Morrison in N1030 considered he had damaged a U-boat bombed from 700 feet. Operations continued on the lines of coastal anti-submarine activities in Home Waters.

In July 1917 the Sopwith Works were requested by the Air Board to concentrate on spares, while the sub-contractors supplied the complete Sopwith airframes in quantity. The reasons were obvious. The sub-contractors, by virtue of large orders, were having new erecting sheds made, but the Canbury Park Road Works of Sopwith, admirably fitted out with shops for metal and wood-working, did not have the floorspace for lines of erecting bays, or large doping shops. A chance came in September 1917 when there was reversal of Air Board policy on distributing work to a large number of small firms, in favour of centralising production.

A National Aircraft Factory scheme was launched with premises and equipment financed by the Government, but with management in the hands of proven industrial organisations. Factories were planned at Waddon near Croydon, Heaton Chapel near Stockport, Aintree near Liverpool and at Ham near Richmond within two miles of the Sopwith works.

T. O. M. Sopwith was one of the industrialists initially interested in the scheme, but he soon had misgivings; not without good reason, for it proved an expensive failure. Before committing himself too far, he opted out of the national scheme and made a bid for the new factory being erected at Ham and was offered it on lease. This being accepted, ensured the continuation of Sopwith Aviation as a large production centre for aircraft as well as a designing firm.

CHAPTER TWENTY-TWO

Sopwiths in the Aegean, 1917

The seaplane carrier Ark Royal was used as a depot ship at Mudros and carried out major repairs to aircraft and training work with Sopwith floatplanes. In this view a Sopwith 807 floatplane is on deck, having been hoisted aboard by one of the two cranes.

Up to 1917, the main Sopwith aircraft in Eastern waters had been floatplanes, but in 1917, following the success of the new Sopwith fighters on the Western Front, a few were spared for the so-called minor theatres.

In January 1917 the R.N.A.S. units at the Aegean bases of Thasos, Thermi, Imbros and Stavros, known as Flights 'A' to 'D' respectively, were raised to Squadron status. With the base at Mudros which included *Ark Royal* they composed No. 2 Wing, R.N.A.S. The Flight at Thasos was the first to use 1½ Strutters in Eastern waters. Insufficient in numbers to carry out formation raids on Western Front scales, the normal operating procedure was for a single bomber 1½ Strutter to set out with a fighter version as escort. Typical was the raid on January 9th; N5110 piloted by Flt. Sub-Lt. H. R. Aird, escorted by Flt. Lt. G. A. Magor piloting N5086 with Midshipman R. J. Dashwood, R.N.R. as observer, left Thasos Island at 10.45 a.m. They arrived over Tatar Bazarjik, the target, just over a hundred miles distant, at noon. Four 65 lb. bombs were dropped, two falling between the railway station and the bridge over the Maritza River. Another bomb fell on some buildings near the bridge and the fourth close to a large camp of 300-400 tents. Photographs were also taken and the two Sopwiths returned safely unmolested by aircraft or ground fire.

It was different next time Magor and Dashwood set out on N5086. Officially their mission was a reconnaissance of Xanthi. When near their task area, they observed in the distance two trains leaving a station in opposite directions. Chasing one, they arrived over it about at 900 feet some three miles south of Xanthi, whereupon the train stopped and troops rushed out into the fields and fired at the aircraft. Dashwood retaliated with machine-gun fire, observing several men dropping before they had time to take cover.

By February ten 1½ Strutters had reached the Aegean, eight Type 9400S fighters and two Type 9700 bombers (N5110 and N5119). In transit, shipped late in 1916, were a further five fighters and seven bombers. Additionally, on January 6th 1917 a single Sopwith Triplane, N5431 the only Triplane ever to serve the R.N.A.S. in the East, was shipped out.

A few of the limited numbers of 1½ Strutters were sent to 'C' Squadron at Imbros. One of these, a bomber, set out on February 17th, escorted by a Bristol Scout, to attack an enemy gunboat reported by a reconnaissance earlier that day. The pilot aimed two bombs at the boat and one at a seaplane hangar at Kusa Burna and returned, but the Bristol Scout disappeared and was lost.

A more ambitious raid was made on the enemy's main seaplane base at Gereviz on February 27th, by 'A' Squadron. While still dark, three Henri Farmans, two 1½ Strutter bombers and a fighter, and a Bristol Scout, set out to attack at dawn. Due to poor weather the crews could not see the results of their work but the Sopwith fighter was involved in a spirited fight. A German seaplane, flying low over the lakeside base, was attacked by the Bristol, supported by the Sopwith fighter. The enemy pilot, realising he could not escape in the air, landed on the water and taxied for the shore where both he and his observer ran for cover.

Flt. Sub-Lt. J. N. Ingham in the Sopwith circled the enemy machine at 200 feet while his observer, Flt. Sub-Lt. N. H. Starbuck, R.N.V.R., fired three trays of ammunition into it, and made a few diving passes.

On March 20th, shortly before midnight, a Zeppelin raided Mudros and a Sopwith Schneider took off into the night in an attempt to intercept, but lost sight of the airship in the dark.

The wisdom of the paired 1½ Strutter bomber and fighter operating method was proven three days later. A Halberstadt dived on to a bomber, riddling the fabric and the clothes of Sqn. Cdr. J. R. W. Smythe-Pigott, D.S.O. with bullet holes and damaging the controls. Fortunately at this stage the fighter was able to intervene and drove the enemy machine away.

Inevitably, the first loss came to an Eastern-based Sopwith 1½ Strutter; N5223 with Flt. Sub-Lt. J. M. Ingham and Sub-Lt. J. E. Maxwell, R.N.V.R., was lost in a battle with a leading ace in the East, Leutnant Rudolph von Eschwege, on March 30th. Sopwith Baby N1018 was reported lost five days later on a delivery flight by air from *Ark Royal* at Mudros to Thasos. A search by another Baby found it stranded on the sea with engine trouble and it was later towed into Thasos Seaplane Base by a patrol launch. The same day three Sopwith fighters from Stavros had engagements with enemy aircraft.

In the same way that the R.N.A.S. on the Belgian Coast sent reinforcements to the R.F.C., so in the Aegean; and again Sopwith aircraft were involved. To assist the R.F.C. in Salonika, an 'E' Squadron was formed with the precious few Sopwith 1½ Strutter fighters and sent to Hadji Janos. Their first engagement was on April 5th when twelve twin-engined German bombers attacked Karasuli. For forty-five minutes four Sopwiths pursued and harried the raiders, but without success. However, in an engagement three days later, a Friedrichshafen aircraft was forced down and its crew of three captured.

On April 26th, another mass attack by 16 enemy machines crossed the path of an R.F.C. squadron escorted by the R.N.A.S. machines and another fight on Western Front proportions took place, with the difference that the Germans rarely risked day raids with large aircraft in the West. One twin-engined Freidrichshafen was sent down, the bulk of it falling in the enemy lines, but the rudder became detached and fluttered down into the British lines.

Every serviceable Sopwith of 'E' Squadron left the ground on April 28th when six twin-engined bombers, escorted by two Halberstadt scouts appeared. Flt. Sub-Lt. E. P. Hicks took N5099 to a position 150 yards below the first two, permitting his gunner, Acting Air Mechanic (1st Class) Albert King to fire a drum of Lewis gun ammunition into each. Then, climbing slowly, getting into position 30 yards behind a third, another Lewis drum was emptied. Moving on to a similar position behind a fourth, a further two drums were expended, all to no avail as the German formation swept on serenely.

On May 2nd 1½ Strutter 9748 plummeted to earth from 500 feet, killing Flt. Sub-Lt. H. L. Gaskill and his observer, 2nd Lt. J. Watts, Royal Scots, seconded R.F.C. The machine had been badly damaged by anti-aircraft fire and was being nursed back. As soon as the nose was put down to descend at Hadji Janos the wings had folded back.

While No. 2 Wing's 1½ Strutter fighters had been depleted to form 'E' Squadron, their 1½ Strutter bombers were now drawn upon to form 'F' Squadron which flew out from Stavros to Amberkoj on April 29th, to operate under R.F.C. control. Their first raid, on a storage dump, was made next day and the day following they made morning and afternoon raids on dumps dropping a total of 2,500

The No. 2 Wing 1½ Strutters were not the only aircraft of their type in Eastern Europe as this example, operated by the Austro-Hungarian Air Force, shows. It was apparently captured from the French.

lb. of bombs. Operating almost every day they visited Hudovo, base of the twin-engined German bombers on May 10th, but found their tented hangars had been removed.

A combined operation was attempted on May 20th when the monitors M.29 and M.33 bombarded the Customs House, Post Office and harbour shipping at Kavalla. While 'A' Squadron from Thasos spotted for the monitors, 'E' and 'F' Squadrons attacked the Kavalla forts.

Such were the calls made on 'E' Squadron aircraft by other stations, that it was reduced early in May to only two pilots, apart from the commanding officer. Then, on May 27th, just as the squadron were bombing up for their last operation on the Struma front, prior to returning to Thasos to refit for operations against Bulgaria, the squadron aircraft were wiped out in a second. An accidental explosion of a bomb, setting up sympathetic detonations, wrecked ten machines, killed four men and wounded five.

However, 'F' Squadron, then at Marian, was having marked success. On May 25th they twice attacked a large dump in the Kreshna Defile; three bombs made direct hits on a cluster of sheds causing a fire on one side of the dump and another fire was started on the other side. One bomb, bursting among fifty houses, caused a stampede. Later the sheds were seen to be gutted. Next day they located the enemy bombers' new camp near Livunovo, with ten hangars clustered close together in a gully with the landing ground on the other side of the road. 'F' Squadron were withdrawn to Thasos on June 3rd and re-equipped with newly arrived Sopwiths. Its new task was to join with 'A' Squadron in destroying enemy crops, interspersed with attacks on Gereviz Seaplane Base.

Triplane N5431 in Admiralty records, this aircraft was the Triplane in the Aegean area, being the only British Service triplane to be used outside the Western European theatre of operations.

A Sopwith opponent. A Friedrichshafen FF33F floatplane captured intact in the Aegean, and brought to Mudros for detailed examination.

On June 7th, with a strong wind, the first crop-burning attacks were made; fifteen fires were started and four were observed still alight several hours later. Altogether, about 400 acres of crops were estimated as destroyed. The attacks continued, using chiefly petrol bombs; special incendiary bombs sent from England having proved ineffective.

On June 12th a 1½ Strutter, repeatedly attacked by a Halberstadt, was forced down into the sea only 1,000 yards from the Bulgarian coast. Both occupants were good swimmers and set out, under fire from the shore, for a friendly island. Unfortunately Air Mechanic A. E. King suddenly disappeared, probably hit by a bullet. The pilot, Flt. Lt. E. P. Hicks, was picked up by a French motor boat.

The Triplane shipped out early in the year had been kept at 'C' Squadron, Imbros, where it had made a number of interceptions, but without decisive results. On June 9th it was flown back to Mudros for adjustments as it had warped out of true. The Depot there was exceedingly busy on Sopwith aircraft that June. In the first week three 1½ Strutter bombers were converted into fighters, presumably to make up some of the losses sustained by the explosion at 'E' Squadron. Sopwith Pups were beginning to arrive in numbers and two were erected within 24 hours of receipt, one even being tested within that time. Then 'F' Squadron arrived from Thasos on the 18th and their 1½ Strutter bombers needed both engine and airframe overhauls. The Triplane's reconditioning being completed the day of 'F' Squadron's arrival, it was flown this time to 'B' Squadron at Thermi, while one of the new Pups replaced it with 'C' Squadron at the aerodrome on the island of Imbros.

The few Pups arriving in the Aegean bases were a great improvement over the Baby seaplanes, hitherto used on standby for intercepting intruders, but which did not have the necessary speed. While it was British policy to put the latest and best aircraft on the Western Front, the Germans allotted a proportion of their best to the various fronts. In consequence for operations over the Aegean bases, the German aircraft that outpace the best of their adversaries.

The Pup in mid-1917 was still a star in the East and its limited numbers are reflected in the reports of the time; 'C' Squadron at Imbros for example recorded the activities of *the* Sopwith Pup. A follow-up attack of the night bombing of Smyrna Harbour on July 7th was reminiscent of the attacks of the war on the Western Front, by the varied Sopwith types participating. On this occasion a Sopwith 1½ Strutter bomber, accompanied by a similar fighter version, was escorted by a Triplane.

By mid-July 'F' Squadron were back into their stride and for an attack on an enemy camp near Suvla Salt Lake on July 15th they set out from Mudros nine strong—five bomber and two fighter 1½ Strutters and two Pups. All their bombs dropped in the camp area and all returned to base.

That month a Camel arrived, the first 'out East' and made a dramatic début. A large enemy seaplane approached Lemnos early on the morning of July 27th. The four aircraft on standby took off in pursuit. Flt. Lt. J. W. Alcock, later to win fame with Whitten Brown in the first non-stop Atlantic flight, ran to the single Camel, erected the day previous but not yet tested, and took off. Flying towards Tenedos he observed the seaplane below with two smaller floatplanes above, evidently escorts. With the Camel's superior speed and climb he managed to deliver an unexpected attack on the tail of one of the escorts which nose-dived into the sea off Seddu Bahr. The other machines dived for the coast to seek the cover of their own side and Alcock flew on to Imbros to refuel before return to Mudros.

By August there were sufficient Sopwiths to equip 'Z' (Greek) Squadron co-operating with the British and operating under R.N.A.S. command. On August 11th Sopwith bombers from the Greek squadron attacked Gereviz. When starting to escort them, the pilot of Baby N1117 tested his guns, to find the top Lewis gun jammed. While hammering away at the gun with his hands, the Baby side-slipped and then dived into the sea; fortunately the pilot escaped injury, but it was the end of the floatplane. Another Baby was attacked by three Halberstadts which put 50 holes into the fuselage and floats and wounded the pilot who dived down

When the first Camels were received in the Aegean they were allotted one or two each to the various squadrons at Mudros, Stavros, Imbros and Thermi. Unlike aircraft on the Western Front they did not have squadron markings.

Blackburn - built Sopwith Baby floatplanes as well as reaching the Aegean, were also being exported to Norway in 1917. This example is shown in the markings of the Royal Norwegian Naval Air Service.

and landed on the sea alongside a patrol launch.

More Camels arrived in August with Hamble Baby Seaplanes to supplement the Baby Seaplanes. The lone Triplane continued to fly escort patrols. During early September intelligence sources reported that 10 enemy machines had been redeployed for the defence of Smyrna and three Camels were moved from Mudros to 'B' and 'F' Squadrons at Thermi on the 10th. Two days later a Camel from 'D' Squadron, Stavros, that went up to intercept a large enemy machine escorted by a scout, was shot down. The enemy press attributed this victory to Eschwege.

The tempo of work slackened in the second half of the year. Apart from lack of pressure from enemy aircraft the R.N.A.S. personnel were debilitated by the climate. In early October only one naval pilot at Stavros was fit for duty and two R.F.C. pilots were sent on loan. Prior to evacuating Thermi 'B' Squadron was run down to four 1½ Strutters (2 bombers and 2 fighters) by flying its Bristol Scout, a 1½ Strutter bomber and a Nieuport to Mudros and dismantling its entire Camel strength—four—for dispatch by sea to Mudros. On September 30th when enemy seaplanes were sighted off Mudros they were engaged by three fighters—a Pup, a Camel and the single Triplane. Two of the enemy were shot down, but the Camel was forced to retire with a broken inlet valve.

Camels were fitted for the carriage of light bombs in October and bombed Chanak on the 18th.

On December 2nd, a Pup from Imbros had one of the strangest combats of the war. To quote the official account —'A captured Bristol Scout which came up from the vicinity of Chanak was engaged. The hostile machine made a nose-on attack firing a short burst. The Pup side-slipped to avoid the fire and, turning on the enemy, fired on him as he continued his nose-dive away'.

While there was little wrong with the basic Sopwith designs, their power units, armament and equipment constantly bedevilled their activities in the East as on the Western Front. In the Aegean difficulty was experienced with the disintegrating ammunition belts for the twin Vickers, and in mid-November one Camel was flown to R.F.C. units in Salonika to try and gain the benefit of their operating experience with the belts. The Camels had frequent encounters with enemy aircraft but rarely with any decisive result. One Albatros stalled and crashed evading a Camel and two others were sent on their way trailing smoke. The report of Flt. Lt. P. Fowler in B6254 on December 9th was typical of many 'Owing to guns jamming the attack could not be pressed and while the jams were being cleared the hostile machine made off flying towards Galata Aerodrome'.

It was not just jams. On December 12th one of two 1½ Strutters attempting to bomb Galata airfield was forced back because the muzzle attachment of the Vickers came adrift and fell into the propeller which split with the impact.

Nevertheless, the Sopwiths were the mainstay of the R.N.A.S. in the Aegean in 1917, the 1½ Strutters being superior to the earlier and still existing Henri Farmans as bombers until D.H.4s arrived in October. As interceptors, the Baby seaplanes, later the Pups and the few Camels— and not forgetting the single Triplane still flying in December 1917—made the enemy aviators extremely wary of the islands which had airfields. On anti-submarine work the surviving Schneiders and Baby floatplanes did admirable patrol and sea search work. As a Minor Theatre the Aegean received little publicity and little has been written of aerial activity there—but of that activity in 1917, it was certainly Sopwith aircraft that performed the major part in operations.

A Mishap on Mudros? Most of the Clayton & Shuttleworth-built Camels numbered B5676 to B5700 were shipped to Mudros in 1917. By the end of the war a total of seventy Camels were serving on the Ægean island bases, all with Clerget engines.

CHAPTER TWENTY-THREE

Camel Crisis

The prototype Camel at Brooklands in the Winter of 1916-1917. This machine differed from production models by having the top wing made in one piece instead of the wings attached to a top-wing centre-section.

No other aircraft has achieved the fame of the Camel with its record of shooting down more enemy aircraft than any other British fighter in the 1914-1918 War. But there was a time in 1917, after it had been committed to production, when there were misgivings. Once again a Sopwith aircraft was bedevilled with troubles from engines and guns.

The Camel had evolved first from an Admiralty request for an improved Baby floatplane and Contract AS26088 was issued for two prototypes N4 and N5. Its design was another brainchild of the trinity—Sopwith, Sigrist and Hawker, but its construction was undoubtedly traced out by Herbert Smith. N4 was wrecked on an early flight in March 1917 and N5 was produced with a wheeled undercarriage to become the prototype 2F1 Camel shipborne fighter, which is dealt with in a later chapter.

This was a time when the Army's needs for fighters were far greater than the Navy's and three prototypes were built to Army requirements to a new standard. These aircraft became the protoype Camel F.1s and they were supplemented by an Admiralty order for two to a similar standard. The record of their development is given in the Type-by-Type Review near the end of this book.

It has become a misconception that the R.F.C. Camels were F.1s, while those for the R.N.A.S. were 2F.1s, but the designations applied respectively to landplane and shipborne Camels and both the Navy and Army used the F.1 in the land battles on the Western Front.

The first order for Camels in January 1917 was placed by the Navy with the Sopwith Company, and Army orders were placed the following May with both Sopwith and sub-contractors. On May 7th Sopwith commenced delivery of their Navy Camels and in June 1917 the R.N.A.S. had No. 4 (Naval) Squadron, stationed at Bray Dunes, operational.

With most Sopwith types, it was either engines or armament that caused most trouble, but with the Camel it was both. Its revolutionary feature was its two Vickers guns, synchronised with the engine to fire through the propeller arc, whereas all previous fighters had only one such machine-gun.

The Vickers guns in the 1½ Strutter and Pup had been fed by a canvas belt of ammunition carrying up to 500 rounds. Unfortunately, the canvas not only deteriorated quickly but it would freeze stiff in the upper air after becoming damp in the lower regions. The canvas belt was replaced by

Another view of the prototype Camel, but apparently from film that gave different tonal values to the national colours, since the blue outer ring of the roundel and leading rudder stripe have been "lost". The contract price of a Camel airframe without guns or instruments was £874.10.0 A B.R.1. engine added another £650 to the cost.

a belt of metal links that disintegrated after firing, and were ejected with the case of each expended round. With two guns there was little space to permit ejection into the normal chutes. Additionally, the Vickers guns all had the normal right-hand feed and not until later in 1917 were Vickers guns produced with left hand feed blocks for Camels.

Apart from frustrating gun stoppages, the Camel did get rather a bad name as a dangerous machine, by a tendency to spin. This was a combination of the sensitiveness of the machine in responding to control and the propeller torque which caused the nose to rise in a turn to port and drop in a turn to starboard. If the turn was too tight the Camel would drop into a spin without warning. Flt. Sub-Lt. Ellis of No. 4 (Naval) Squadron in N6337, was probably its first victim. Shortly after leaving the ground on July 12th 1917 he turned, spun down and was killed in the wreck. At that stage of the war, the spin had been mastered and a

Squadron at Martlesham Heath reviewed their own assessments. They recalled that when B3751, the first production Camel, had been tested in March it had been extremely cold and they wondered if instruments had been affected. Captain R. H. Carr was sent out to France to find out for himself. At G.H.Q., R.F.C. in the Field he was taken to St. Omer airfield where Camels were put at his disposal for testing. A few had been delivered new from England and there were three from No. 9 (Naval) Squadron, then serving under R.F.C. command. During that August the squadrons had been involved in 29 combats, not one of which was decisive and they had lost one pilot, Flt. Sub-Lt. Woodhouse, who had been shot down in our front-line trenches on the 9th. Yet in the previous month they had driven nine enemy machines down out of control for the loss of one pilot.

Captain Carr set to work. In all he tested seven Camels at St. Omer and his recorded figures showed a wide diver-

Camels of B Flight, No. 10 (Naval) Squadron. This squadron's aircraft had black, red and blue striping, alternating with white; for A, B & C Flights respectively. Additionally, wheel discs were patterned for individual aircraft identification.

Camel would come out after neutralising controls, but in poor Ellis's case he had not the necessary height. In experienced hands the Camel was potentially the finest fighter of 1917, but it took its toll of pilots under training.

In the naval squadrons, the Camel had by early August replaced the Pups in Nos. 3 and 4 Squadrons, while No. 6 had switched from Nieuports to Camels. They were followed by Nos. 8, 9 and 10 which gave up their Triplanes for Camels.

By that time the R.F.C. had the Camel in service, No. 70 Squadron after operating with 1½ Strutters for over a year, gladly exchanged their Sopwith 2-seaters for single-seaters. Their first victory with a Camel appears to be July 27th 1917 when Captain C. F. Collett flying B3756 shot down a German fighter.

Steadily the R.F.C. continued to build up squadrons on the Western Front, logically replacing Pups with Camels. But whereas in June and July there had been general praise for the Camel, in August and September came complaints. The performance of some of the Camels was dropping alarmingly. This had widespread effects since in any patrols by formations, the maximum speed of the formation was, of course, that of the slowest Camel. One Camel was sent home in disgrace! No. 70 Squadron returned B3862 to the St. Omer Aircraft Depot as unsuitable and from there it was sent back to England as having a performance too poor for operations.

Since the Camel was committed to large-scale production the authorities were gravely concerned. The Testing

gence in performance. To give the reader an indication, without boring him with statistics, just the time to reach 10,000 feet is quoted for the seven Camels tested:

Minutes	Camel	Engine	Condition
10¼	N518	B.R.1	Engine 25 hrs.
13¾	B3905	Clerget (RP)	New machine
10¼	B6206	Clerget (RP)	New machine
12¼	B2338	Clerget (RP)	New machine
12½	B6204	Clerget (Gwynne)	4 hrs. 9(N) Sqn.
16	B3898	Clerget (Gwynne)	48 hrs. 9(N) Sqn.
13	B3881	Clerget (French)	64 hrs. 9(N) Sqn.

There had been ugly rumours of sub-contracted Camels being sub-standard, but these figures, with other tests refuted any idea of construction being at fault. It became evident that the engines were the cause of the trouble and that performances were dropping after they had run for some time; thus explaining why the faults were materialising at this late stage. There were indications that the French built engines, when new, were of considerably greater effectiveness at height than English-built Clergets.

To an extent, all Clerget engines were prone to lose power after running for some time. Their valves had to be perpetually adjusted and retimed due to their actuating mechanism wearing through case-hardened parts. It did not take much wear to cause a 2½ h.p. loss which, on a Camel, corresponded to a 50 feet per minute climbing loss at around 15,000 feet and a loss of 1,000 feet on its ceiling.

The flying officers of a Camel Squadron forming up to go overseas, No. 80 Squadron at Beverley, Yorkshire, January 1918. Immediately beneath the Camel's twin Vickers is the squadron commander, Major V. D. Bell. The officer standing third from the right, is Captain H. A. Whistler who became a Camel ace.

Remedial measures were forced through. The ideal solution was to substitute the B.R.1 engine as the valve actuating mechanism, working on roller bearings, covered a wider area and was not subject to the same wear, and thereby loss of power. However, although the Air Board whose job it was to organise and co-ordinate the supply of material, thus preventing competition between the two Departments (Admiralty and War Office), had control over airframes, they did not control the contracts for engines. Thus, Camels for the Royal Naval Air Service had B.R.1 engines and those for the Royal Flying Corps 130 h.p. Clergets. Other engines, such as the 110 h.p. Le Rhone, were being fitted on Camels for training units.

A temporary measure, to compensate for the inevitable loss of power, was to push ahead with experiments already underway to modify the Clerget with a longer piston stroke and higher compression, which raised the horse-power to 140. This modification was ordered into production.

Meanwhile, in the squadrons, mechanics were doing their best to keep the engines tuned, with considerable success. On September 3rd, the Official Communique ran: 'While on an offensive patrol Sopwith Camels of No. 45 Squadron engaged 14 Albatross [*sic*] Scouts, flown by pilots who appeared to be more experienced than the majority.

Captain Harris (later Marshal of the R.A.F. Sir Arthur Harris) dived at one of the scouts which was attacking the rear Camel and shot it down in flames. Another German Scout, however, followed Captain Harris, so Lieutenant Moore dived at it and shot it down out of control. Lieutenant McMaking also drove one down out of control. After the fight three enemy machines were seen wrecked and burning on the ground. Another patrol of the squadron encountered 11 enemy aircraft and in the fighting 2/Lt. Clarke shot down an Albatross Scout which fell in flames, and 2/Lt. Frew shot down one out of control. In an engagement earlier in the day 2/Lt. Macmillan (now Wing Commander Norman Macmillan, O.B.E., M.C., A.F.C.) flew to the assistance of R.Es (R.E.8s), which were attacked by Albatross Scouts. He shot down one of these machines which was seen by an R.E. pilot to burst into flames.'

This report typified the aggressive way the Camels were being handled and later that September a new role was imposed in which rate of climb and ceiling was of little import—ground strafing. Fitted with racks under the fuselage for four 20 lb. bombs, No. 70 Squadron R.F.C. and No. 10 Squadron R.N.A.S. appear to have been the first to employ the Camel operationally as a bomber. This was on September 19th, eve of the Third Battle of Ypres.

Cold start. The Camels of No 8 (Naval) Squadron being started for an offensive patrol during the winter of 1917-1918, at St. Eloi where the squadron was based from May 1917 to March 1918.

The four Camels of No. 70 Squadron that set out on this first bombing mission each carried two 20 lb. bombs, and all four were piloted by Second Lieutenants, youngsters who revelled in the opportunity for low flying. Young Stuart used up his ammunition by repeatedly diving on a body of about 600 troops and then dropped his bombs on targets in Houthulst Forest. F. G. Quigley, a Canadian who finally ranked among the top 20 British aces, got rid of his bombs on strongpoints and then tried to shoot up enemy communication trenches. Dalton used both bombs and bullets against bodies of troops and Michie caused a fire with one of his bombs. Four pilots from No. 10 (Naval) Squadron did similar work.

The following day, with the opening of the Battle, Camel squadrons were employed almost exclusively on ground strafing work. In this they were joined by the remaining Pups, also fitted for dropping 20 lb. bombs. From No. 66 Squadron, 2/Lt. Warnock flew along the canal in a Pup from Armentieres under low cloud. Coming to some railway sidings he dropped his bombs among 30 to 40 trucks, and then fired seventy rounds at transport, lorries and a team of six horses drawing a wagon. Lt. H. Lascelles, of the same squadron crossed the lines at Messines and made for Roulers at 1,000 feet. Coming down to 200 feet he fired a hundred rounds at a car and bicycle on the Ypres-Menin road, and glimpsed the occupants of the car racing for the shelter of a wood. His bombs he reserved for a battery which he had observed firing and then had his attention caught by a crowd of men towing a barge along the canal. He had fired 150 rounds when his gun jammed, so he re-crossed the lines and landed at Bailleul to have it cleared. Rearmed, he returned to stop the further progress of the barge and found it unattended drifting crossways down the canal.

No. 66 Squadron's Pups also went out on the night of October 2nd to drop 20 lb. bombs on enemy aerodromes. 2/Lt. W. Pritt, bombing Waeregham airfield started a fire behind a hangar and then with his gun attacked the men who were trying to put it out. No. 66 gave up its Pups for Camels later in the month.

Another Pup squadron, No. 46 changed to Camels in time to participate in the Battle of Cambrai. On the opening day of the battle, November 20th, 2/Lt. J. Cooper dropped a bomb on a wagon which he saw blow up. Either he was too close to the explosion or he was hit by ground fire for a main spar of a lower wing collapsed, yet he managed to bring the Camel back to his airfield.

In his intriguing book *No Parachute*, Air Vice-Marshal A. S. C. Lee gives a vivid account of his adventures with his Camel on November 24th. This is endorsed in the words of the official day's communique: 'Capt. Lee, No. 46 Squadron, assisted in driving down an enemy two-seater north of Bourlon Wood. He dropped four bombs into the wood, but the bursts were difficult to observe owing to shell fire. He then fired at hostile troops at Fontaine. While retiring temporarily to correct jams in his guns a shell burst underneath him, and he made a forced landing so close to the Germans he had to run 100 yards under machine-gun fire, and was consequently unable to destroy the machine. While

Flight Lieutenant A. Frauenfelder with the Camel he used to instruct instructors at No. 2 School of Special Flying, Redcar, Yorkshire. Camels were also allocated to the No. 1 School of Special Flying at Gosport, Hampshire.

walking back to Cantaing with a guide a shell burst on a house he was passing and blew him across the road. On the way back to get into communication with the squadron he assisted stretcher-bearers to bring in wounded.' Air Vice-Marshal Lee mentions in his book that shortly after this he took seven pilots to collect new Camels from Candas (No. 2 Aircraft Depot) where fifty aircraft a day were being issued to make good losses.

Fortunately there was by then a good supply of Camels reaching France, both from the Sopwith works and sub-contractors. The acceptance park at Brooklands had been closed and after construction at Kingston, Camels were being dismantled and transported to Brooklands for re-erection. Up to October 1917 they were flown to No. 7 Aircraft Acceptance Park at Kenley. However, after some 300 Camels were delivered there, and coinciding with T. O. M. Sopwith's acquisition of the Ham works, Brooklands reopened as No. 10 Aircraft Acceptance Park while No. 7 at Kenley from then onwards dealt almost exclusively with Dolphins. Camels fitted with B.R.1 engines for the R.N.A.S. were delivered direct by road to Hendon for acceptance.

Camels from Boulton & Paul, Portholme, Clayton & Shuttleworth and Ruston Proctor were being flown to France from Nos. 3 and 4 Acceptance Parks at Norwich and Lincoln respectively while four other firms were about to make their initial deliveries when 1917 came to a close. They were to be needed when operations were intensified in the Spring of 1918 with the launching of the great German offensive.

CHAPTER TWENTY-FOUR

Bentley and his Rotaries

The newly commissioned Lieutenant W. O. Bentley, Royal Naval Volunteer Reserve (Engineer Officer) photographed in 1915.

The successes of the R.N.A.S. Camels and the promise of the Snipe was undoubtedly due to their power units, B.R.1 and B.R.2 engines respectively. Originally the B.R.1 had been called the A.R.1 (Admiralty Rotary No. 1) but in deference to the designer, it was renamed the B.R.1 (Bentley Rotary No. 1) in July 1917.

Lieutenant W. O. Bentley, R.N.V.R., was *the* Bentley whose cars became world famous. Having experimented with aluminium pistons before the war, he offered his ideas to the Admiralty. There, Commander Briggs was sufficiently impressed to send Bentley to Gieves for a naval officer's uniform, with two wavy gold rings, denoting Lieutenant's rank in the Royal Naval Reserve. Next day he was sent to Derby where Rolls-Royce were making Renault engines under licence, and later to the Sunbeam factory at Wolverhampton. It was not until his assignment as naval liaison technical officer at Gwynnes, Chiswick, that his association with rotary engines and Sopwith aircraft commenced.

Since Sopwith's concentrated on fighters, rotary engines were an obvious choice; with no radiator and a small flat crankcase, they had a lower weight to power ratio than stationary engines, on average about 2·5 lb/h.p. to 3·5 lb/h.p. respectively. Also, being flat, their main weight was close to the centre of gravity of the aircraft and facilitated manoeuvrability. These considerations, for the class of aircraft in which Sopwith specialised, outweighed their disadvantages—the gyroscopic effect of a 350 lb. engine rotating at 1,250 r.p.m., and the difficulties in regulating petrol and air feed making slow running unreliable thereby necessitating 'blipping' to cut off the ignition for periods when taxiing or making a landing approach.

When Bentley was assigned to Gwynnes, he had a staff of two, Petty Officer Aslin and a naval artificer. The firm, well-known as makers of marine and water pumps, had obtained the British licence rights of the Clerget rotary engine. Superior to the Gnôme, and to any British engine in its class, it was needed in numbers for the Sopwith 1½ Strutter. It had, however, one grave failing; the heat generated by combustion was not dissipated evenly. The front of the cylinders, getting the full benefit of the airflow were cooled better than the rear of the cylinders and the walls of the cylinders became distorted. To compensate for this, 'Obturator' piston rings were fitted because they gave extra flexibility by adjusting their form to fill the distorted shape.

While obturator rings reduced the frequency of engine failures, they were far from satisfactory. Bentley, who had visited the Western Front and made a few flights, realised the consequences of engine failure. Having managed to get Gwynnes to adopt the aluminium piston for the Clerget, he set out to get them to also adopt a new lined cylinder and to improve on the obturator rings. At this stage he met opposition from Gwynnes who did not wish to depart from the engine for which they held licence rights.

Bentley referred the matter to Cdr. Briggs at the Admiralty who decided to give Bentley his head and design an engine of his own. He was sent to Humber's at Coventry whose head designer, F. T. Burgess, was an engineer of Bentley's calibre. Aided again by Petty Officer Aslin and others in the Humber works, a design was soon presented to the Admiralty. In effect, it was the engine that was later to become the B.R.2 that Bentley advocated, but the Admiralty thought that by presenting a 200 h.p. project, Bentley was running before he had learned to walk and he received the go-ahead for a much more modest project. But this proved the project. An engine the size of the 130 h.p. Clerget developed more power and went into production.

The Camel soon proved the worth of the engine. Tests of a B.R.1-powered Camel (or A.R.1 as it was then known) were completed in early April 1917 and production engines were being fitted next month—B.R.1 No. 13 was put into Camel N6345 erected at Dover on May 30th 1917 and on the 25th of next month, flown by Flt. Sub-Lt. Chadwick of No. 4 (Naval) Squadron it shot down an enemy aircraft—one of the first of many to fall to B.R.1-engined aircraft.

As a naval sponsored venture it was natural that the first B.R.1s should go to R.N.A.S. Camels, but the effects of Lloyd George's newly instituted Ministry of Munitions, providing a procurement agency similar to that given today by the Ministry of Technology to the Services, saw to it that the R.F.C. had their quota.

Rivalling the new B.R.1 was a new long-stroke version of the Clerget, known as the Clerget LS. The first of the Ruston Proctor built Camels, B3751, originally powered with French Clerget No. 20521 when it arrived for test at Martlesham Heath on June 15th 1917, had this exchanged for No. 1 Ruston Proctor-built Clerget LS the following September. Not to be outdone, Bentley had B.R.1 No. 19 (ex-Camel N6336), with the compression ratio varied, fitted into B3751.

During this same time Clerget LS No. 2 was fitted in

Camel B3835 which Captain G. W. Gathercole, the Martlesham test pilot, reported as 'vibrating horribly'. A B.R.1 with a 5·3 : 1 compression ratio was then tried in this Camel, before being passed on, in turn, to N518 and B3751 for further tests.

All three engines were eventually used and it is significant that under the official nomenclature the standard 130 h.p. Clerget, with the piston modification as Clerget LS was renamed the 140 h.p. Clerget—whereas it was the 150 h.p. B.R.1. The official tag of superiority meant little to Bentley who had tried to give a 200 h.p. engine at the outset. While being engaged in consultation for a series of modifications to the B.R.1, some of which were introduced in production, he had further plans. Having won the confidence of the Admiralty with the B.R.1 he was given his head for his original venture which appeared as the B.R.2.

The B.R.2 was an engineering work of art. The crankcase started as a lump of metal weighing a hundredweight and after machining to fine limits weighed only 28 lbs. Similarly, each detachable cylinder head was machined from a solid 20 lb block that, after milling and drilling for its fins, holes and parts in fifty different operations weighed only 5 lb. 5 oz.

The aluminium cylinder, for lightness, yet with adequate strength and consideration for maximum cooling had a steel sleeve. The inner surface of cylinders was machined to an accuracy of 100th part of a millimetre and such delicate parts had special cases built for transit in manufacture to avoid damage.

From March 7th 1918 there were fortnightly conferences to discuss B.R.2 engine production while development proceeded on cylinder heads and carburettors using the early hand-made experimental, and early production, engines. By June the R.A.E. were concluding tests of a B.R.2 in a Snipe with a higher compression, and elsewhere Snipe B9965 and Salamander E5931 fitted with the Badin

Bentley's 'masterpiece'—the B.R.2 rotary engine.

petrol system did tests of jet and needle sizes with Bloctube carburettors; in August a Bentley Bloctube carburettor was on running tests. There were anxious moments when production B.R.2s were put on 50-hour running endurance test. B.R.2 No. 14 had only done $13\frac{1}{2}$ hours when the Snipe, in which it was installed, crashed. However, the engine was little damaged and was transferred to Snipe E8044, but after only a further $5\frac{3}{4}$ hours running a fault developed in the petrol system. Eventually, the test was completed and after stripping down and examination all was found to be well. There was considerable trouble at first with the Badin petrol feed system and rumours that failures were caused by the freezing up of certain tubes were so persistent that the R.A.E. were forced to conduct a series of high altitude tests. They proved the failures to be due to faulty installation.

The choice of an efficient propeller is largely related to engine running speeds and in September 1918 Snipes E7987 and E8006 did comparative tests with three different propeller types. One of the engines used was the hand-built second experimental model (known as B.R.2 Ex.2) which later was used for tests with a higher compression ratio.

Many were the tests and experiments to which the B.R.2 was subject, but the basic design was considered sufficiently sound for orders to be placed for 30,000 and several thousand were built before contracts were cut following the Armistice. The B.R.2 remained a standard R.A.F. engine until 1928 when it was declared obsolete for service purposes.

The official *A Short History of the R.A.F.* sums up the whole range of rotary engines with the words: 'As far as the rotary engines were concerned, finality in design would appear to have been attained as a result of their intensive development during the war'. It was Bentley that took the rotary engine to its ultimate, and mainly as the power unit for Sopwith aircraft.

The morning after Zeppelin L33 was forced down at Little Wigborough, during the night of 23rd/24th September 1916, Lt. W. O. Bentley (left) examined the propeller drive from one of the six 240 h.p. Maybach ASLu engines.

CHAPTER TWENTY-FIVE

Sopwith Defenders

A Sopwith Pup at a home station, equipped for home defence flying with a Lewis machine-gun mounted on the top wing. It also has a non-standard headrest fairing and is inscribed with the name 'Monkey'.

Because the spectacular successes of home defence were achieved against Zeppelins in B.Es., it is often forgotten that many other types served in Home Defence in the 1914–1918 War, and that more enemy aircraft were shot down by Sopwith aircraft in defence of the United Kingdom during those years, than by B.Es. or any other design. In the early war years Sopwith Tabloids, Spinning Jennies and Schneiders of the Royal Naval Air Service rose to meet marauding enemy aircraft, but paradoxically Sopwith aircraft of the R.N.A.S. became far more involved in home defence after responsibility for Home Defence had passed to the Army.

On February 16th 1916 the War Office took over responsibility from the Admiralty for the air defence of the United Kingdom. This was a time when all the Sopwith aircraft in service were operating with the Royal Naval Air Service and thereby were no longer directly involved in defence. Indirectly, activities continued much as before. The Admiralty were conscious that the Zeppelins operating for the German Navy, had a fleet reconnaissance role. Their pilots were as eager to destroy the enemy's aircraft as the R.F.C. were to bring down the raiders. Sopwith Baby floatplanes, in particular, continued to fly against Zeppelins operating off the British coast, and incidents involving them have been mentioned in Chapter 15.

Names such as Leefe Robinson stand out in Home Defence operations because of his spectacular success in bringing down an enemy airship in flames that was witnessed by Londoners. But probably no pilot flew more persistently against the raiders of his homeland than Squadron Commander C. H. Butler, D.S.O., D.S.C. He was posted to Manston, a sub-station of Westgate on the South-East coast of Britain on March 14th 1916. He already had night flying experience and at every opportunity he took the air by day or night. In early 1917, Triplane N5424 was allotted to the War School at Manston and Butler had this kept ready for his use.

The Triplane's debut in the Home Defence role was at 5.35 a.m. on March 16th 1917, five minutes after a German seaplane had flown over Westgate dropping 13 bombs. Butler, flying the Triplane, was followed two minutes later by a Bristol Scout, but after patrolling from the North Goodwin Lightship to the North Foreland, without sighting the enemy machine, he returned to Manston.

When some sixteen enemy aircraft attacked Shorncliffe Camp on May 25th, dropping bombs on Tontine Street, Folkestone that killed many civilian shoppers, they were intercepted on their return by only one aircraft from a British base—the first prototype Pup which came up from Dover. The Sopwith Tabloid had been the first ever prototype to be used operationally and the Pup was the first prototype aircraft to fire its guns in anger. Piloted by Flt. Lt. Leslie, 150 rounds were fired at one of the enemy machines. Leslie saw tracers enter the fuselage and black smoke came from the engine but he was forced to break off the attack as other aircraft were attacking him from behind. While manoeuvring to avoid his attackers, he went into a spin, came out but, engine trouble forced a return to Dover.

But Dover had already signalled Dunkirk and from the other side of the Channel there were production Pups to

The new Sopwith L.R.T.Tr., a long range, three seat, triplane which was deemed suitable for home defence by virtue of the top nacelle gun post.

oppose the raiders, five from No. 4 (Naval) and four from No. 9 (Naval) Squadrons. No. 4's Pups sighted a formation of five German aircraft when some thirty miles out to sea. Setting out in pursuit, they attacked a large twin-engined biplane which was seen to explode in mid-air and the wreckage dropped in the sea some fifteen miles off Westende. Of No. 9's Pups, Flt. Sub-Lt. Tanner attacked an L.V.G. off Zeebrugge which spun down; another machine at which Flt. Sub-Lt. Le Boutillier fired nearly 100 rounds, was seen to go down out of control.

On June 5th, 22 bombers of the German group *Kagohl* 3, under Hauptmann Ernst Brandenburg, set out to bomb British bases. His leading aircraft was first sighted at 6.25 p.m. approaching Sheerness where five houses were demolished and a store in the Dockyard was set on fire. Sqn. Cdr. Butler in his Triplane successfully intercepted this time and chased two machines far out to sea, driving both down. He landed in France from where Pups had also been in action.

While on a routine offensive patrol, the pilots of three Pups of No. 4 (Naval) Squadron off Ostend, spotted about sixteen enemy aircraft flying northwards towards England. They started in pursuit but only one pilot managed to engage and he had to break off when his Vickers gun jammed. By then they were too far from Dunkirk to return and so made for Manston. Three more Pups from the squadron went up to intercept the bombers on their return and both Flt. Cdrs. A. M. Shook and A. J. Enstone claimed victims. No. 9 (Naval) Squadron also sent Pups up to intercept and a number of combats took place.

Eight days later came the first of the big daylight attacks on London by enemy aeroplanes. Bombs were also dropped at Margate. In all 162 people were killed and nearly 320 injured. Two bombs actually fell in the Tower of London, but both failed to explode. One of the few British aircraft that managed to get within range of the intruders was Pup 9940, piloted by Flt. Lt. Fox, who attacked one of the rearmost machines at 15,000 feet. After firing a tray of ammunition from his upward firing Lewis gun, he had to withdraw from the fire of other enemy machines to change a tray. But after returning to the fight a tracer round became wedged in the barrel of his gun and he was forced to return to Manston. Again No. 9 (Naval) Squadron joined up from the French side; at this time they were in the process of changing their Pups for Triplanes.

At night, the Zeppelins still attempted to raid England. Five were ordered to attack Southern England on the night of June 16th/17th, but due to cross winds L44 and L47 could not leave their sheds at Ahlhorn and L45 had repeated engine failures and limped back on one engine. This left only L42 and L48 to raid England and the latter was shot down by Lt. L. P. Watkins in a B.E.12 from No. 37 Squadron which, next month, was partly re-equipped with 1½ Strutters fitted out for night fighting. The remaining Zeppelin L42 had a narrow escape; it was pursued by several aircraft including Pups but only Flt. Sub-Lt. G. H. Bittles succeeded in getting within range—and he was flying a Sopwith Baby. Coaxing N1064 up to 11,000 feet Bittles fired a tray of ammunition at L42 from about 100 feet below with his elevated Lewis-gun. It apparently had some effect, for the airship quickly dropped ballast and rose to 15,000.

Fourteen enemy aeroplanes attacked Felixstowe and Harwich in the early morning of July 4th, destroying one Curtiss flying boat and seriously damaging another. Pups and Triplanes rose from Yarmouth, Covehithe, Dover and Walmer but as the raiders did not penetrate inland and made off out to sea, they escaped the defending aircraft—but not the Pup pilots of No. 4 (Naval) Squadron from the other side of the Channel who, being forewarned, set out in their Pups and engaged the returning raiders.

On Saturday morning July 7th the Gothas returned to London in force and nearly a hundred aircraft took off to intercept. One of the first to attack was one of No. 37 Squadron's new 1½ Strutters which ascended from Rochford near Southend. Bravely flying into a formation of eight Gothas, the pilot and his mechanic air gunner were killed by the intensive defence fire and their machine fell in the Thames Estuary.

Butler, still using Triplane N5424, had been up at the first alarm but had failed to sight the enemy and landed for information. Learning that a major raid on London was in progress, he set out again on N5424 with two other Triplanes, Flt. Sub-Lt. R. H. Daly in N5382 and Flt. Sub-Lt.

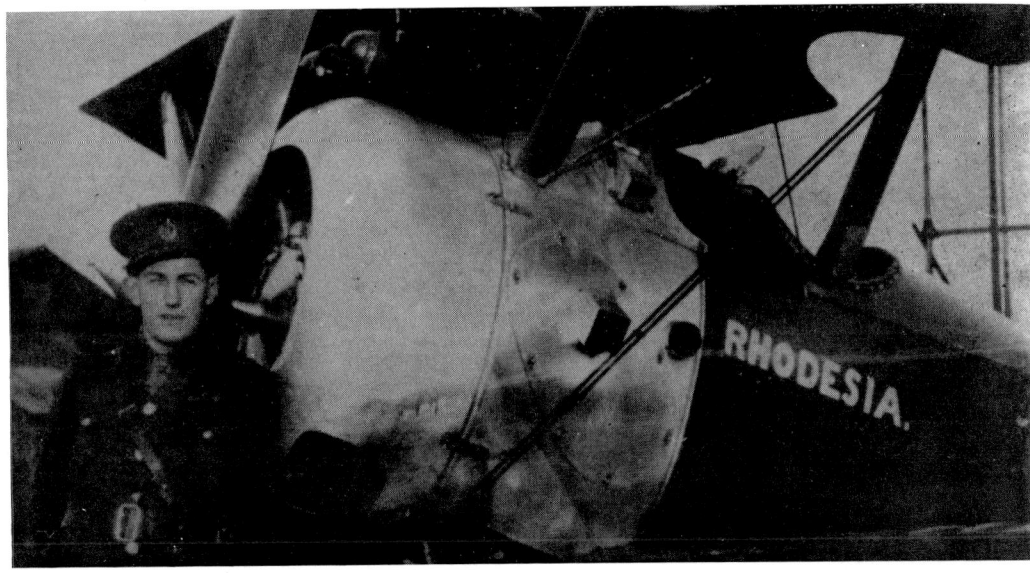

Lt. D. G. Lewis, who became von Richthofen's 80th and last victim when he was shot down in Camel B7393, is here shown when serving with No. 78 Squadron at Suttons Farm on Home Defence duty. His Camel was presented by Rhodesia, a country in which he is still living.

A. H. Lofft in N509. All three attacked Gothas and Lofft followed one raider to Walcheren Island where both his guns jammed. On returning his engine failed within sight of the coast but he glided in to a safe landing.

One Camel pilot, Flt. Lt. Scott, flying B3774, sighted one of the raiders at 8,000 feet some 35 miles out over the sea. Firing 475 rounds he saw the enemy machine nose-dive into the sea from which the tail and a wingtip subsequently surfaced. Although according to German records, this was the only enemy machine destroyed, the Triplanes' guns had a telling effect, as four of the raiders crashed on landing.

In London raids were having their effect. A public outcry galvanised the military authorities into taking action to contain the raiders and new defensive measures, in the main, concerned Sopwith aircraft. No. 46 Squadron of Pups was withdrawn from Bruay in France to Sutton's Farm in Essex and No. 66, another Pup squadron, was re-deployed from Liettres on the Western Front to Calais on the Channel coast. The one other squadron concerned was No. 56 with S.E.5s. However, no more large-scale daylight raids occurred on London and in late August all three squadrons returned to the Western Front.

The next German raid in force, on August 12th, had more limited objectives at the South-Eastern tip of England. Again Sopwith aircraft were in the forefront of the defending aircraft, but again it was gun troubles that stopped them from exacting a heavy toll of the raiders and no less than 16 pilots reported gun stoppages. To quote from a summary of the action:

'Sopwith F.1 (Camel) B3798 engaged one hostile machine, into which the pilot fired a considerable number of rounds, apparently wounding the gunlayer and damaging the machine. During the attack the Sopwith's engine was hit by a bullet.
'Sopwith F.1 (Camel) N6333 engaged an enemy machine until compelled to break off owing to gun jamming. Sopwith's engine was hit in several places by bullets.
'Bristol Scout N5391 engaged enemy machine over the coast, but had to retire owing to the gun jamming four times.
'Sopwith F.1 (Camel) B3844 engaged one Gotha off North Foreland and after firing 30 rounds both guns jammed and owing to the centre-section wire snapping (probably hit) the pilot was forced to land one mile north-east of aerodrome.
'Sopwith F.1 (Camel) B3925 sighted a Gotha off North Foreland and went in pursuit as far as West Hinder and engaged the hostile aircraft. After firing 180 rounds the pilot was compelled to break off the engagement owing to gun jams.
'Sopwith Triplane N5382 attacked a Gotha over North Foreland and attacked five times between there and the Scheldt. The pilot fired 350 rounds and observed tracer entering the enemy machine. When off the Scheldt the hostile aircraft appeared to dive apparently under control. The Sopwith broke off the engagement owing to having no ammunition left, and on return journey passed eight hostile aircraft, which from 500 yards fired on the Sopwith machine with small shell.
'Sopwith F.1 (Camel) B3761 attacked a hostile machine 20 miles east of North Foreland four times, firing in all 420 rounds. Tracers appeared to enter the fuselage on each attack, but the machine seemed well under control. The Sopwith followed the enemy machine until north of Zeebrugge, until one gun jammed, the other being empty; the Sopwith was hit four times by gunfire.

The Sopwith Triplane was particularly successful in the home defence role due to its good rate of climb, but due to large Western Front demands, less than a dozen were available at R.N.A.S. home stations.

'Sopwith Triplane N535 engaged an enemy formation of ten machines over the mouth of the Thames, and after firing 250 rounds both guns jammed. The pilot cleared one gun and again attacked the hostile formation, until his guns jammed a second time, after which he was compelled to break off the engagement.
'Sopwith (Pup) 9947 attacked two Gothas off Holland but had to break off the engagement owing to gun jams.'

This report shows clearly the preponderance of Sopwith aircraft engaged on home defence work in the second half of 1917, and indicates what might have been achieved had the Sopwiths carried reliable weapons.

Camels were not yet allocated to Home Defence units of the Royal Flying Corps, but Camels at acceptance parks were used by ferry pilots whenever there was a raid alarm. Units on the Western Front had priority of issue, but units stationed on the French coast joined in, and in the raid of August 12th, related above, the first sustained attack by Camels against the raiders occurred.

It was quite by chance that a formation of nine Gothas winging their way to England came within sight of five

First used on home defence work at Dover, the Sopwith Pup made by Beardmores, No. 9901, was used for ditching trials. It is seen landing in water off the Isle of Grain, near the mouth of the Thames.

Camels of 'A' Flight, No. 3 (Naval) Squadron at Furnes out on a fleet patrol duty flight. Setting out in pursuit, one pilot was forced back with engine trouble and another failed to make contact, leaving three pilots to pursue the raiders all the way to England.

Flt. Lt. H. F. Beamish, leading the Flight, caught up with the bombers at 15,000 feet about 15 miles from the English coast. But as soon as he opened fire the petrol in his Camel's pressure tank gave out. After switching over to the gravity tank without any response he decided to come down in the sea near shipping. However, giving the engine one more try, it picked up and he set course for Rochford. His fuel gave out just as he landed.

Flt. Sub-Lt. Hayne caught them up ten miles off Southend. He dived three times on the rearmost bombers, but had to abandon the attack with gun jams and as his propeller was shot through and badly splintered, he landed at Manston to have it changed.

The third pilot was Flt. Sub-Lt. G. S. Harrower, whose epic 1½ Strutter flight is related in Chapter 19. He caught up with the bombers at the mouth of the Thames, dived at them several times, using up all his ammunition, and landed at Eastchurch. Incidentally, the only other Harrower in the R.N.A.S. was Warrant Officer P. Harrower at that time attached to the Sopwith works since January 1917 for Admiralty inspectional duties.

No. 4 (Naval) Squadron sent up another flight to catch the raiders on the way back, but of the four Camels that got within range—all had to break off attacks with gun troubles. One Gotha was shot down and 4 crashed on landing.

The Pups built by Beardmores had the Admiralty designation Type 9901 from the serial number of the first—which was stationed at Dover on August 22nd 1917 when the Germans next made an air raid on Britain. Flt. Cdr. G. E. Hervey took this machine up and met ten bombers at 12,000 feet off the North Foreland. He fired 100 rounds at one aircraft from 100 yards range and watched it go into a slow spin and fall in the sea about half a mile off Margate.

But this victory was also claimed by Flt. Lt. H. S. Kerby who took Pup N6440 up from Walmer (later this machine was stationed at Dover), and sighted the enemy formation approaching Broadstairs. Attacking one of the machines, he also watched it fall.

Flt. Sub-Lt. E. B. Drake from Westgate, flying Camel B3844 attacked one bomber at 15,500 feet between Herne Bay and Canterbury and saw several tracers enter the fuselage. Keeping on the enemy's tail he attacked again over Deal until his port gun jammed and he broke off to clear it, before attacking again off Dover.

Squadron Commander C. H. Butler was in the thick of the fighting as usual but he had discarded his Triplane in favour of a Camel. On this occasion he had met his first bomber over Canterbury but as he closed up to make sure of his aim, enemy fire cut the wires controlling the firing of the twin Vickers guns. Butler, breaking away, tried reaching over to press the gun levers and finding that this worked he attacked another machine over Dover, firing at close range until one gun was empty and the other jammed. With one aileron control wire shot almost through—it was holding by one strand—and a bullet embedded in a cylinder, Butler returned, ran to Camel B3843, and took off again.

This time Butler, meeting up with enemy machines over Ramsgate, again experienced yet another stoppage, but again cleared his guns and carried on. Just as he was about to attack a Gotha he observed Camel B3834 dive into the attack, hit the starboard engine of the bomber and set it on fire. The machine then burst into flames to crash between Westgate and Manston. The pilot was Flt. Lt. A. F. Brandon who had been up making his first practice flight on a Camel when he had noticed the 'Readiness' ground signal laid out on his airfield.

Altogether, three Gothas had been shot down and others had been damaged. The Germans decided to restrict their activities to nights when targets were more difficult to locate, but on the night of 3rd/4th September at Chatham 107 naval ratings were killed and 86 injured in one barracks, apart from other service and civilian casualties.

Few of the Sopwith single-seaters were equipped for night flying at that time and it fell to the crew of a two-seater, 1½ Strutter N5617, to make the next sighting which they did at ten minutes to midnight on October 31st, and again in the early hours of the following morning. But in spite of expending all their ammunition, they did not appear to disable the enemy machines.

The pattern was much the same 23 years later in the Battle of Britain when the Germans, having failed by day turned to night attacks. If this analogy is continued, then there is no doubt that in this first Battle over Britain, the Hurricanes were the Sopwiths.

Flying boat protection—an escorting Camel photographed from a flying boat.

CHAPTER TWENTY-SIX

Shipboard Sopwiths

Once holder of the coveted Blue Riband of the Atlantic, the 13,000 ton Cunard liner Campania *was converted to a carrier in 1915. She is seen here* circa *1917 with a Ships Strutter flying from the forward ramp.*

Fleet requirements were for ships to carry aircraft for scouting purposes or for attacking, or at least warding off reconnoitring Zeppelins. With the early floatplanes it had been necessary for ships to stop down to hoist out the floatplanes and then, if there was a swell, take-off was a risky matter. The trials and tribulations of the Schneider floatplanes have already been chronicled in earlier chapters.

A special decking on a carrier ship was the ideal, and that it was possible had been proven pre-war both in America and Britain. But the *Ark Royal*, the first of the carriers modified with a clearway for flying-off, just could not make the speed to assist the aircraft in rising from the deck. However, H.M.S. *Campania*, originally laid down as a liner and commissioned by the Navy in 1915, was capable of 22 knots, over double that of the 'Ark'. In July 1915 she was committed to experiments with aircraft deck take-offs.

During the afternoon of August 6th 1915 a Sopwith Schneider was placed at the top of the forward seaplane hatch of H.M.S. *Campania*, its tail close up against the fore bridge supports. The floatplane's wheels were 152 feet from the bows where the flying deck ended. While mechanics checked over the engine, Flight Lieutenant W. L. Welsh prepared for the first wartime take-off from a ship.

The Schneider had been fitted with a jettisonable wheeled dolly under its floats to facilitate running down the decking. Welsh ran the engine of the 100 h.p. Gnome Monosoupape at a steady 1,050 revs, while the ship steamed at 17 knots into a wind of 13 knots force.

The six seconds from releasing the wheeled floatplane to its leaving the deck, seemed an age to those watching. The engine, taking the strain, spluttered momentarily—but it was a characteristic of this power unit and the Schneider, after a slight check in its forward motion, forged ahead. After some thirty feet the tail was seen to lift, but the machine veered to port; the pilot brought it back on to the centre-line chalked on the decking. With 39 feet of deck left, the wheels lifted clear. Welsh dropped the wheels near a motor boat belonging to his parent ship and landed on the water, to receive a signal of congratulation from the Commander-in-Chief, Grand Fleet.

It was a start, but by no means expedient for operations. The Schneider had been lightened to 1,268 lb., including the 18 lb. wheel attachment, whereas 1,700 lb. was the normal loaded weight for operations. Further development, it was realised, would be best pursued with an aeroplane that would not have the weight and drag of floats. But an aeroplane could not land safely on water, which led to two possibilities: providing a deck on which aircraft could land back, or facilitating ditching. Both were tried, but two years had elapsed before any significant steps had been made.

Experiments were continued with H.M.S. *Manxman* newly commissioned in 1916 as a carrier, but its speed proved insufficient to fly off Sopwith Babies except against a strong wind. However, with Pups it was a different story. The Grand Fleet Committee on Air Requirements recommended on February 5th 1917 that although the Baby floatplanes were admirably suited for anti-Zeppelin work, they should be replaced by Pups on *Campania* and *Manxman*, as this would greatly increase the number of occasions when anti-Zeppelin type aircraft could be flown from these ships. As a result, throughout 1917, there was a general move to replace seaplanes by aeroplanes on ships of the Fleet.

The technique of landing back on ships had not then been perfected and the best that could be done for the Pups was the installation of an airbag in the fuselage to assist flotation. However, when on April 29th Flt. Cdr. F. J. Rutland had to ditch a few miles off the Danish coast, after taking off from H.M.S. *Manxman* for an anti-Zeppelin patrol, his Pup only floated for twenty minutes. Following this ditching, trials were conducted at the Isle of Grain and flotation bags successfully tested on Pup No. 9901 became the Mk. 1 Flotation Bag standard.

Sqn. Cdr. Dunning making his historic landing back on to H.M.S. Furious *in a Sopwith Pup, August 2nd 1917. The aircraft was flown alongside the ship, then brought crab-wise over the deck where a landing crew waited to grab toggle-ropes to secure it on deck.*

Ship's Pups on the fore-deck of H.M.S. Furious, 1917. A collapsible palisade acted as a wind shield, as necessary, to protect the Pups from cross-winds. The wheeled dolly-runners in the foreground were used for launching floatplanes and were pushed clear when Pups were launched.

Some six weeks later the first serious attempts were made with H.M.S. *Furious* to land back aircraft on to the ship. Five Pups were carried, and only with an aircraft of such superb controllability could the experiments be tried. As it was, it was August 2nd before Sqn. Cdr. E. H. Dunning, flying alongside *Furious* and 'crabbing' inboard, brought his Pup back over the deck to be secured by a handling party. On a later attempt the engine choked and the aircraft stalled and fell over the side; Dunning was drowned before his aircraft could be salvaged.

Landing on, being still very much at an experimental stage, was not immediately envisaged for operations. When a Zeppelin was sighted from *Furious* at 06.55 hrs on September 11th and a Pup was hoisted out there was little question of it landing back. There was also little chance of catching the Zeppelin as it took thirty minutes to prepare the Pup for its adversary sighted at an estimated thirty miles distance. Nevertheless, Flt. Cdr. W. G. Moore took off and conducted a search. On returning, he landed on the sea ahead of a destroyer. The pilot was picked up safely but the Pup broke up. A fortnight later, Flt. Cdr. B. D. Kilner left the deck of H.M.S. *Vindex* in a Pup to attack a Zeppelin seen passing overhead, and was never seen again.

Earlier an extemporised platform had been fitted to the weather deck of H.M.S. *Yarmouth* from which Flt. Cdr. F. J. Rutland had successfully flown off in a Pup during June 1917. The following August 21st the platform was used operationally. H.M.S. *Yarmouth* in company with

Routine by 1918, Flt. Sub-Lt. H. L. P. Lester takes off from H.M.S. Pegasus *in a Sopwith Pup during March that year after receiving his training on H.M.S.* Furious.

the 1st Light Cruiser Squadron was off the Danish coast when Flt. Sub-Lt. B. A. Smart was sent up in a Pup to attack a Zeppelin shadowing the cruiser squadron.

Climbing steadily to 7,000 feet, Smart set course for the Zeppelin some fifteen miles away. Catching up, Smart climbed above the Zeppelin to avoid return fire from the gondola and then dived, aiming at the stern of the airship with his Lewis gun. As this had no apparent effect he made a wide circling and climbing turn, and attacked again, this time coming within twenty yards of the tail of the L27. Suddenly the airship burst into flames and Smart had difficulty in avoiding the aft section which in a second had become a roaring furnace. Within a minute the Zeppelin had crumpled and fallen burning into the sea. Smart returned to the cruisers and landed in the water near H.M.S. *Prince* less than an hour after his first take-off from a ship.

Following this success, plans were made to fit all light cruisers with similar platforms and for the larger ships experiments were already underway with a platform fitted on the 'B' turret of the battle cruiser H.M.S. *Repulse*. On October 1st, Sqn. Cdr. Rutland took off from this turret which was trained to head to wind. A take-off was also made from the 'Y' turret, aft, nine days later. So far only Pups had been envisaged as ship-borne aeroplanes, but with the possibility of carrying two aircraft, it was considered desirable that the second should be a two-seater, carrying a naval observer for fleet observation work.

To design a new two-seater would take a considerable time. It was suggested that since single-seat Sopwiths were successful in taking off from ramps, why not Sopwith two-seaters. Thus it came about that the 1½ Strutter, obsolete for operations on the Western Front, was given a new lease of life with the Fleet. The first attempt, on March 5th 1918, of getting a 1½ Strutter off from one of *Repulse*'s turrets was a failure, but on April 4th Flt. Cdr. F. M. Fox made the first successful take-off in a 1½ Strutter, from H.M.S. *Australia*. Following this, the Admiralty issued instructions for all battle cruisers and then battleships, to be fitted with a lengthened ramp above the forward turret for the launching of 1½ Strutters, while aft a ramp would be fitted for the launching initially of a Pup and subsequently of a 2F.1 Camel, then coming into service. This was a sphere of operations on which the percentage of Sopwith aircraft involved was to be precisely 100 per cent.

CHAPTER TWENTY-SEVEN

Discourse on the Dolphin

Grooming the first prototype Dolphin for its maiden flight at Brooklands in May 1917. The civilian extreme right is T. O. M. Sopwith, who at this time arranged for test pilots to fly production aircraft so as to leave Hawker free for prototype testing.

Speedier than a Camel, more manoeuvrable than an S.E.5 and produced in quantity, the Dolphin was the world's first multi-gun single-seat fighter in service. But like so many other British aircraft including Sopwith types, it was bedevilled by engine trouble. Also, by its very appearance, service pilots were prejudiced against the Dolphin for it was one of the few Sopwith aeroplanes that, by a backward staggered wing, just did not look right.

A backward staggered wing was unusual, but not unique. The D.H.5, then in service, had such a feature and around it grew, without much justification, a reputation of being a dangerous aircraft. Also, because the Dolphin's low top wing, pilots were afraid that in the event of the aircraft turning over they would be crushed by the structure.

Harry Hawker, working closely with the design staff—the Dolphin was largely Herbert Smith's brain child—must have had similar views. Yet the design staff persisted in their beliefs and not without good reason. The configuration of the Dolphin resulted from Herbert Smith's appreciation of the basic need in air fighting—an unobstructed view of the sky around and above—by the level of the pilot's eyes being above the centre of the top wing. By so fixing the wing and pilot's position, the centre of gravity and associated aerodynamic factors decreed the position of the lower wing in its backward staggered position.

The first prototype, distinguished by a large car-like radiator, was tested by Harry Hawker shortly after being passed out by the Sopwith Design Department on May 23rd 1917. In June it was forwarded to Martlesham Heath, followed closely by a second machine which was reported upon rather unfavourably. The first, flown to France on June 13th by Captain H. T. Tizard (later Sir Henry Tizard) for evaluation by G.H.Q., R.F.C. in the Field, was greeted by fire from our own anti-aircraft gunners to whom its outline was unfamiliar.

Making the Dolphin a multi-gun fighter was a joint Sopwith/Service venture. Herbert Smith had discussed the possibility during a routine visit of armament officers on July 17th 1917 and a general meeting on the subject was convened to take place on the 19th to which both Army and Navy representatives were invited. Strangely, the Army, the customers, sent only two R.F.C. armament lieutenants, while the Navy turned up in force. The Armament Captain of the Air Department, Admiralty, Captain The Honourable A. Stopford, R.N., drove up with Wing Commander J. L. Forbes head of the Air Department's N3 Armament

Evidence of the world's first multi-gun fighter, the armament of the Dolphin—twin fixed Vickers machine guns synchronised to fire through the propeller arc and two elevated Lewis machine-guns with Norman vane sights. This photograph, taken February 15th, 1918, also permits a glimpse of the reduction-gear drive from the 200 h.p. Hispano - Suiza engine to the propeller shaft.

A prototype Dolphin modified with new tail unit and horn-balanced rudder. This photo was obviously taken at Brooklands as the banked racing track can be clearly seen in the background.

Section, while Engineer Lieutenant F. W. Scarff, R.N., designer of Scarff gun ring, joined, and perhaps rather dominated, the R.F.C. officers in the discussions.

The third prototype, being erected in the Kingston shops, was used for planning the gun arrangement and the party then moved on to Brooklands to see the second prototype in flight conditions. However, the aircraft was having radiator troubles and was being returned to Kingston for a new tailplane and radiator to be fitted.

Radiator troubles and engine modifications continued to delay development, but so impressed were the Expeditionary Force with the potentialities of the Dolphin that the Ministry of Munitions gave Sopwith's their largest order to date, for 500 Dolphins, on June 29th 1917. Production deliveries commenced in November 1917.

In the first month of 1918 Herbert Smith was beset by problems both from the tests of the prototypes and the early production machines, added to which important decisions had to be made on its future development. With production well under way, it was a case of both the Service and Sopwith making every effort to overcome each problem as it arose—and many did arise.

That January, Herbert Smith advised T. O. M. Sopwith that the Ministry of Munitions suggestion to adapt the Dolphin for a B.R.2 engine as an alternative could not be met. Sopwith accepted Smith's views, not only on account of his confidence in his designer, but because several other promising designs on the stocks were being designed around the B.R.2.

The Dolphin was committed to the Hispano engine of which the most serious defect was a spate of connecting rod failures. The Hispano-Suiza Company, who had long since passed the design, averred that the fault lay with Sopwith Aviation in piping the oil from its tank, placed to suit the Dolphin's layout, and that the failures were a natural result of oil starvation. Added to the con-rod problem, excessive vibration was causing the cracking of the oil tanks themselves.

These were not the only troubles. Magnetos had been a problem since the supply from Germany had failed on the outbreak of war. Dixie magnetos had to be imported from America to supplement British manufactured models. It was found that the distributor of the Dixie magnetos could not be removed for servicing or adjustments unless the curve of the Dolphin's top longeron was altered. As Sopwith's already had fittings for the first hundred under way, a decision was made to fit only B.T.H. and S.E.V. magnetos on this first order, but have the curve of the longerons altered on later and all sub-contracted aircraft.

Apart from the engine troubles, there were protracted teething troubles. After trials at Orfordness of one of the prototypes, recommendations were made for the re-positioning of instruments. There were also difficulties with spent cartridges hitting the radiators for which wire netting protection was suggested. While not initially a Sopwith Aviation responsibility, they immediately became one, for such matters were discussed with the design staff for modifications to be incorporated on production aircraft. Even a recommendation for the use of an extra clip on an oil pipe line meant drawing changes and notices to contractors.

All these matters were being dealt with in the Sopwith Works during January 1918. At the same time the Engine Branch of the Ministry of Munitions discussed with Herbert Smith the final form the radiators should take on production Dolphins. The choice was narrowed to three alternatives and the Ministry's officers agreed to have these checked on machines at Martlesham Heath and advise Smith.

Yet another important matter that month required Smith's attention. By the end of 1917 nearly twenty Dol-

For night fighting a special cabane was fitted which could take one or two Lewis guns. By having the gun mounting tube raised, the pilot could swing the gun to full elevation to fire directly upwards in the anti-Zeppelin role.

A Dolphin built up from salvage at Farnborough shown serving with No. 79 Squadron. This aircraft was flown by Lt. H. E. Snyder of 'B' Flight.

phins had reached France. These were issued to No. 19 Squadron whose Lightnings today bear a Dolphin insignia in commemoration of operating this aircraft type. Major E. R. Pretyman, the next Officer Commanding designate for the squadron, saw Herbert Smith to discuss using the Dolphin for night fighting.

In selecting the first squadron to re-equip with Dolphins, No. 19 was an obvious choice. Their Spad VIIs were due for replacement and for their mechanics it was merely a change of one type of Hispano engine for another. The Squadron received their first Dolphins in January 1918 and they achieved their first victory with Captain Irving flying C3799 on March 8th. The other R.F.C. Spad squadron at the Front, No. 23, was similarly re-equipping with Dolphins when the great German offensive was launched. This attack affected the Dolphins in several ways; No. 19 Squadron had to shelve their ideas of night fighting and their chance of making a sizeable score in day fighting was limited by a commitment to low-level strafing in an attempt to stay the advancing enemy troops.

More squadrons had been mobilising in England and No. 79 at Beaulieu and No. 87 at Hounslow were preparing for the Front. However, reports from Beaulieu of bursting radiators, due to a restriction in the overflow pipe, and from Hounslow of a spate of forced landings due to siphoning in the petrol system, brought further troubles to the design staff who were so overburdened that the conversion of six Dolphins with special cabanes and gun mountings for Home Defence, had to be carried out at Farnborough.

Barely had No. 79 Squadron reached France and was settling down to line patrols, when the German onslaught came and they too were embroiled with ground strafing. No. 87 Squadron, still in England was affected in a rather different way. Ordered to fly to St. Omer, their Dolphins

Mishap to C4147 of No. 23 Squadron. This squadron exchanged its Spads in March 1918 for Dolphins which were used for the rest of the war.

were immediately commandeered to provide urgently-needed replacements for the other three Dolphin Squadrons. The pilots were sent back to Hounslow and the squadron re-mobilised and moved to France again on April 25th. After two days at St. Omer, where the pilots were apprehensive that again they were to be parted from their aircraft, the squadron moved as a corporate unit to Petite Synthe for line and offensive patrols. Shortly after this the squadron was moved to Estrees-les-Crecy for interception of German high-flying photographic aircraft.

So far the Dolphin had largely been denied a chance to show its worth; first by the unreliability of its engine, and then the contingencies of war whereby a machine designed for fighting enemy airmen, with a ceiling superior to most of its contemporaries, had been switched to low-level attacks against enemy troops. But the odd chance had come its way and in capable hands it proved a formidable fighter.

The great exponent of the Dolphin was Captain Albert Desbrisay Carter, D.S.O. & Bar, M.C., Croix de Guerre (Belgium), who achieved his first victory with a Dolphin on March 15th 1918 and was frequently in combat during the following eight weeks. On May 15th, the day of the Official Communiques mentioned the existence of the Dolphin, it was reported that in C4132 he 'got on the tail of an enemy triplane which was attacking another Dolphin; he fired a long burst into the E.A. (enemy aircraft) which dived. Pieces of material were seen flying from the E.A., the right-hand plane fell off and it went spinning to the ground ... he also drove one E.A. down O.C.C. (out of control)'.

The pace at which Carter fought could not last. Using C4017 he returned from patrol on May 17th with his propeller boss burnt. Next day he returned with the propeller split. He insisted on flying the same machine, so it was repaired and used by him the following day. On that occasion, he did not return. He was seen to go down some 10–15 miles behind the German lines.

The appearance of the Dolphin at the Front brought one more problem that Smith, in designing it, could hardly have foreseen. Its blunt nose and large radiator bestowed on it the characteristics of a German biplane—'a Hunnish appearance'—in the jargon of the pilots of the time. Indeed, there was once a tussle between some Breguet 14s of a Belgian squadron and a pilot of No. 87 Squadron.

Pilots were growing to like the Dolphin. With two Vickers guns and a Lewis gun it had that extra fire-power. They also proved very sturdy aircraft. C4129 of No. 19 Squadron was successfully landed at St. Omer after the port lower wing had been badly gashed by anti-aircraft fire. Dolphins were frequently used for the most dangerous of ventures—balloon attacks. On June 7th Lt. H. N. Compton on C4130 forced down an enemy balloon after firing 150 rounds into it at close range.

One of the most gallant, but little-known fights, occurred with a Dolphin of No. 23 Squadron. It was confirmed by an Australian private at Villers Bretonneux who saw the officer concerned fall at 10.30 a.m. on May 20th after a fight with three German aircraft. To use his own words 'The Germans appeared to have the best of the fight, when Lt. Crysler dived and rammed the enemy's plane which had flown underneath him. The German plane dropped like a stone, blazing, in the enemy's lines. Lt. Crysler circled and again engaged the other two planes, but we could see that his machine was on fire. He attempted to come down to earth, but his plane was then a mass of flames. He jumped from the plane at about 500 feet landing in our own lines.' The Dolphin was C3870.

At a time when parachutes were not standard equipment, the greatest fear a pilot had was of being burnt alive before his aircraft could be landed. Sopwiths had not experimented with parachutes, but they approached the fire hazard from a new and novel way—a detachable fuel tank. Using Dolphin D3747 at Brooklands, in conjunction with R.F.C. officers, Sopwith Company members watched on June 25th as a pilot jettisoned his Lloyd-Lot tank at 1,500 feet over the airfield. For the purpose of the experiment, the tank was artificially weighted to avoid a fire risk. Sopwith's were prepared to effect design changes to incorporate the tank, but the delays which would be incurred could not be accepted by the authorities.

However, before the war ended, the Calthrop (Guardian Angel) Type A1 parachute had been selected both for the Dolphin and Snipe. This was before the days of seat parachutes and the top decking had to be cut away to accommodate the parachute and a Snipe and Dolphin, so modified,

The fourth and final Dolphin prototype with lower cut fuselage sides and fuselage decking to the production standard recommended by the Testing Squadron, Martlesham Heath. This prototype left the Sopwith works in October 1917 and was flown to France.

were flown to France for a demonstration to the G.O.C., R.A.F. in the Field.

The radiator troubles still persisted and not until September 1918 was a standard form of radiator agreed between the Expeditionary Force and the Air Board. Until then the Sopwith Company were constantly issuing changes of drawings on the Dolphin's cooling system.

By that time, with engine troubles still prevalent, a decision had been taken to use either the ungeared 200 h.p. Hispano or the Sunbeam Arab. But before the Arab was fitted, instructions were issued for converting the geared Hispanos to direct drive. Initial tests were completed in October on Dolphin E8914 and from then on contractors were instructed to fit only ungeared engines which necessitated a new cowling.

The French who had more faith in the Hispanos than the British were interested in utilising the Dolphin airframe for the new more powerful 300 h.p. Hispano Suiza. An example of this engine (No. 205002) had been acquired by the Royal Aircraft Establishment in June 1918 for examination and evaluation. After being dismantled and re-assembled, it was given a propeller for the normal 50 hour running test of experimental engines—but after only seven hours a failure wrecked the crank case. Fortunately a second example (No. 205005) was held in reserve and within a month, the French Government allotted 250 to the R.A.F.

In the closing stages of the war the Dolphins did good work. By virtue of their high ceiling, they were often deployed for intercepting high-flying German photographic reconnaissance aircraft. As stated earlier, the Dolphin was the first of the multi-gun fighters. There was provision for four guns, two fixed Vickers machine guns synchronised to fire through the propeller arc and two Lewis machine guns mounted on the centre section set to fire upward and forward at a mean elevation of 30 degrees. A three-position ratchet allowed a limited elevation or depression. So rarely was an opportunity presented to fire the Lewis guns in this setting that some pilots had them removed, or flew with only one. Then a Flight Commander advocated mounting the Lewis guns on the lower wings, just outside the radius of the propeller blades, firing forward to converge at a hundred yards, while the Vickers were set to converge at fifty yards. This gave a good spread of fire-power all aimed by the one Aldis sight. The arrangement had its advantages and disadvantages; the Lewis gun drums could not be changed, limiting 97 rounds to each gun, but it did allow the use of special ammunition for balloon attacks without fear of incendiary ammunition, which tended to leave the Vickers guns at a slightly slower speed, hitting the propeller.

Captain L. N. Hollinghurst, D.F.C. who retired from the R.A.F. in 1952 as Air Chief Marshal Sir Leslie Hollinghurst, G.B.E., K.C.B., D.F.C., had this to say of the Dolphin just after the war . . . 'As a machine the Dolphin was very easy to fly and was strong and manoeuvrable. The pilot had an excellent view and in my opinion, it was one of the best war machines ever built. Unfortunately, it stalled rather quickly and that, and the peculiar shape of nose which prevented the inexperienced pilot keeping the nose on the horizon, increased pupils' dislike of the type.

'Human nature being what it is, it is the simplest thing in the world to give a machine a bad name. When one considers the immense waste in getting a machine to the production stage only to have it damned by those who have to fly it, this becomes a serious consideration. It is worth considering whether a little judicious propaganda in the right quarters as to the splendid qualities and lack of bad habits of a new type would not be advisable. The ante-room technicist

E4708 bears the marking of No. 35 Company, Royal Army Medical Corps of the Millbank Hospital, London, whose personnel had saved National War Bonds to the value of a Dolphin—a nominal £2,000.

would not be able to condemn the type until he had at least flown it, while the nervous pilot would start with confidence in his new mount. Above all, the machine would start with a fair chance—but the propaganda would have to be very judicious or it would defeat its own object. I have gone somewhat deeply into the merits of the Sopwith Dolphin because I think this is a typical instance where a type has got an undeservedly bad name through a variety of circumstances.'

The Dolphin was used for strafing, intercepting, line patrols, balloon bursting and as bomber escorts. It was on escort duty that the Dolphin's last and major day fight occurred, on October 30th 1918. Nine Dolphins of No. 19 Squadron, led by Captain J. D. Hardman, were escorting D.H.9s of No. 98 Squadron returning from an attack near Mons, when they were attacked by some thirty enemy scouts. Captain Hardman shot one enemy machine down in flames just as it had aligned onto the tail of a D.H.9 and another fell in flames to his guns before the lines had been re-crossed. Four of the bombers were lost, and five Dolphins went down but three of the pilots were uninjured. Four German planes were seen to go down in flames and four others to crash, two more being driven down out of control.

In the last few days of the war Captain F. W. Gillett, an American serving with No. 79 Squadron, scored his 17th victory with a Dolphin and Captain R. B. Bannerman of the same squadron, his 15th. In all the squadron's Dolphins accounted for 64 enemy aeroplanes and 8 enemy kite balloons. No. 87 Squadron, which had also used only Dolphins at the Front, scored a total of 89 victories.

Surprisingly, the Dolphin squadrons at the Front were never increased, and Nos. 19, 23, 79 and 87 remained the only squadrons at the Front in spite of a thousand Dolphins having been delivered to the R.A.F. before the Armistice. No. 90 Squadron did form up in England with some Dolphins but was disbanded in August 1918 to reform as a bomber unit. For Home Defence No. 141 Squadron was equipped with Dolphins, but they were replaced by Bristol Fighters. Finally, No. 81 Squadron at Upper Heyford was forming with Canadian personnel and in November 1918 it became No. 1 Squadron of the newly formed Canadian Air Force—just too late for the fighting.

After the war, early in 1919, Nos. 19, 23 and 87 handed in their Dolphins and returned as cadres to the United Kingdom. No. 79 Squadron, however, took its Dolphins to Bickendorf, near Cologne, with the Army of Occupation and after this squadron disbanded in June 1919 the Dolphin, which was serving in training units, was withdrawn.

Starting a Dolphin in the winter of 1918 - 1919 at Bickendorf, where No. 79 Squadron were the sole Dolphin squadron to be stationed in Germany with the Army of Occupation after the Armistice. During the war the squadron had destroyed 64 enemy aircraft and 8 enemy kite balloons with its Dolphins.

CHAPTER TWENTY-EIGHT

The Camel helps to avert a Crisis

Steadily throughout 1917 the R.F.C. continued to build up strength on the Western Front, aided by naval Camel squadrons operating under R.F.C. control. It has been related in the previous chapter how Dolphins were flung into battle during the great German offensive in the spring of 1918; the main tasks of ground strafing had fallen naturally to the Camel squadrons which had been regularly exercised in this kind of work for some months.

However, Camel strength had already been depleted on the Western Front as a result of severe reverses suffered by the Italians at Caporetto in October 1917. The Allies had sent reinforcements and Britain's contribution had included the VIIth Brigade R.F.C. which, apart from two Squadrons of R.E.8s and two balloon companies, consisted of three Camel squadrons—Nos. 28, 45 and 66. The Camels achieved many successes there against the Austro-Hungarian Air Service and from November 29th until the end of the year, destroyed 13 enemy aeroplanes and two kite balloons, and drove down six other enemy machines for the loss of two aircraft.

It was on this Front that the sole Victoria Cross for an exploit in a Camel was awarded. To quote an official account—'on the 30th March a gallant performance earned the Victoria Cross for Lieutenant Alan Jerrard, a pilot of No. 66 Squadron who had been distinguishing himself for some time previously. While on patrol (in B5648) with Captain P. Carpenter and Lieutenant H. Eycott-Martin, a formation of four Albatros D.IIIs was attacked. The enemy scouts made for their aerodrome, and Carpenter shot one down, while Jerrard attacked another and followed it down to 100 feet where it burst into flames. He was next seen flying over the aerodrome at 50 feet firing into sheds and attacking, one after another, aeroplanes which were trying to take off. Soon there were about 19 enemy aeroplanes in the air, of which Jerrard was engaging six. Lieutenant Martin went to his assistance, and crashed a D.III which was on his tail. Jerrard then destroyed his second enemy aeroplane, and continued to "shoot up" the aerodrome; a D.III then climbed on Martin's tail and was shot down and crashed by Jerrard. The latter immediately resumed his one-sided combat with six D.IIIs, the other two Camels being driven off when they endeavoured to go to his assistance. He then tried to regain the formation, but was flying very weakly, as though wounded. He was followed and attacked by 10 D.IIIs, so he turned and engaged them, firing repeated bursts at each in turn until he finally crashed four miles from the hostile aerodrome, where he was made prisoner'—He escaped later.

One of the Aces of this Front was Captain W. G. Barker, who was to meet T. O. M. Sopwith the following year. He moved to Italy with B6313, the Camel he had flown in France, and then had it transferred to No. 66 Squadron when he was posted from No. 28 Squadron. But on being promoted Major to command No. 139, a Bristol Fighter squadron, he was not permitted to take a Camel with him; however, by having the machine sent back to 'Z' Aircraft Depot, and from there 'temporarily attached' to No. 139 Squadron, the same effect was achieved.

Lieutenant A. Jerrard, the Camel pilot who received the Victoria Cross for his gallant exploits in Italy.

Captain K. R. May (above) whom the German ace von Richthofen was following when he was shot down; and (below) Captain John Trollope of No. 43 (Camel) Squadron who, on March 24th 1918, shot down six enemy aircraft with his Camel.

Line-up of No. 208 Squadron's Camels on the Western Front. The aircraft at the extreme left has a non-standard spinner fitted over the propeller boss. The squadron was commanded by Major C. Draper, D.S.C.

Meanwhile, at home, two Camels were pioneering radio telephony (R/T) at the Wireless Experimental Establishment, Biggin Hill. On March 22nd, B6303 fitted with a propeller-driven generator on the undercarriage powering a transmitter mounted behind the petrol tank, transmitted on 430 metres to C1614 fitted with a receiver. Messages were audible up to three miles with the transmitting aircraft trailing 150 feet of aerial wire and the receiving Camel, 199 feet. However, not for another ten years was R/T introduced in single-seat fighters. At this time wireless telegraphy (W/T) was in general use, i.e. Morse signalling of a tone as apart from speech, but did not apply to Camel F.1s except for Clerget-engined machines at Turnhouse, B5581 and B5739, used for fleet co-operation work.

In France at that time, only guns and bombs mattered with aircraft. The German attack had come on March 21st. Their advance on a broad front was cloaked by mist, but such was the desperate situation that machines took the air, descended through the fog, and expended their bombs and ammunition on sighting the enemy, then returned to refuel and re-arm. There was no lack of targets—marching troops, horse-drawn batteries and supply columns were seen on a scale never before encountered.

As the British troops gave ground so airfields were evacuated and on occasions patrols would return to their airfields to find them already evacuated. Several times pilots flew back to their airfields under fire, and in searching for the new location would be forced to come down in fields and perhaps wait days before petrol could be acquired.

Such was the state of the country-side, pock-marked with shell holes, that all but the troops had to advance along the roads. With front-line troops overrun and reinforcements having to march long distances, and artillery withdrawing and re-registering in new target areas, it fell to the aircraft to harass the invaders. Along the Bapaume-Cambrai road Camels flew in relays, blowing craters in the road to slow down vehicles, killing and maiming horses with machine-gun fire to block the road with wagons, ditching lorries and diving down and firing in the hope of killing the driver, or unnerving him to swing his vehicle off the road—and once off it was often a salvage operation to get back. Marching troops scattered as the Camels swooped and if it was open land, it became a killing area as the Camel pilots repeatedly traversed the area on such rewarding targets—for their Army had, as Field Marshal Haig put it in his famous order, their 'backs to the wall'.

It is difficult to assess just how effective was the spirited defence put up by the strafing fighters in this dire crisis, but one German regimental history alone, at this time, reported the loss of 8 officers and 135 men in one attack from the air lasting only a few seconds on March 25th. That evening the armament officer of No. 4 (A.F.C.) Squadron indented for 15,000 rounds of ·303 ammunition to replenish stocks used that day by his squadron's Camels.

Altogether, there were ten Camel Squadrons on the Western Front, representing 41 per cent of the total fighter strength. They were:

Squadron	Commander	Base
No. 3	Major R. Raymond-Barker, M.C.	Warloy
No. 3 (Naval)	Sqn. Cdr. R. Collishaw, D.S.O., D.S.C.	Mont St. Eloi
No. 4 (A.F.C.)	Major W. A. McCaughry, M.C.	Bruay
No. 43	Major C. C. Miles, M.C.	La Gorgue
No. 46	Major R. H. S. Mealing	Le Hameau
No. 54	Major R. S. Maxwell, M.C.	Flez
No. 65	Major J. A. Cunningham	Droglandt
No. 70	Major H. B. R. Grey-Edwards	Marieux
No. 73	Major T. O'B Hubbard, M.C.	Champien
No. 80	Major V. D. Bell	Champien

The locations changed as squadrons withdrew to airfields further back or were switched to meet new advances. Following the initial attack on the St. Quentin Area, came further attacks on new sectors. Squadrons were flying from dawn to dusk. Replacement pilots were hurried out from England, machines in store were withdrawn and production was further urged. At La Gorgue sixteen Camels, one of them just flown in on its delivery flight from Clayton & Shuttleworth's factory at Lincoln, were set on fire to prevent them falling in enemy hands. The two great Aircraft Depots and associated Supply Depots were choked with repair and issue work. No. 1 at St. Omer and No. 2 at Candas repaired and supplied Clerget-engined Camels to Squadrons according to the particular Wing in which they served, while No. 4 Aircraft Supply Depot at Dunkirk dealt with all Bentley-engined Camels. During April, both Nos. 1 and 2 were threatened by the German advance and substitute camps were hastily erected at Vron, Bahot and Marquise to take elements of these two depots which employed some 150 officers and 3,250 men. They were, of course, concerned with all types of aircraft, but at that time the attrition rate of Camels exceeded that of all other aircraft types due to the numbers engaged.

On April 21st 1918, the leading ace of the war, Manfred von Richthofen, was shot down following a combat with the B.R.1-engined Camels of No. 209 Squadron, recently recalled to the Front. They were led by the same Squadron

Commander C. H. Butler who had distinguished himself on home defence work as related earlier. Whatever controversy there is as to who shot down von Richthofen—and there is good reason for controversy*—it was *officially* credited to Captain Roy Brown flying Camel B7270. The official R.A.F. Communique stated 'Captain A. R. Brown, No. 209 Squadron, dived on a red triplane which was attacking one of our machines. He fired a long burst at the E.A. which went down vertically and was seen to crash on our side of the lines by two other pilots of No. 209 Squadron.' Brown, let it be said, never claimed to have shot down Richthofen. He claimed to have shot at a red triplane which went down vertically and two other officers who did not witness the fight, saw a Fokker Triplane crash. It was Brown's superiors who assessed that the red triplane he shot at, was the one wrecked near Corbie.

That same day, the redoubtable Captain R. A. Little was in yet another fight. Shooting down his first enemy aircraft with a Pup, he became an ace on Triplanes and ranked eighth in the British list of aces. On this occasion, using a Camel, he attacked the rear machine of a formation of twelve and as he watched it fall out of control he was attacked by six of the enemy. Spinning down with controls shot away, the machine dived to within a 100 feet of the ground where it flattened out with a jerk, breaking the fuselage just under the pilot's seat. Little, having undone his belt against prevailing advice, was thrown clear as the machine struck the ground. Two enemy machines, following Little down, raked the wreck with fire to which Little replied with his revolver until reinforced by some British infantry who drove them off with their machine-gun fire.

Commanding Little's squadron, No. 203, was the famous Raymond Collishaw who, like Little, was a skilled exponent of the Triplane and Camel, and the highest scoring surviving ace.

Another famous Camel pilot at this time was the balloon-bursting Captain H. W. Woollett of No. 43 who on April 12th achieved six victories on the one day. Ten days later the communique gave a typical report on his special activity 'fired about 60 rounds into a balloon which fell in flames, the two observers jumping out. He then climbed through the clouds and flew north and dived on another balloon which he had previously seen. Forty-five rounds were fired into this balloon which fell in flames.'

Another ace who was rapidly increasing his score was

* See *Von Richthofen and the Flying Circus*, Harleyford.

D. R. MacLaren who rose to command No. 46 Squadron which he had joined in February. He kept flying and fighting until shortly before the Armistice, when he broke his leg wrestling. His final placing was sixth on the list of high-scoring aces of the British Services.

The superiority of the Camel was now apparent to the American Forces who were learning quickly from their Allies. The deeds of the United States Air Service in 1918 will for ever be associated with Nieuports and Spads, but a determined effort was made to get Camels. T. O. M. Sopwith was asked by the Air Board to consider granting a licence for French production of the Camel, from 200 to 500 were initially suggested, to be fitted with Gnome Monosoupape engines made in France. French industry, however, was committed to large-scale production of Spads and eventually the Americans decided to take Spads from the French and Camels from the British, but the latter powered with French-built 150 h.p. Gnome Monosoupape engines.

Since there was some delay in proving the Camel with the 150 h.p. Mono (see Type Review) it was decided to equip one squadron, as an interim measure, with readily available 130 h.p. Clerget-powered Camels. It says much for the terrible attrition rate of fighter aircraft, that to arrange to equip this one squadron of 24 Camels, for that short period, 104 were requested—24 initial equipment and a wastage of 80 in just 5 months! These figures were based on British squadron experience. In the event, the Americans formed two Camel squadrons, which remained under R.A.F. command until November 1st. These were the 17th formed on June 20th and the 148th on July 1st.

In May 1918 the Camel had been introduced to a new front—Salonika, by forming a 'C' Flight of Camels in No. 150 Squadron which operated S.E.5As and Bristol Monoplanes. On May 4th Lt. C. B. Green in C1587 had the first combat involving the Camel in that theatre, and Lt. G. C. Gardiner, flying C1598 on June 6th, made the first Camel victory in this theatre. Flying D6549 on September 3rd, he also scored the last Camel victory in the Macedonian area.

In France, the tide of battle turned and in August the Allies were on the offensive. But this did not mean a respite for the Camel squadrons, but intensified ground strafing, this time of retreating troops. Air Chief Marshal Sir Robert Foster, whose account of Sopwith Pups was given earlier, had been recalled to the Front in the earlier crisis and was still flying Camels in the Autumn. He wrote:

Camel B6230 of No. 209 Squadron. The marking of large fuselage bands, where the roundel normally appeared, was a characteristic of the aircraft of this squadron.

Camel B9268 at a training unit in England with a camera gun fitted above the engine cowling in place of the usual twin Vickers guns. This was typical of the Camels at the four Fighting Schools in Britain in 1918.

'The Camel, when one knew it well, was a really splendid fighting machine: one could always turn inside the enemy, but we always were inferior in speed both in level flights and in the dive. In his last months of 1918, Camels were involved in a great deal of ground attack work. In this type of operation, the wise thing to do after the initial dive in to the attack was to stay at ground level both to complete the mission and to get home. The great handiness of the Camel enabled one to jink round houses and trees, go out of sight down valleys and do steep turns at no height. Of course it was impossible to avoid all the concentrated fire from the ground: but without the Camel's exceptional manoeuvrability our casualties would have been more severe.

'As evidence of the structural strength of the Camel, the following is an extract from my Log-Book.

23.8.18 Sop Camel C61 25 mins. Raid on Canton aerodrome. Gibbons collided with me picking up formation. Got down safely with buckled top plane and centre-section.

'This was one of the jolly operations thought out by our belligerent air staff. A squadron of Camels, each with 4 by 20 lb. bombs on a rack below the fuselage and a high proportion of incendiary rounds in their ammunition, would go 20 miles or so over the lines—quite a distance in those days—and do a low attack on a German airfield, hoping to catch their aircraft on the ground. There was a fighter escort overhead, while the Camels dived down from 10,000 feet to ground level and attacked aircraft, hangars and anything else in sight. The Camels then retired, hedge-hopping all the way back to the lines, leaving the escort to fight their way home as best they could. Five days before, we had had an outstandingly good raid on Phalempia airfield; but the air staff could not leave well alone, and we were told to repeat the operation on Cantin. As I had led the Phalempia outing, I was detailed to do the same for Cantin.

'There were fifteen aircraft in three flights of five, and we climbed to 8,000 feet and headed for Arras where we were to rendezvous with our Bristol Fighter escort. I had just spotted our escort and was doing a gentle turn towards them, when I felt and heard an ugly crunch. I looked up to see four 20 lb. bombs just above my head, a red-painted wheel sticking through my port upper plane and another red wheel firmly pressed on to my centre section. My left-hand man, instead of watching me, had been looking out for our escort. As luck would have it, we both of us did the right thing—I shut my engine, hoping to drop clear, while Gibbons, instinctively, I suppose, did a flat left turn. The wheel, which was sticking through my top plane freed itself by turning in the hole it had made, while the one on the centre section was already clear. Unfortunately Gibbons' left turn brought his tail skid with a sweeping arc right across my right top plane, ripping the canvas and away it went. Then my revolving airscrew took a large chunk out of his tail plane as he got clear—but, anyway, we were separated.

'For me, there was only one thing to do—no parachutes in those days—to get down if I could before the aircraft broke up: it didn't look too good with the broken mainspar sticking through the canvas on the port wing, the centre section obviously badly damaged, though the cross-bracing wires were not broken, the canvas ripping off the starboard plane as we went along, and the engine vibrating badly from losing a sizeable part of one propeller blade. The first thing to do was to edge the aircraft round so as to face west and away from the lines, then to find the optimum gliding speed and hope for the best. In the event all did go well and I landed safely, although I nearly botched it by doing a slow flat turn to get into Izel le Hameau aerodrome, which I did. Gibbons also got down safely with half his tail-plane and elevator missing. My mechanics were astounded that the aircraft had held together as it did—but I was flying it again only a few days afterwards, and did many more operational sorties with C61.'

In September a rudder of enlarged area was fitted to a standard Clerget-engined Camel and gave a distinct improvement in directional control at low altitudes. This was then recommended for all Camels engaged on low flying duties and examples were sent to France the following month. However, it was December before the modification was incorporated on production aircraft.

Also during September tests were carried out at Farnborough to investigate the behaviour of the Camel when spinning and the effect of different motions on the controls, in bringing the aeroplane back to level flight. It was found that the Camel spun smoothly to the left, but in a series of jerks to the right, but the period was about the same i.e. two seconds per complete turn. The B2312 was committed to these trials with a Clerget engine, and D1965 with a B.R.1.

As mentioned in the previous chapter, a Lloyd-Lott detachable tank was dropped from a Dolphin. This was also considered for the Camel, but with modifications. The turning rod method of release was liable to jam—as it had on its initial Dolphin test and a bomb dropping release mechanism was mooted. As a further precaution it was thought the tank should be covered with rubber to effect a measure of self-sealing in the event of being shot through.

A self-sealing tank was fitted on Camel H7343 in the Hooper works during October 1918 while the Sopwith works experimented with an alternative method of fitting self-sealing tanks. But as before, the parachute was still thought to provide the best safety measure in the case of fire in the air. A well-known Camel pilot wrote a book in recent years in which he highlighted the fact that pilots in the 1914–1918 War did not have parachutes which is true enough – but the Camel did play a small part in the development of the Service parachute.

The Parachute Committee had 14 sets of parachute harness sent to Orfordness for test on October 17th. Trying one out, Major Oliver Stewart reported two days later, 'I flew a Camel wearing a Guardian Angel parachute harness and found that the movements of my body which would be required when fighting were restricted and that on turning my head to look backwards and downwards, my chin came against the rings which go round the shoulders. The top part of the rings cut into my shoulders, causing me considerable discomfort, which was increased by the movement of the arm, such as that necessitated by altering the throttle or fine adjustment openings.'

His views were endorsed by a report by Captain R. M. Charley next day who flew for an hour in a Camel with a harness over his flying jacket. He found it impeded movement and that in a long flight as a scout pilot continually turning to scan the sky it 'would become extremely uncomfortable—in fact, impossible'. The station medical officer, Captain J. E. Rawlings, M.C., of the U.S. Army, giving a summary on the reports stated: 'It is my opinion that any harness for receiving the weight of the body in a parachute descent should receive and distribute that weight to the parts of the body most suitable for receiving shocks automatically, that is, first, the hips and thighs and secondly the chest and armpits. Any system that departs from this principle will be deficient and fail to perform the function intended.' The question of parachutes in Camels was not pursued, but experiments were continued with the Snipe.

Within a week of the Armistice the Camel reached its peak in active combat by an engagement on November 4th which an official historian described as 'the record combat of the War in the Air.' Led by Captain J. L. M. White a patrol of Camels from Nos. 65 and 204 Squadrons met forty enemy scouts. In the ensuing fight, nine enemy machines were claimed as destroyed, six driven down out of control and two others driven down, making a total of 17 for one pilot missing and one taken prisoner with wounds. Since moving to France on October 24th 1917, No. 65 Squadron had used only Camels and by the time of the Armistice it was credited with 219 victories of which 136 enemy machines were recorded as destroyed.

When the war ended on November 11th 1918 squadrons using Sopwith aircraft on the Western Front were:

Sqn.	Type	Commanders	Base
3	Camel	Major R. St. Clair McClintoch	Inchy
4 AFC	Snipe	Major W. A. McClaughry, M.C. D.F.C.	Ennetieres
19	Dolphin	Major H. W. G. Jones, M.C.	Abscon
23	Dolphin	Major C. E. Bryant, D.S.O.	Bertry
43	Snipe	Major C. C. Miles, M.C.	Bouvincourt
45	Camel	Major A. M. Miller	Bettoncourt
46	Camel	Major G. Allen	Busigny
54	Camel	Major R. S. Maxwell, M.C.	Merchin
65	Camel	Major H. V. Champion de Crespigny, M.C., D.F.C.	Bisseghem
70	Camel	Major G. W. M. Green, D.S.O., M.C.	Halluin
73	Camel	Major M. LeBlanc-Smith, D.F.C.	Malincourt
79	Dolphin	Major A. R. Arnold, D.S.C.	Reckem
80	Camel	Major D. V. Bell	Bertry
87	Dolphin	Major C. J. W. Darwin, D.S.O.	Boussieres
151	Camel	Major C. J. Q. Brand, D.S.O., M.C., D.F.C.	Bancourt
152	Camel	Major E. Henty	Carvin
201	Camel	Major C. M. Leman	La Targette
203	Camel	Major T. F. Hazell, D.S.O., M.C., D.F.C.	Bruille
204	Camel	Major E. W. Norton, D.S.C.	Courtrai
208	Snipe	Major C. Draper, D.S.C.	Maretz
209	Camel	Major T. F. W. Gerrard, D.S.C.	Bruille
210	Camel	Major B. C. Bell, D.S.O., D.S.C.	Boussieres

Sopwith aircraft composed just 50% of the total fighter strength on the Western Front.

An innovation late in the war — the two-seat Camel dual-control conversion for training purposes. The original proposal for a two-seat Camel was rejected and not until the last month of the war was the project approved and the drawings passed. As a result only a few Camels were converted to this standard.

CHAPTER TWENTY-NINE

Trench Fighters

The downward firing guns of the Camel T.F.1. This Sopwith photograph, dated February 19th 1918, is titled Sopwith Camel F.1 Armoured Trench Fighter.

General Headquarters of the British Expeditionary Force were well aware in late 1917 that the Germans would launch a massive offensive in 1918 in an attempt to crush the Allies before the Americans entered the Field in force. Among the requirements tabled, was E.F. Type No. 2 for an Armoured Trench Fighter. The Ministry of Munitions referred this requirement personally to Sopwith, who discussed it with his design staff.

The initial proposal was to arm a Camel with downward firing guns, protected by 700 lb. of armour and powered by a 110 h.p. Le Rhône which, being lighter than a Clerget and taking less fuel, would compensate to a degree for the additional weight. Nevertheless, a new set of strengthened wings was thought to be necessary.

At the works, the project was designated T.F.1 (Trench Fighter No. 1), but already the lines had been cleared of Camels as Snipe production was getting well under way. Boulton & Paul were ordered to despatch two Camel airframes on December 10th 1917, but as these had not arrived some days later, presumed temporarily lost on the railway, a machine was allocated from Martlesham Heath.

The first armoured Camel flown at Brooklands on February 15th 1918, had 5 gauge armour plate with ballast to make up the weight of 11 mm. plate ordered for the next two. After brief trials it was flown to France on March 7th. Meanwhile the Sopwith design staff were already working on a new trench fighter, based on the Snipe to the E. F. Type 2 requirement. Designated T.F.2 it became known as the Salamander. Before January was out the first was being erected. Originally plans had been to restrict the normal forward-firing Vickers to one, and have two downward-firing Lewis guns with a variable depression of 35–55 degrees, but this was changed. As Wing Commander Alec Ogilvie of the Ministry of Munitions Design Branch explained in writing to the Controller of Technical Development on March 8th 1918—'It has been provisionally arranged that three machines shall have two synchronised Vickers gun mounted in the ordinary way for forward firing instead of two Lewis guns arranged for downward firing, so that in the event of E.F. deciding that the downward-firing Lewis guns on the Armoured Sopwith Camels are unsuitable, there will not be any delay in producing an alternative.'

It was a wise precaution, and Ogilvie had probably taken into account the simple fact that the Expeditionary Force authorities endorsed, when they returned the T.F.1 Camel to Brooklands in mid-March, that a machine firing forward at a small angle of depression, covers a far greater killing area than a machine in level flight with acutely depressed guns.

The German offensive was launched in March 21st and the British Army, hard-pressed, was fighting desperately. Some idea of the critical situation became obvious at the works as all available aircraft were collected from the Brooklands sheds. Emphasising the situation was the num-

Salamander E5429, the first prototype, awaiting its turn to commence its trials at Brooklands on April 27th 1918. The machine on the right is an S.E.5A.

Modification TF2/28 called for a special camouflage for Salamanders from October 15th 1918. On this Salamander women workers have effected this new finish which, unlike all other camouflage schemes, had the division of colours in the disruptive pattern of light (grey/green) and dark (green/purple) shades, marked with a thick black line.

ber of components stamped 'PP'—'Passed for the sake of Production' that left the works, a concession that the Aeronautical Inspection Directorate used liberally only once per world war—1918 and in 1940.

The T.F.2, now named Salamander, was passed out on April 26th and that same day was taken over to Brooklands for flying trials to commence the following day. Meanwhile the engine for the second prototype was being shunted into sidings at Kingston Station, a matter of 200 yards from where the aircraft was being erected.

Tests on the first Salamander were hastened, but just before the final trial flight, prior to going to France for evaluation, a piston fractured. A new engine was fitted and the aircraft eventually flew overseas on May 9th and was tested at No. 1 Air Supply Depot two days later. Favourable reports having been received, it was ordered into production early in June, a large contract being awarded to the Sopwith works.

In both the T.F. projects the firm worked closely with Firth's as armour plates, initially ordered, had to be cut to size at the works, fitted, dismantled and returned for hardening, and then back to Kingston for final embodiment. For the production models, with proper drawings prepared, annealed plates were delivered to size.

The prototype E5431 underwent official Ministry trials in late June, after the type had been committed to production! It was found to be heavy on lateral control and thereby considered tiring to fly. The firm did try to remedy this, but their main attention was diverted by the Snipe and this feature remained a flying characteristic; nevertheless the aircraft was quite sensitive fore and aft, and in spite of its weight could be easily looped, spun or rolled.

Understandably, it gathered speed quickly in a dive and had to be eased out gently. A steep gliding angle made landing difficult under the existing landing technique, as for the Snipe, of cutting the engine out at 200 feet and gliding in. Often this had to be punctuated with quick bursts from the engine to prevent the aircraft sinking too quickly. On tests with the prototype the duralumin axle bent in a not abnormal landing and on all subsequent machines this was considerably strengthened.

As a machine scheduled for large scale production, the use of Grade B timber was authorised necessitating an increase in the thickness of top wing spars and entailing the revision of many detail drawings. Nevertheless, the Kingston drawing office had the complete production drawings out during August 1918 and this essential medium for production coincided with an operation in France that was to open up new possibilities in the use of the type.

The former naval squadrons at the extreme left of the Western Front, operating on the Channel coasts, maintained unrelenting bombing attacks on the German-held Belgian ports. In August the pattern of these attacks changed. On receiving reports through intelligence sources that the important German airfield of Varssanaere had received reinforcements, a full scale attack was ordered by No. 5 Group Headquarters. It was planned that the bomber escort squadrons, Nos. 210 and 213 Squadrons and the 17th U.S. Aero Squadron, all flying Camels, would make a low-level strafing attack on the 13th.

Setting out at dawn, some fifty Camels formated over Dunkirk, and flew up the coast well out to sea, at about 5,000 ft. At a Very light signal from the formation leader, Lt. W. E. Gray, they turned inland diving on airfields where three flights of Fokkers were being warmed up for their morning patrols.

Dropping incendiary bombs and making diving attacks firing their twin Vickers, the Camels of No. 210 and the 17th wrought great damage. Machines and sheds were wrecked and an adjacent chateau known to be used as a Mess was riddled with bullets. Meanwhile the Camels of No. 204 arrived, escorting the D.H.9s of Nos. 211 and 218 Squadrons which added 112 lb. bombs to the confusion. On the way back, flying direct at low altitude, the Camels shot up transport on the ground with their remaining ammunition and the escort Camels, having conserved theirs, shot down the single enemy aeroplane encountered. All our aircraft returned safely but from the riddled appearance of some, it was realised it was only by considerable luck and if such attacks were to be repeated an armoured machine was preferable to the unprotected Camels.

Within two days, on August 15th, Captain J. W. Pinder of No. 213 Squadron was at Brooklands to test the Salamander for operations in No. 61 Wing which was attached to the Dover–Dunkirk Command for work with the Royal Navy. He received Harry Hawker's own performance reports on E5432, E5433 and another Salamander, referred to as No. 240, which cannot be identified. With armour and armament, full fuel load but no ammunition their best was 125 m.p.h. at 3,000 feet and a climb to 6,000 feet in under six minutes.

Captain Pinder after flying the Salamander himself, reported: 'The machine is considerably heavier on the controls than a Camel by reason of its weight—the loading is nearly 9. Manoeuvrability is about the same as a Bristol Fighter and it is capable of being looped and half rolled and turns fairly fast. Below 10,000 feet it could almost be used for fighting an Albatross Scout. In dives a great speed is obtained in a short distance but the machine answers well to controls all the while. It is also easily manageable flying along close to the ground with engine at full revolutions. The visibility is somewhat poor.

'I ascertained from the department which has to do with armour plate that the plates are tested for each machine. They are capable of stopping German Armour Piercing Bullets at 150 feet range, except at the sides: these plates (at the side) will stop any bullet hitting at an angle of over 15 degrees from the vertical and any of the plates will stop shrapnel from A/A fire. There are two fixed machine guns firing forward with tanks *each* holding 750 rounds of ammunition. The Salamander would carry the same load of bombs as a Camel—4 × 20 Coopers or 1 × 112-lb. bomb.

'I was given to understand that the primary object of the Salamander was for trench gunning and Kite Balloons but I consider that this machine would be an admirable one in which to raid Zeebrugge, Ostende, etc., performing under the same conditions as the daylight low bombing Camels. Practically the only way in which the machine could be brought down would be by explosive bullets in a main spar—two flying wires shot away—or a direct hit from Anti-Aircraft gunfire.'

Both the Snipe and Salamander were urgently required on the Western Front and Sopwith had set up production lines for both in the new Ham works. Both had their inevitable spate of modifications. For the Salamander it was revised drawings to cover balanced ailerons, modified fin and rudder in September, a new cowling strengthened with stays in November and in the last month of the year the Admiralty Compass Department asked that the dashboard be re-arranged and the position of the instruments altered.

The Salamander had an automatic fuel feed from the armoured tank under the pilot's seat to the gravity feed tank just behind the engine. An indication of the fuel feed was given to the pilot by an oscillating diaphragm. If this should stop movement, the pilot operated a Weymann hand pump fitted on the left of the cockpit. With the weight of armour restricting range to well under two hours, a stopwatch was another important instrument on the dashboard.

One other change just before the Armistice altered the aircraft's complete appearance. It was the only machine to have the 'Observation Machine' camouflage bestowed as a production finish. This was a disruptive pattern, as illustrated, of Dark Purple, Green, Light Earth, Light Green Grey (fuselage sides only) and black lines of 2–4 inches dividing the different colour areas.

With the Armistice, the new standing AMA doping scheme replaced the special camouflage while the contracts were run down. Unlike the Camel F1, the Salamander was not declared obsolete at the end of 1918 and although it never reached squadron service, some fifty were kept stored until the early 'twenties. It is believed that F6524 went to the French for evaluation and it is certain that F6533 went to the U.S.A., for it is recorded as being at McCook Field bearing the marking 'This machine is not to be flown'.

It is something of a paradox that the Salamander, designed as a trench fighter and regarded essentially as an Army support aircraft, would in fact have been a naval support aircraft had the war continued.

The third prototype Salamander at Martlesham Heath for trials in June 1918. Basically a Snipe, the Salamander differed by armour, streamlined headrest and slightly staggered Vickers guns. The airframe contract price was £1,138 plus £800 for a B.R.2 engine.

CHAPTER THIRTY

Quest for Superiority

A busy scene at Brooklands, April 27th 1918. From front to rear, a Snail with a monocoque fuselage, Snipe B9967 and a prototype Salamander await test by Harry Hawker. T. O. M. Sopwith, back to camera at far left, talks to an R.A.F. officer on his left; a Camel can be seen behind them.

Towards the end of 1917 a method of initiating, co-ordinating and progressing aircraft designs was inaugurated by the Air Board. Based on military requirements, a series of specifications for design requirements was prepared, and issued to the new, and still rapidly expanding, aircraft industry with an invitation to tender. It was the start of the system that appertains today. Then it was the Controller of the Technical Department, Department of Aircraft Production of the Ministry of Munitions—a post that could be traced through to a Controller General today at the Ministry of Technology.

At South Kensington, the newly-opened offices of the Air Board's Technical Department scheduled the progress of all prototypes and officers examined the reports as each aircraft was checked out by the Testing Squadron at Martlesham Heath. Among the specifications issued in late 1917 was Spec. A.1a for a single-seat fighter. To meet this specification there were nine promising designs—two of them Sopwith types, the 7F.1 Snipe and 8F.1 Snail.

The Air Board had to decide which fighter to place in large scale production and a process of elimination began. By the beginning of 1918 it had been decided, irrespective of modifications pending, that the B.A.T. F.K.22/2 Bantam would be relegated to an advanced fighter trainer, while the engine troubles of the 200 h.p. Hispano precluded the Vickers F.B.26. Three serious competitors to the 7F.1 Snipe remained, the Austin Triplane, Boulton & Paul Hawk and British Nieuport B.N.1, all utilising Bentley's proven B.R.2 engine and all expected to be ready for test in January 1918.

The Sopwith 8F.1 was in a rather different class, being one of three types with the A.B.C. Wasp radial engine to be considered—the other two were the F.K.22/1 Bantam and the Westland Weasel. Sopwith had been invited to build a fighter to utilise the Wasp engine in October 1917. A radial engine, as apart from a rotary engine in which the whole unit revolved, was then an innovation. Unfortunately there were difficulties in producing this new type of engine and testing was held up, but another innovation had already been introduced. Sopwiths were asked to try their hand at a monocoque fuselage.

This was quite revolutionary. A wooden box-like frame was considered the basis of any fuselage, with a plywood or fabric covering. For the 8F.1 it was the intention that the strength of the fuselage should be in the plywood covering itself, shaped by nailing pliable wood into multiple hoops with a number of bulk-heads. This was the beginning of stressed-skin construction and here again, Sopwith's were one of the pioneers. It was this shell-like construction that brought the 8F.1 its name of Snail.

Monocoque construction had yet to prove itself and of the six 8F.1's ordered, only two had monocoque fuselages, but the name Snail was taken to apply to both types.

Early in 1918, complementary to plans for the amalgamation of the naval and military air services into the Royal Air Force, the Specification designations were changed to meet the overall force requirements. Fighter requirements were to two specifications:

A.F. Type I Single-seat Fighter (High Altitude)
A.F. Type II Single-seat Fighter (Ground Targets)

The A.F.II requirement was being met by this time solely by Sopwith with the Air Board torn between choosing an armoured Camel (T.F.1) or the new Sopwith T.F.2 which, as related in the previous chapter, became the Salamander.

For the A.F.I requirement there were eight contenders. The Snipe with its three B.R.2-engined rivals, to which a fourth, the Armstrong Whitworth F.M.4 Armadillo, had been added, and the Wasp-engined set of which Sopwith's Snails vied with Westland and B.A.T. types.

By early March all of the B.R.2 contenders, except the Armadillo, were at Martlesham Heath under test. The Air

Board, pressing for a decision, watched the preliminary reports closely; each type had its pros and cons.

As a final test, service pilots were invited to Suttons Farm to examine the new fighters and give their opinions. This was the nearest the British came in a counterpart of the famous fighter competition the Germans had held at Adlershof that January to select their superiority fighter, for which such pilots as Richthofen were invited to attend and adjudge.

Among the pilots invited to Suttons Farm was Captain J. B. McCudden, by then victor in 57 aerial combats. He flew the Snipe for fifteen minutes and was full of praise of its capabilities. During the day's trials the Nieuport B.N.1 caught fire in the air and was destroyed. Three days later the Snipe, B9965, was flown to France for further examination; the other B.R.2 contenders were returned to their makers. In general their faults had been a lack of manoeuvrability, an important characteristic in air fighting and one for which Sopwith aircraft were famed. The Ministry of Munitions placed contracts in March for no less than 1,700 Snipes.

T. O. M. Sopwith himself, although as General Manager was busily engaged in controlling his still expanding organisation, continued to identify himself very closely with the experimental work. When on January 7th 1918 he was made a Commander of the Order of the British Empire for 'services in connection with the War', his status was described in the *London Gazette* as Chairman and Test and Experimental Manager of the Sopwith Aviation Company. By coincidence, on the same day a similar honour was conferred on John Davenport Siddeley. The following year, Mr. Siddeley formed the Armstrong Siddeley Development Company which was purchased by Mr. Sopwith in 1935 and so linking the names Hawker and Siddeley.

While ideas continued to flow from the famous trio—Sopwith the overall controller, Sigrist the trained engineer and Hawker the test pilot, a permanent design staff drafted out working drawings to meet specifications and modifications called for by the Services. Not that it was called a design staff at the time; the distinctions between the various members of the drawing office—and there were some fifty—was in status and salary rather than by title.

Corresponding to Chief Designer was Herbert Smith who had joined the firm in March 1914. He had taken a Diploma in Engineering at the Bradford Technical College and had served with the British & Colonial Aeroplane Company at Bristol as a draughtsman. From 1915 onwards he had charge of all the draughtsmen working on experimental designs. The chief draughtsman was R. J. Ashfield (See Chapter 6) who was also responsible for most of the stress work. One other member, steadily proving his worth as a designer was W. G. Carter. In the ex-skating rink experiment shop itself, Jack Pollard was in charge of the construction of the experimental airframes.

The development of the Snipe, and all that it entailed, represented but a fraction of the experimental and development work in 1918. At the beginning of the year in the experimental shop the first Snipes were being prepared for Martlesham Heath. At the same time the first Snail was ready for its engine and scheduled to fly and the second was also nearing completion. The Hippo built to the specification for a two-seat fighter was already at Martlesham on trials together with the Rhino bomber triplane to Spec A.2b for a light bomber. A series of reports about them, preceded by rumours, was being fed back to the design staff. At the same time a new set of wings was being designed and built for the Bulldog.

When the new Sopwith factory opened at Ham, the experimental shops remained at Kingston in the ex-skating rink and the general and drawing offices remained in buildings that are now (1970) occupied by the Kingston Technical College. The rest of the factory was largely reorganised. Camel production was tailed off and left to sub-contractors and the Kingston works concentrated on the Dolphin while the new Ham Works tooled up for the Snipes.

The Bulldog represented a requirement for a two-seat fighter. Fitted with a 200 h.p. Clerget engine it flew makers' trials at Brooklands in early November 1917. A new pair of wings were fitted on November 13th and a different cowling was quickly produced to prepare it for official tests at Martlesham Heath in December, where it was in competition with the Avro 530 to meet the Expeditionary Force A.2a Specification for a two-seat fighter. Early in the new year the wings were again changed on the

Another scene on April 27th when Hawker, taking off in a Snail shown at the top of the opposite page, attracts attention. T. O. M. Sopwith is standing back to camera, the light figure in the middle of a trio at the right.

Bulldog and it is not clear from documents if it was at this stage, or during the earlier trials, that a change was made from single to two-bay wings.

The earlier Hippo, designed as a 1½ Strutter replacement in a fighter reconnaissance role was also considered to the A.2a Specification, and there arose the unusual situation in January 1918 of the one firm grooming two aircraft to rival each other. However their period of competition was brief as favour soon swung from the Hippo to the Bulldog.

Such was the acute shortage of certain engine types at this time, that while the Bulldog was having its new wings fitted its engine was taken out for another machine. Once modified, it was the Hippo that had to give up its engine to permit the Bulldog to continue trials in its revised form. It was all to no avail; not that it was outpaced by the Avro, but that the Expeditionary Force revised its requirements. The Bulldog was then officially classed as an 'experimental machine possessing features likely to be of use to the Royal Air Force'. It was not altogether the end for the Bulldog as it was considered later as a contact patrol aircraft—but as a fighter it was out of the running. The R.A.F. already had the Bristol Fighter in service and it remained supreme in the two-seater fighter role. It is possible that the Sopwith had the edge over the Bristol on certain aspects, but for a machine to be committed for production it would have had to show very considerable improvements.

The Ham factory, as related, was built for the Ministry of Munitions under the National Aircraft Factories scheme, but instead of functioning as an N.A.F. it had been leased to Sopwith. Not only was production underway before it had commenced in the National factories, but the contract price for airframes remained roughly the same instead of trebling as in the national factories.

H. C. Miller, a fellow Australian for whom Harry Hawker had obtained employment at the Sopwith works before the war, revisited the works in 1917 having left them on the outbreak of war to join up. He wrote 'Before leaving for the front I got leave to visit Sopwith's and found it was like entering a new world. Thousands of workers, about half of them women, bustled about the huge new factories turning out row upon row of shining fighter machines. Harry Kauper, working under terrific pressure, showed me quickly around and wished me well. Harry Hawker, now married (he married Miss Muriel Penty, November 14th 1917) was testing aircraft by the hundred. He found time, however, to slip down with me to Brooklands...'. On the way back to Kingston, Hawker's advice to Miller, who was a pilot in the Australian Flying Corps, was 'Always climb. You will out-climb the German, you cannot out-dive him'.

Hawker, by this time, no longer test-flew normal production; there was a spate of prototypes of the Sopwith 'Zoo', such as Rhinos and Hippos, for attention. Built in the old skating rink, dismantled and re-erected at Brooklands, where they were there assiduously flight-tested by Hawker and reported on to T. O. M. Sopwith's satisfaction, they were handed over to a Service pilot for test at Martlesham Heath. But while the quest went on unremittingly for a superiority fighter, it soon became evident that the new-formed R.A.F. had found this in the Sopwith Snipe.

The prototype of the superiority fighter of 1918, the Snipe, in the early stages of its development. Having a slab-sided fuselage, top wing cut-out and single-bay wings, the early prototype Snipes were very different to the final production versions.

CHAPTER THIRTY-ONE

Superiority with the Snipe

In its original form the Snipe had a Camel-type fin and was powered by a 150 h.p. B.R.1 engine.

While some of the Sopwith designs for fighters were unorthodox—triplanes, negatively staggered wings and monocoque fuselages—the Snipe represented a straightforward improvement on the orthodox form of the day. An enhanced Camel it might be said, and its armament of twin Vickers machine-guns firing forward through the propeller arc was precisely the same; but whereas the Camel had various engines and was at its best with the 150 h.p. B.R.1, the 230 h.p. B.R.2 was the standard power unit for all production Snipes. It was, once again, the successful combination of a Sopwith-designed airframe, with a Bentley-designed engine.

Like the Camel, the origins of the Snipe are obscure. The first documentary evidence of its inception is a report at the Air Board about August 1917—'The Drawing Office of Messrs. Sopwith is now engaged, and will be so for a week or ten days, on a single-seater for the same engines [*sic*] as the Camel. This machine will be essentially a Camel with the view of the Dolphin and will probably be a better fighting machine than either. If successful, the change from the Camel will be an easy matter from the point of view of production.'

The first Snipe, which probably flew in September 1917, had like a Camel a B.R.1 engine and a slab-sided fuselage. In October a Snipe was reported flying with a B.R.2 engine and modified wings, but a sticking relief valve caused a fuel tank to burst damaging the machine which went back to Kingston for repair. The performance had been disappointing and on re-building there were alterations to the fuselage and the wing union connections.

A report issued in November, presumably on this prototype after repair, stated 'The machine has been flying maker's trials and appears satisfactory. It appears to be a great improvement on the original Snipe, B.R.1 engine. It was stated earlier that the machine is in balance fore and aft, but that it behaves in a similar manner to the Sopwith Camel, i.e., tail heavy when climbing and nose heavy when diving. Since this preliminary report, the machine has been flown continually and it is thought that this defect is more noticeable in the B.R.1 than in the B.R.2 Snipe and when diving at excessive speeds (up to 200 m.p.h.) the machine becomes uncomfortably nose heavy, and a considerable pressure has to be exerted on the stick in order to pull it out, this is not so noticeable in the B.R.1 and it is probably accounted for by the extra weight of the engine. The firm are going into the question of overcoming this difficulty. It is not thought that moving the top plane further forward or altering the angle of the tail plane would be feasible, as the machine is already rather tail heavy when climbing. The machine has since been crashed'. The 'crash' was evidently an incident in the late afternoon of November 16th when, on its third flight that day, a ground fog came up quickly and the Snipe crashed on landing.

An order for six prototypes (B9962-9967) was placed by the Air Board and a concentrated development programme was flown throughout the winter of 1917/1918 to iron out the inevitable difficulties that arise with any new type. A new Expeditionary Force requirement necessitated modifications to increase fuel capacity which meant more weight and revised strength calculations. A new set of wings to the revised specification was fitted to a Snipe which was reported being prepared for flight to Martlesham Heath on January 10th 1918 but was, in fact, still on Sopwith trials at Brooklands on the 23rd. Between then and March 20th this and following prototypes must have engendered enormous confidence, for on that date 1,700 were ordered from seven different firms.

The question is often asked just how much better was the Snipe than the Camel and, almost as if the Testing Squadron at Martlesham Heath presaged the question, tests in August 1918 between production Snipe E8006 and production B.R.1-engined Camel B3835 were conducted. The time to climb to 10,000 feet was precisely 9.4 minutes by both aircraft, but at that height the Snipe made 121 m.p.h. to the Camel's 115 m.p.h. Also in the Snipe's favour was extra fuel which bestowed a greater range and it carried over double the weight of ammunition. However, as with the Camel, performance deteriorated with service and E8006, after 24 hours flying, took a minute longer to reach 10,000 feet and lost 6 m.p.h. at that height.

Only in one vital factor in air fighting did the Snipe fall behind the Camel—in manoeuvrability. With a view to improving the Snipe in this respect, particularly lateral and directional control, B9966 was fitted in August with

The Snipe's second stage, with a 200 h.p. B.R.2 and modified fin, bearing the characteristic Sopwith testing pitot tube on the wing struts, clear of the propeller, which gave a visual indication of the air speed by a counter on the bulge.

118

An early Snipe under construction at Kingston. At this stage only one of the two Vickers guns has been fitted. The airframe, excluding armament and instruments cost £946 and the engine another £880.

balanced top plane ailerons and increased fin and rudder area. Service pilots were invited to fly the machine and venture their opinions. So favourably did they report that the modification was hastened through the Sopwith drawing office as a Class 1 modification, which meant that where possible retrospective action would be taken. No sooner had this been arranged than it was decided to modify the tailplane and elevators to increase the sensitivity of fore and aft control and improve longitudinal stability. This too was then tested by service pilots who reported favourably, but this modification was not embodied.

Detail improvements were being effected all the summer. A new knife-edged tail skid was introduced to give better control when taxying and a guard had to be put on the dashboard to prevent the control column hitting the instruments when hurriedly put into a dive. An innovation, seemingly a good idea, had been the introduction of celluloid inspection panels on the wings, but as they were liable to blow out they were replaced by aluminium plates.

The first squadron to receive the Snipe was No. 43 who conducted their first operational patrol with the type on September 23rd. In the middle of October No. 4 (Australian Flying Corps) Squadron changed its Camels for Snipes. On the 22nd, a full patrol of nine of their Snipes, flying at 14,000 feet over Tournai, encountered fifteen Fokker D.VIIs. Manoeuvring to approach from the sun, Captain T. C. R. Baker led one flight of Snipes into the attack and engaged the enemy leader himself. Then both his guns jammed—but he managed to clear them in time to shoot another Fokker down that was on the tail of another Snipe. Meanwhile Lt. T. H. Barkell, leading the other flight, destroyed one Fokker in flames and sent another down out of control. The fight went on for some minutes and died out over a wide area after two more Fokkers had been driven down. The only casualty was Barkell who landed near Peronne with leg wounds.

Five days later came one of the most gallant fights of the war. Major W. G. Barker, who was on a refresher course from England, went on patrol in Snipe E8102 with the Camels of No. 201 Squadron. Over the Foret de Mormal

In its third progressive form, the Snipe had its characteristic round-sectioned fuselage and two-bay wings. This Sopwith photo was taken March 11th 1918.

he attacked an enemy two-seater from which one of the crew parachuted as it broke up in the air. He was then fired at from below and, wounded in the thigh, spun down only to pull out in the middle of an enemy formation of fifteen Fokkers which he proceeded to attack in turn. The first two escaped from his fire but a third, on which he fired from an estimated thirty feet, went down in flames. At this stage he was again wounded and fainted; his machine fell out of control. After regaining consciousness, he was attacked by a whole formation and after shooting an aircraft down in flames he was *again* hit; a bullet this time shattered his elbow. Fainting again, losing control and being followed down, he *again* recovered in time to regain control. Noticing smoke coming from his machine, he believed it to be on fire and after trying to ram a Fokker, opened fire on it at close range sending it down in flames. Diving in his disabled machine to escape, he found that his retreat was cut off by eight enemy scouts. Successfully evading the enemy machines, yet without the use of his legs, and with one arm disabled, he brought his Snipe back flying a few feet above the ground. Finally, he crashed near a balloon winch where he quickly received attention. This was the last Victoria Cross to be awarded in the 1914-1918 War for a gallant deed in the air.

Undoubtedly, the Snipes were superior in every way to the best the Germans could offer. Snipe pilots would accept battle even when they had not the advantage of height. In the battles of the closing weeks of the war the Snipe squadrons scored quickly and had few losses. It was soon noticeable that the enemy often chose to give Snipe formations a wide berth. No. 28 Squadron was the next to receive Snipes and No. 208 was receiving them when the Armistice came.

It was said that when No. 4 (A.F.C.) Squadron after the Armistice took its Snipes to Cologne and showed their manoeuvring powers to some German airmen, they expressed first their astonishment and then their gratification that they personally had not met them in action.

From performance figures and test reports that Martinsyde F.4 Buzzard is often quoted as the best fighter of the 1914-1918 War period. That assessment is not disputed. However, the Martinsyde Buzzard did not enter squadron service and thereby few can dispute the statement that the Snipe was the finest fighter, from any country, to *operate* in the 1914-1918 War.

CHAPTER THIRTY-TWO

Sopwith Night Fighters

Night fighters. Pilots of No. 37 Squadron which until October 1918 had operated from Woodham Mortimer.

Such was the reception of the German raiders in daylight at the hands of pilots in Pups, 1½ Strutters, Triplanes and Camels, that towards the end of 1917 the Germans restricted their aeroplane raids to the hours of darkness.

Hitherto, the Camel had been regarded essentially as a day fighter and, indeed, was even considered dangerous to fly at night; No. 44 Squadron proved otherwise. This squadron had been formed with 1½ Strutters for Home Defence duty on July 24th 1917. To improve their capability to meet the day raiders, they had been re-equipped with Le Rhone-engined Camels. But when German raiders operated on the night of September 3rd/4th, three of the squadron pilots insisted on taking-off. They proved conclusively that the Camel could be safely flown at night. General E. B. Ashmore, responsible for home defence, wrote of this as "perhaps the most important event in the history of air defence".

The squadron commander of No. 44, Major G. W. Murlis-Green, scored the Camel's first victory on the night of December 18th. Patrolling at 11,000 feet over Essex, he perceived a small glow and as he approached to investigate a searchlight illuminated a Gotha. He managed to get in a few bursts of fire before the Gotha took evasive action to shake off the searchlight. Once that had happened, Murlis-Green had little hope of again picking out the dull glow of exhausts after operating in the blinding shaft of light from the ground. However, he had achieved his object for the bomber, after releasing its bombs over Bermondsey, lost height steadily and finally came down in the sea near Folkestone.

Two No. 44 Squadron members, Captain G. H. Hackmill and Lt. C. C. Banks, had an extended combat over Essex during the night of January 28th/29th and finally despatched a Gotha in flames. An official history quotes this as marking the first unqualified victory at night in a combat between aeroplanes.

From No. 44 Squadron's experiences a special night fighting version of the Camel was suggested and eventually produced. The modifications were reported as follows: "Alternatives consisted of removing both Vickers guns with their chutes and magazines, and covering in the front part of the fuselage from cowl ring to cockpit with 3-ply decking, and of cutting away the top centre section for a central width of $30\frac{1}{4}$ inches and mounting two Lewis guns on the top centre section on S.E.5 gun mountings, together with the removal of the main and gravity petrol tanks, and fitting in their place an 18-gallon B.E. petrol tank on special tank bearers fitted across the fuselage in the rear of the oil tank. The pilot's seat, cockpit, dashboard and all controls have been moved back a distance of 18 inches which adjusts the upward view. The alternatives appear to produce a machine suitable for the purpose aimed at, i.e. a night flyer for short patrols." Already, luminous dials and Holt landing flares were standard equipment for night-flying Camels.

At the same time a special cabane was being fitted to six Dolphins for home defence use. From tests of Dolphin C3778 at Orfordness it was found that the cabane made no significant difference to performance and it was believed to make the machine safer in the event of its turning over. Major Pretyman discussed with Sopwith's the possibility of making it suitable for night flying by an increase in dihedral, larger tail area, an adjustable tailplane, altered Lewis gun mountings and hooped bars to permit a pilot to crawl clear in the event of overturning.

In January 1918 the first Dolphin for home defence work reached No. 141 Squadron at Biggin Hill where it was crashed soon afterwards. Sufficient Dolphins were issued to equip a flight, but it was not liked for night work and was discarded by No. 141 Squadron in favour of Bristol Fighters. In other units, however, the Camel was finding increasing favour for night work. No. 112 Squadron, which had originally formed with Pups for home defence on July 3rd 1917, had its first and only victory of this war with a Camel on May 18th 1918 when the Commanding Officer, Major C. J. Q. Brand, M.C., shot down a Gotha over Leysdown.

Due to the activity of German bombers behind the Allied lines on the Western Front, it had been decided to form night fighter squadrons, as apart from home defence squadrons; the first to be for the defence of the Abberville area. Flights were drawn from three Home Defence squadrons, Nos. 44, 78 and 112, all equipped with Camels, to form No. 151 Squadron at Hainault Farm in June 1918.

Under the command of Major G. Murlis-Green, the squadron arrived in France on June 21st. Within a week, 'A' Flight was detached to No. 101 Squadron which operated F.E.2b pusher night bombers. The idea was for the Camels to accompany the F.E.2bs at night and shoot up or bomb searchlight bases revealed when the lights functioned in search of the bombers. The remainder of the squadron had defence zones, but this was as unsuccessful in 1918 as in 1940.

Encounters were by chance. On the night of June 29th Captain D. V. Armstrong, noted for his daring unofficial exhibitions of flying in a Camel, sighted an enemy machine in the beam of a searchlight. After firing several hundred rounds at the enemy machine, it went down vertically; but in the darkness, after the dazzling light, Armstrong could not discern its fate. Captain A. B. Yuille achieved the first confirmed night victory over the Front at 00.45 hrs on

July 25th when he made two attacks on a Gotha over Etaples. Like Armstrong he lost it in the darkness, but on this occasion the machine was so damaged it landed in the Allied lines. The crew, a squadron commander, his adjutant and equipment officer, were taken prisoner. They said that Yuille's fire had first wounded one of them and then put each engine out of action in turn.

Shortly after this the squadron's tactics were changed to fit into an integrated defence system. The fixed defences of guns and searchlights were placed around the centres of communication and depots and the Camels, instead of patrolling the enemy's target areas, were moved forward to follow the raiders back to their bases.

That summer the Germans committed their *Riesenflugzeuge* (Giant) bombers to attacks behind the Allied Front. On the night of August 10th, the first of the three Staaken R.XIVs built, the R43, was coned by several searchlights, drawing several Camels to the spot, among them one piloted by Captain Yuille. Getting close in behind the bomber, Yuille fired some 75 rounds and was engaged by one of the Staaken's gunners. Firing close in from a different angle of approach, he succeeded in getting the machine on fire and it went down. Five crew members attempted to use their parachutes, but all nine crew were killed. A Camel had shot down the largest aeroplane ever to be destroyed in aerial combat.

The Camels soon became the scourge of the German night bombers and twice in September they shot down two in one night. In all, during five months at the Front, No. 151 Squadron had 20 confirmed victories of which for 16 there was evidence in wreckage on the Allied side of the lines, and six other claims which could not be confirmed. Of these claims, two concerned five-engined "Giants", 22 twin-engined Gotha, A.E.G. or Friedrichshafens and two single-engined two-seaters.

Such was their success that further Camel night fighter squadrons were planned. No. 152 was formed at Rochford in October and moved to France that same month and No. 153 was forming up when the war ended.

The United States Air Service also decided to do something about the marauding German bombers and they too chose Camels for their 185th (Night) Pursuit Squadron. Their first operation was on the night of October 22nd when Major H. Hartney took a black-painted Monoengined Camel up to intercept a Gotha. Catching up behind the enemy bomber he opened fire, but in his excite-

General J. F. A. Higgins inspecting a Le Rhône-engined home defence Camel at Stow Maries.

ment his gloved hand switched off the ignition. The nose of the Camel dropped as the engine stopped and by the time he completed his checks and found the trouble it was too late for further pursuit. So disgusted was Hartney with his clumsiness, that he did not even put in a combat report. However, three weeks later, after the Armistice, advancing Allied troops found the wreckage of a Gotha in a wood in the vicinity of a Hartney's fight. Thus the A.E.F.'s only night victory was by a Camel and was confirmed to Hartney's satisfaction, but never *officially* since there was no official report of a combat!

Towards the end of the war, some of the day fighter Camel squadrons commenced night strafing attacks, while a few aircraft of No. 151 Squadron were regularly detached to picket enemy airfields to catch raiders taking off or landing. Within a few days of the Armistice, the first of the Snipes was delivered for Home Defence work and No. 112 Squadron was re-equipped with Snipes in the early part of 1919.

Although the Camel was almost immediately withdrawn from night flying, it had become the most effective night and day fighter of the 1914-1918 War.

The most sophisticated of the 1914-1918 War night fighters, a Snipe at Biggin Hill in 1919. This aircraft, F2390, has navigation lights and flare brackets fitted to the lower wing. During standby, flares were kept handy in containers slung under the wings. On take-off, mechanics would remove the containers and affix the flares, which were ignited electrically from the cockpit, to the brackets.

CHAPTER THIRTY-THREE

Anti-Sub Sopwiths

A Parnall-built Fairey Hamble Baby of basic Sopwith Baby design, which was used in an anti-submarine role from coastal stations around the United Kingdom.

If the Sopwith floatplanes had failed as fleet aircraft in the early war years, at least it can be recorded that they had a measure of success in the last years of the war. By the end of 1916 less than 30 Schneiders or early Babies with 100 h.p. Monosoupape engines remained. A larger number of 110 h.p. powered Sopwith Babies were at coastal stations and Blackburn-built models with 130 h.p. Clergets were arriving early in 1917. All these were very urgently needed.

Germany, having been forced to a declaration of unrestricted submarine warfare in an effort to break the stranglehold of the British blockade, changed tactics from surface attacks by gun and torpedo, to submerged torpedo attacks without warning. The effect was soon felt and sinkings for April 1917 reached an unprecedented high level. Part of the counter-measures, of new coastal stations and increased coastal patrol activity, affected the Sopwiths. Twice that Spring Baby floatplanes from Newlyn sighted U-boats and attacked with single 65 lb bombs. Two such bombs were all they could carry, and in 1918 this was reduced to one to increase their fuel and thereby range. Barely sufficient to cause damage let alone sink a U-boat—nevertheless sufficient to scare off U-boat commanders from coastal shipping.

The pilots of the Baby floatplanes faced the same hazards as before. Sheltered waters were essential for landing and taking off in these light craft. At Hornsea on the Yorkshire Coast the Baby pilots wisely used the Mere inland in preference to the sea. In heavy winds they could not be expected to fly, but in the still air they often flew in mists and collisions were not uncommon. The Baby pilot that hit a ship's mast has been related earlier and there is in existence a photograph of the Baby that hit and stuck to a 300 foot mast.* On July 17th 1917 Baby N1102 flew into a Curtiss H12 flying boat; the pilot was reported missing, but the Curtiss, although damaged, made a safe landing.

Patrols from the French side were apt to be contested. On July 19th Short 184 Seaplane No. 9057 set out on a dawn anti-submarine patrol escorted by Babies N1015 and N1016. At 5.30 a.m. three enemy single-seat floatplanes set on the formation. The Short was forced down on the water and N1015, piloted by Flt. Sub-Lt. Potvin, was shot down and fell sideways into the water. Flt. Lt. Graham in the other Baby shot one enemy machine down and then flew on and landed by a French destroyer, to instruct it to go to the rescue of his comrades. A French trawler then took his Baby in tow back to Dunkirk. Shortly after this, Baby floatplanes were withdrawn from bases in Belgium and northern France.

Together with the Hamble Babies, the Fairey-built version of the Sopwith Baby, the Sopwith floatplanes came into their own operating around Britain's shores from the autumn of 1917. On September 29th Blackburn Baby N1433 from Plymouth bombed an enemy submarine. Next month, it was the turn of Hamble Baby N1203 on patrol from Westgate, whose pilot sighted a periscope and wash of a submarine 25-30 miles east of the North Foreland. Two 65 lb bombs were dropped and the area was searched but a strong wind bore the machine off-course and the pilot was forced to land at Calais.

Another Hamble Baby distinguished itself on November 8th. N1469's pilot patrolling a few miles east of Scarborough saw a streak in the water which he estimated to

---

*See p. 19 HARLEYFORD'S "Marine Aircraft of the 1914-1918 War".

be caused by a periscope going through the water at about 12 knots. As the machine turned towards it, losing height down to 500 feet, the vessel appeared to stop. Two 65 lb bombs were dropped with a 2-second interval and the Baby circled the area for five minutes when, fuel being low, the pilot returned to his Hornsea base to report. Ten days later the pilot of Baby N1446 saw a submarine in almost the same area. A single 65 lb bomb was dropped and appeared to hit within three yards of the target. The pilot fired Very lights which brought the Armed Yacht *Jason II* on the scene to carry on the search.

Yet again that November Babies were harassing the U-boats. On the 23rd Hamble Baby N1199 set out from Fishguard following sighting reports of an enemy submarine. Its pilot observed a wake 15 miles N.N.W. of Strumbles Head. Diving from 1,000 to 500 feet the machine dropped 2 x 65 lb bombs of which the first failed to explode. The second burst 10 to 15 feet ahead, but after the water from the explosion had stopped swirling, no further sign of the U-boat could be detected.

Due to deteriorating weather and smaller numbers in service, the triple attacks of November by Baby seaplanes was not repeated and not until the spring did the Sopwiths join in with the Short 184s, and increasing numbers of D.H.4 and D.H.6 landplanes on coastal patrol, in attacking U-boats. While they did not achieve sinkings, they provided valuable sighting reports and their attacks discouraged bold action on the part of U-boat commanders, forcing them, literally, to keep their heads below water. They also did valuable auxiliary work; sea-searches for survivors, occasional carrier work to ships anchored out and mine-watching.

After a winter of hard work, but of no actual contact with the enemy, Spring brought a resumption of actions in the campaign between German U-boats and the Sopwith Baby floatplanes. Patrolling from Fishguard on March 22nd both Short 184 N1683 and Baby N1127 sighted a submarine and each dropped one bomb apiece.

Three days later, Flight Sub-Lieutenant E. F. Waring in Baby N2101 from the Tees station, dropped a 65 lb bomb astern of a submarine sighted 15 miles east of Hartlepool. Next day, from Hornsea, Flight Sub-Lieutenant G. F. Hyams, in N2087, spotted what could have been the same submarine 10 miles off Scarborough. When he fired a recognition signal, the submarine dived, so he went in with his single 65 lb. bomb dropping it about 70 yards ahead and slightly to one side. No results were seen.

On April 6th another Baby from the Tees station, N2110, was involved. Lieutenant E. J. Addis left forty minutes after noon to patrol southwards where an enemy submarine had been reported. At 1.15 p.m., when five miles off Seaham, a periscope was sighted. Knowing that none of our submarines were in the vicinity, Addis went into the attack with his 65 lb bomb dropped from 800 feet. Immediately after the explosion oil came up. A destroyer raced to the scene and after dropping depth-charges, caused more oil to come up from the depths.

Baby N2101 mentioned earlier had been moved to Seaton Carew in April and on the first day of May, a submarine was sighted south-east of Sunderland by the pilot, Lt. R. R. Richardson, who was at the limit of his patrol. After dropping a 65 lb bomb the pilot was forced to return to base. Ten days later the same pilot, using N2109, caught sight of a periscope four miles south-east of Skinningrove. Again a 65 lb bomb was dropped and Baby N2107 was sent out to continue patrolling the area.

Aircraft at Seaton Carew saw plenty of action that May, and not least the few Baby seaplanes on station. Lt. Addis at the controls of N2111 on May 30th and Lt. Clutterbuck flying it next day, both released their 65 lb bomb over periscopes.

In June, not until the end of that month did Baby seaplane pilots see action. On the 30th Lt. Addis, in N2067 during a convoy escort, sighted a periscope; first he released his bomb over the spot and then followed up with Very lights to indicate to motor launches the sub-

Blackburn - built Sopwith Baby with 65 lb. bomb for anti - submarine work at Hornsea Mere, a lake in Yorkshire used for its sheltered waters and proximity to the coast. Unusual is the Lewis gun firing forward through the propeller arc in addition to the Lewis gun mounted on the top wing centre-section.

Menace over. First of a line of surrendered German U-boats, escorted by British coastal motor boats, approaches Harwich, November 28th 1918.

marine's position. The enemy U-boat was not seen again.

The next action for a Baby was July 26th involving the same aircraft and pilot. Again a bomb was dropped and Very lights set off, but without effect. Next day Lt. Clutterbuck, on N2107, saw a ship actually torpedoed and sink two miles off Whitby. Circling the area, he sighted a periscope a half mile from the stricken vessel and dropped his bomb. Following his action, two destroyers, five motor launches and several drifters combed the area, but without success.

On August 9th it was the turn of a Hornsea-based Baby, N2095 piloted by Lt. Sherwood, to bomb a suspected U-boat, but rough seas allowed no further possibility of sightings. Three days later Lt. Taylor on N2109 from Seaton Carew, out on early morning patrol, aimed a bomb at a periscope. This same Baby was involved in the last recorded bombing attack by a Sopwith floatplane;

Lt. Clutterbuck observing the outline of a submerged submarine immediately below, dived down to 400 feet and dropped his bomb which fell twenty-odd feet to the starboard bow. Red Very lights were then let off and succeeded in attracting two motor launches and some drifters to the scene. This was on August 27th; the drifter dropped depth charges on the suspected area, giving people of Scarborough the impression that a naval battle was being fought out at sea.

In the final analysis in the last six months of the war, some fifty Baby floatplanes were held for anti-submarine patrol from U.K. bases of which about twenty were daily serviceable. In that time they sighted 13 U-boats and attacked ten of these. In the course of their patrols averaging $1\frac{3}{4}$ hours each, their sightings worked out at one per 236 hours which is roughly one per 16,500 miles of flying. In these respects its sighting rate was better than any other type of aircraft engaged—D.H.4, D.H.6 and D.H.9 landplanes, Short 184 floatplanes or airships. Yet the Baby, unlike the Short or D.H.4 and D.H.9, did not have the advantage of an observer in addition to the pilot. It has been suggested that its success was due to being less conspicuous and less noisy than the others.

Patrols continued for some days after the Armistice from the six stations still using Baby floatplanes: Dundee, Hornsea, Seaton Carew, South Shields, Westgate and Yarmouth. This was lest the surrender messages had not been received by some U-boats at sea. But before November was out all Baby type floatplanes were declared obsolete and the remaining 51 Sopwith Babies (all Blackburn-built) and 12 Hamble Babies (Fairey-built) at home and overseas were dispersed. The last Sopwith-built Baby, No. 8127 had been dispersed the previous month. It was the end of the Sopwith floatplanes in the British Services, but in the Norwegian Naval Air Service several were to survive until the 'thirties.

Ten Blackburn-built Babies were delivered to Norway in 1917. Although it was not normal policy to sell aircraft to neutral countries during the war, it was considered expedient in this case since they were required for neutrality patrols that would keep a look-out for German U-boats using Norwegian fiords as bases in their anti-shipping campaign against the Allies. Fitted with floats or skis according to season, they operated from water and ice. Some were fitted with wireless and one was converted to a two-seater.

New methods at sea. A Camel fitted with special flotation gear shown ditched and at left a lighter for Camel take-offs at high speed when towed by destroyers.

CHAPTER THIRTY-FOUR

The Torpedoplane Transact

The prototype Sopwith T.1 numbered N74 with 200 h.p. Hispano-Suiza engine, showing wing-folding arrangement for ship stowage. Not until post-war was the name Cuckoo used for this type.

Strictly speaking, there was not a Sopwith Cuckoo of the 1914-1918 War for the name was not bestowed on the Sopwith torpedo-carrying aeroplane until after the end of the war. However, it is an indisputable fact that Sopwith Aviation, specialists in fighter aircraft, designed the only service shipborne torpedo-carrying aeroplane of the war—the Sopwith T.I.

Torpedo-carrying aircraft in the early years of the war had been floatplanes adapted to carry torpedoes. The combined weight and drag of both weapon and floats, and the need for still water for take-off, so reduced the effectiveness of such aircraft that their operations were greatly limited and few successes were achieved. When the Admiralty eventually realised that a torpedo-carrying aeroplane could offer greater possibilities, they knew to whom they could turn.

Marked 'Most Secret', and delivered by an Admiralty courier, T. O. M. Sopwith received a letter from the Admiralty dated October 9th 1916, and signed by Commodore Murray F. Sueter, with a request as follows: "Will you please go into the question, with as little delay as possible re-Torpedo-carrying aeroplane with 4 hours fuel and pilot (1) to carry 1 x 1,000 lb locomotive torpedo, (2) to carry 2 x 1,000 lb locomotive torpedoes. Torpedo aeroplane will probably be discharged by catapult, giving machine an acceleration of 90 ft/sec in 60 feet. Details of Short 225 seaplane attached."

T. O. M. Sopwith designated the two projects T.I and T.2. The second project, T.2 for twin-torpedo carrying, was not pursued and, indeed, not until fifty years later did twin torpedo-carrying aircraft become a practical proposition in the Royal Navy. The TI project became a Sopwith-Admiralty transaction.

Unfortunately, when Murray Sueter was posted to command R.N.A.S. units in Italy at the beginning of 1917, Admiralty interest in this type of aeroplane waned. The matter was not referred back by Sopwith, since his attention was now focused on the production of fighters, with the possibility that a successful bomber, the B.1, might result from the original torpedo-bomber scheme.

An essential feature of the T.1 was a split undercarriage to permit the drop of a torpedo without fouling an axle. A prototype airframe had been built with this feature which is said to have caught the eye of Wing Commander A. M. Longmore when visiting the works in February 1917, following his appointment earlier that month to the Admiralty, naval interest was revived.

Nevertheless, it was June 6th before the aircraft, fitted with a 200 h.p. Hispano engine, was passed out by the Sopwith Experimental Department. During July, the T.1 reached the Isle of Grain for official tests which commenced on the 20th of that month. These proved so successful that next month orders were placed for a hundred.

One of the most ambitious plans of the war was then mooted. The Admiralty considered a proposal of Sir David Beatty's for an attack by the Sopwith torpedo aircraft on the German High Seas Fleet at its North Sea bases, sinking the ships that, although languishing in harbour, maintained a perpetual threat and kept the Grand Fleet constantly at short steaming notice. It was not a scheme for early scheduling because the two T.1 contractors, Pegler and Fairfield, had no previous experience in aircraft building and were in difficulties. Blackburn's, whose Baby pro-

Blackburn-built Sopwith Cuckoo with 200 h.p. Sunbeam Arab engine. This aircraft had evidently made a forced landing for which repairs were necessary, as the pilot is evidently resting in the shade of the wings while riggers attend to the machine.

Practising its function. A Blackburn-built Cuckoo of the Torpedo Aeroplane School, East Fortune, drops a torpedo in the Firth of Forth circa October 1918. Later production models had a rudder of increased area as shown, facing. The price of a Cuckoo airframe was £1,613 plus £1,020 for a Sunbeam engine.

duction had tailed off, was then brought in and changed from the production of one Sopwith type to another.

Meanwhile the prototype T.1, N74, after being returned to Kingston for adjustments, had been dismantled and sent back to Grain where early in 1918 it was re-erected for tests with the 18 inch Mk IX torpedo starting on January 19th. It had been intended that the original 200 h.p. Hispano Suiza engine would be replaced by a 170 h.p. geared type but as this was not available the original 200 h.p. engine was temporarily used, and the 200 h.p. Sunbeam Arab engine was selected for production aircraft.

Late January or early February, N74 was flown to Tadcaster and taken by road to Blackburn's works at Leeds for the Arab substitution to be effected. Flying on the way back to Grain there was trouble with the oil system and a forced landing was made at Elmshott in which the propeller was damaged. A new propeller had to be sent before the aircraft could continue its journey. Back at Grain the Cuckoo prototype recommenced tests. It was found that the heavier Arab engine (No. 16236) made the machine nose-heavy, and the tail unit was adjusted to compensate.

For its intended tasks of operating from a carrier's deck, the rate of acceleration to flying speed was of paramount importance and deck lengths required for various take-off weights had to be calculated. These were plotted in an ingenious way. Application was first made to use the decking set out on the airfield, but this was already in use on deck landing trials for Sopwith 1½ Strutters using hooks to pick up sandbags which would cause a rapid deceleration. A new pitch was marked out on the airfield, with cross lines marked to represent each foot of a ship's decking. A string was tied to the tail skid of the T.1 prototype from a freely moving roller on which, fixed concentrically, was a three-foot diameter wooden disc. The disc had a single protruding contact which completed a circuit every revolution and brought into operation an electro-magnetic pen recording on a ribbon. On the same ribbon another pen, actuated by a metronome, measured set time intervals. Thereby the speed/distance graph could be calculated. Thus there was electronic testing in 1918—even though the tracking was done with string!

Tests were conducted with a series of weights. It was felt that an actual torpedo was too valuable to risk—just as well as it proved. More practicable was an underslung wooden box which allowed, by means of lead shot, the weight to be increased for various readings. At a 850 lb load, still some 100 lb below torpedo weight, and on the second day of trials, the tests were disrupted. Shortly after take-off, at 1,000 feet while orbiting over the airfield, the propeller shaft of the T.I. sheared. The propeller, complete with boss, flew off cutting through the port lower wing and damaging a radiator. The pilot glided N74 into

Believed to be the first Cuckoo delivered to a carrier. It was from this connection with carriers that Harry Hawker met Mackensie Grieve, his navigator, in the Atlantic flight attempt.

A Blackburn-built Cuckoo II with Wolseley Viper engine, of No. 210 Squadron, Gosport. In this model the exhaust pipes extend downwards over the torpedo carrier to prevent its mechanism from freezing up in flight. A total of 607 torpedos were delivered for use by such aircraft during the 1918-1914 War.

a nearby field for its second forced landing.

Then came further trouble as the first production aircraft, N6950, took the air. Completed by Blackburn's early in May, it was delivered by air from Tadcaster to Grain later in the month. On arrival it was found that the starboard side undercarriage tubing had fractured—and the machine had not yet borne the weight of a torpedo. Both the tail skid and rudder post had broken in landing and there was considerable backlash on the controls.

Only a month previous, the Sopwith Drawing Office had completed the drawings for Fairfield, Pegler and Blackburn, now there was a further spate of modifications. Strengthening of the rudder post and a re-designed tail post were made essential, but to facilitate production it was arranged that the first ten aircraft would be modified in service and were not to be flown with torpedoes until the changes had been effected.

The Service, preparing to receive the T.1, used East Fortune airfield for training pilots on the new torpedo-bomber and formed a Torpedo Development Squadron at Gosport, with an authorised establishment of six T.1s and two D.H.4s. An important facet of the trials was to determine the ideal dropping height, which was as low as possible consistent with the safety of the aircraft. Under fifty feet the water splashed up over the cockpit.

While Blackburn's delivered Cuckoos by air to Gosport and East Fortune, development work at Grain changed from performance characteristics to operational techniques. Flotation gear was fitted and extension pipes were fitted to the exhaust pipe as a trial method of keeping the torpedo mechanism from freezing up. An aerodynamic matter was referred back to the Sopwith works. Service pilots reported a tendency for the Cuckoo to swing to the right; Sopwith's design staff countered this by designing a larger rudder and slightly off-setting the fin.

The Official Handbook on the T.1, issued in October 1918 summed up its capabilities.

Easy to control and pleasant to fly in general, pilots were warned that in rough weather the aircraft wallowed considerably and was sensitive to bumps so that they should be well strapped in. While it could take off in four seconds, there was no question of a safe landing back on decks. Operationally it would have to be taken close in to its target by carrier and use its four hours endurance to return to friendly land. With no armament apart from its torpedo its only defence was evasive action; while this would be of little avail against enemy shore-based fighters, it had the usual Sopwith qualities of high manoeuvrability. It could be freely stunted, including looping, once it had dispensed with its torpedo.

The Sopwith B.1 evolved from the Cuckoo and, having a Hispano engine, was intended as a bomber for France, to replace the 1½ Strutter on the French production lines.

CHAPTER THIRTY-FIVE

Sopwiths with the Fleet

The German cruisers *Goeben* and *Breslau*, which had sought sanctuary in Constantinople in 1914 and were nominally purchased by the Turks, put to sea on January 20th 1918. They made their presence felt by sinking H.M.S. *Raglan* at anchor and setting Monitor No. 28 on fire.

Although D.H.4s made the main bombing attacks against the enemy vessels, Sopwith aircraft played an important part in the operations. First, W/T equipped 1½ Strutters shadowed the ships, while Camels provided an escort for D.H.4 bombers. Flt. Lt. Moore, flying a Camel on escort, sighted two enemy seaplanes coming up from Helles. Diving on one and opening fire at 200 yards he forced the seaplane to land near its base. Climbing, he tackled the second machine until it too had landed on the water. Flying back to base, he observed the *Breslau* sinking; it had struck a mine. Meanwhile, both ships had been attacked by Flt. Sub-Lt. Murray flying a Sopwith 1½ Strutter bomber escorted by two Camels.

One of the most gallant air actions of the war followed as two Blackburn-built Sopwith Baby floatplanes, each carrying a 65 lb. bomb, set out to bomb the two warships from low level. Piloted by Flt. Sub-Lts. W. Johnson and R. W. Peel they were escorted by Commander A. Moraitinis, a Greek pilot, flying a Camel.

As they neared the ships, the pilots became aware that some ten enemy seaplanes were providing air cover. Cdr. Moraitinis, after driving away two enemy machines, turned to attack a two-seater firing at one of the Babies. He was too late to save Johnson, whose floatplane fell in flames, but Peel saw his chance and flew directly over the *Goeben*—only to have his bomb hang up. In spite of being chased by two enemy two-seaters, he turned to make another bombing run and this time the bomb fell. With his engine damaged and misfiring, he was forced down in the Straits near Cape Helles between a Turkish torpedo boat and the shore. With his engine functioning intermittently, he half taxied and half flew round the Cape to escape from the enemy vessel and eventually reached the coast of Imbros where he remained for two days awaiting new engine parts. The fuel tank of Moraitinis' Camel had been riddled with bullets in a surprise attack from behind and he had been forced to dive out of the fight and land at Imbros—but he was flying again the same day.

A Ship's Camel taking off from H.M.S. Pegasus, *fitted with a flying deck, in 1918.*

Meanwhile the *Goeben*, after developing a list, altered course and ran aground. This gave the aircraft their chance and R.N.A.S. and R.F.C. machines were concentrated at Mudros and Imbros. They attacked, first with two D.H.4s escorted by seven Camels; then followed up at dawn next morning with five bombers escorted by six Camels. Unfortunately, the weather deteriorated.

Three Sopwith Babies from the *Ark Royal* put in a night attack on the *Goeben;* one was reported missing— for a day. The pilot, Flt. Sub-Lt. Bernard Smith, had to glide into a landing on the water with engine trouble. As he descended six shots were fired at him, presumably from a submarine. He was not molested and was found next day by the crew of a Short 184 sent out to search for him.

During the next week, repeated attacks were made. On January 27th a strong wind hampered operations and a Camel being prepared for dawn reconnaissance was blown over on to its wingtip and damaged. A second Camel got off the ground and reported a 90-knot wind. An attack was attempted but the bombers could not make headway and returned. Under cover of the bad weather, the *Goeben's* crew succeeded in refloating their vessel and made good their escape to Constantinople.

At home there had been considerable development in operating aircraft at sea. To avoid swinging, when taking off from a ship's deck, trough-like runners were made to take the undercarriage skids. After land trials at Grain, they were tried at sea by Flt. Ltd. W. G. Moore on 1½ Strutter A6911 operating from H.M.S. *Vindex*.

The Admiral Commanding Aircraft proposed an establishment of 70 two-seat reconnaissance machines for the Grand Fleet on the basis of:

Furious	14 +	14 reserve
Campania	7 +	7 reserve
Cavendish	6 +	6 reserve
Pegasus	4 +	4 reserve
Nairana	4 +	4 reserve

No sooner had the Admiralty approved their establishment on February 13th, than further requests came for additional two-seaters for battle cruisers. To meet these demands only Sopwith 1½ Strutters were available, and few enough of these remained.

Camel with wheels in fore and aft arresting wires on H.M.S. Argus.

A Ships Strutter with skid undercarriage taking off from experimental trough guides on H.M.S. Vindex.

There were hopes by May 1918 of being able to land the two-seaters back on special carriers, other than their parent ships; arrester trials were made at Grain with hooks slung from the skid chassis that picked up ropes attached to weighted sandbags. In June the experiments were extended to 1½ Strutters, Pups and Camels with wheeled undercarriages and fitted with arrestor hooks.

Already the true carrier concept was operational. H.M.S. *Furious*, with a complement of 1½ Strutters and Camels, screened by destroyers and accompanied by the First Cruiser Squadron, was skirting the Skagerrak early in the morning of June 17th. Two 1½ Strutters were about to fly a reconnaissance, when two German seaplanes crossed over the flotilla, dropping bombs from a mere 1,500 feet. The 1½ Strutters on the flying deck were quickly replaced by Camels which took off and chased the raiders, only for their pilots to find that, after a few rounds, their Vickers guns had jammed. After expending their Lewis gun ammunition and landing in the sea, the Camels were salvaged by destroyers.

A Camel also got off from H.M.S. *Galatea* which, sweeping further to the south, was also unsuccessfully bombed; but after only ten rounds being fired the Vickers on this Camel also failed to function. In pouring rain, the pilot could not find his ship and landed on the shore at Fjaltring in Denmark.

Furious was bombed again and another couple of Camels took off. This time an enemy seaplane was forced down on the water and was destroyed by gunfire after the crew had been taken prisoner.

The primary object of the operations, to destroy Zeppelins, had not been achieved and a more ambitious method of using bomb-carrying Camels to attack Zeppelin bases was planned. Two special bombing flights of Camels were formed and trained, and went to sea towards the end June, but severe weather postponed this enterprise.

In mid-July *Furious*, escorted by cruiser squadrons, despatched three Camels early on July 19th, followed within minutes by a further flight of four. Carrying two 50 lb bombs each, the Camels flew towards the Tondern Zeppelin sheds, eight miles distant. Captain B. A. Smart, flying with the second wave, arrived to see that one of the two giant sheds had been destroyed and he made for the other. Missing with his first bomb, he scored a direct hit with his second. Hedge-hopping away to dodge intense machine-gun fire from the ground, he had just enough fuel left to ditch near the fleet. Altogether two Camels were missing, another ditched with engine trouble and three more landed in Denmark, but the object was achieved —two Zeppelins and their sheds were destroyed. Camels had made the first Carrier strike in history.

Furious made further forays to attack enemy seaplanes which were interfering with our anti-submarine flying boat patrols. Meanwhile flotation and ditching experiments continued with Camels and at this time N6341 was deliberately ditched to assess the effect of a new design of hydrofoil, known then as a hydrovane. These experiments were followed by similar trials for the Snipe, using E8068 with a 4-foot 4 inch span wooden hydrovane. This fitting was then modified to make it interchangeable with a 1½ Strutter and effect a standard production gear.

The final operational fleet innovation of the war was the towed lighter take-off. A Camel, with engine running, was released from a platform on a lighter being towed by a destroyer steaming into wind at 30 knots. Colonel C. R. Samson making the first trial, cartwheeled into the water and had the lighter run over the wreck of his machine; forced under water, he extricated himself from the wreckage and was picked up safely. Lt. S. D. Culley made the first successful take-off on July 31st and left on fleet operations shortly afterwards with his Camel N6812 mounted on a lighter. When a Zeppelin was sighted he took off at once. Reporting on this, Rear-Admiral Tyrwhitt said: 'I consider that Lieutenant Culley's effort was one which is worthy of the highest honour as in the first place he flew his machine off a lighter—a feat which, I believe, has only once been successfully effected before. He made his attack in all intents and purposes in the open, as there were few clouds to take cover in, and he single-handed successfullly attacked and destroyed a Zeppelin in a most gallant and businesslike manner'.

This was the last successful action by ship-borne aircraft during the war. Actions which, since 1917 had been effected only by Sopwith aircraft—Pups, 1½ Strutters and Camels. And of the Camel, it had proved itself the premier fleet fighter as well as the most effective day and night fighter of the war.

Resulting from the world's first carrier strike, the Toska hangar at Tondern, housing the Zeppelins L54 and L60, are seen ablaze after an attack by Camels.

CHAPTER THIRTY-SIX

Sopwiths Universal

During the war far more Sopwith aircraft were required than could be produced, and of British production, quite naturally, the needs of the British Services came first. There were, however, a few offsets to foreign governments during the war years, but once the war was over there was a vast surplus of aircraft and plenty for all—and ten different air services chose to operate Sopwiths.

The Dominions had made a magnificent contribution to the war effort with their manpower; many men from the Commonwealth had served in the R.N.A.S., R.F.C. and R.A.F., but there was a desire to form air forces of their own and two of the Dominions had started their own air arms before the war ended. Post-war, the British Government offered a nominal 100 aircraft to each of the Dominions to form a nucleus of equipment for an air arm. This was also a reciprocal gesture for the funds given by organisations and individuals in the Commonwealth to provide aircraft during the war.

Sopwith Baby 8140, stranded with engine failure off Dutch coastal islands on April 27th 1916, was towed into internment in Holland by a Dutch trawler. It is shown being taken by trailer for repair in the Spyker works after which it served in the Royal Dutch Navy bearing the number T1.

An Australian Flying Corps had been formed during the war and by 1918 deployed four squadrons. Two of these were fighter squadrons, No. 2 with S.E.5As and No. 4 with Camels. A leading pilot in the Camel squadron was Captain A. H. Cobby, D.S.O., D.F.C., and 2 bars, whose 29 victories helped considerably to make No. 4 the highest scoring of the Australian squadrons with a total of 76 enemy aircraft destroyed. In September 1918 the Camels were replaced by Snipes. However, when it came to the choosing of a fighter for the post-war Royal Australian Air Force, the S.E.5A was chosen in preference to either the proven Camel or Snipe. But, what is more surprising, there was a requirement for the earlier Pup. This was seen as the ideal advanced trainer for a stage between the Avro 504K primary trainer and the S.E.5As.

About a dozen Pups were eventually shipped to Australia and one, ex-C476 which reached the Australian Civil Register as G-AUCK, was still flying in 1944.

A No. 1 Canadian Air Force Squadron was forming in England when the war ended. The aircraft that had been chosen to equip their first squadron was the Sopwith Dolphin. At that time one Sopwith aircraft was already in Canada, serving with R.A.F. (Canada)—a large training organisation with Canadian-built Curtiss J.N.4s forming 99% of the equipment. For advanced training Camel B3772, originally shipped to the United States, joined the Canadian School of Aerial Fighting on October 2nd 1918, the first and only Camel at the School.

The original Canadian Air Force was disbanded in the run-down following the Armistice, and in forming the post-war Royal Canadian Air Force, at least five Dolphins and two Snipes (E8102 and E8213) were among the aircraft accepted but Canada, like Australia, had decided to standardise on the S.E.5A as a fighter. However, these were little used and in 1924 of the twelve originally supplied nine were in store. One other Snipe, E7649 which arrived via Washington was taken on strength early in 1921 but it crashed on October 22nd 1923. However, by this time seven Camels arrived via America and one survived to mid-1928.

New Zealand in accepting part of the Imperial Gift showed a preference for two-seat aircraft and Sopwith types were not included. India, on the other hand, was given forty Snipes with sixty D.H.9s, but as no plans materialised until many years later for an Indian Air Force, the Snipes, it is understood, deteriorated in store at Karachi.

In Europe ten different nations used Sopwith aircraft in the early post-war years. France, who had built 1½ Strutters in thousands, reduced her holdings to a few hundred by selling or scrapping depending on their condition.

Belgium needed to retain her military aircraft to form the nucleus of a post-war force. When Belgium had been overrun in 1914, a few survivors kept a Belgian *Aviation Militaire* in being. This was re-equipped from British and

1st Lieutenant J. B. Harvey of the United States Field Artillery, the observer of this U.S. Air Service French-built 1½ Strutter, hands photographic plates to a waiting courier at La Valdahan, January 25th 1919.

French industry including 1½ Strutters and Camels. Both types were used in the early 'twenties and an example of each has survived to today in the Musee Royal de l'Armee et d'Histoire Militaire in Brussels.

Holland's method of acquisition of Sopwith aircraft was unique. The Netherlands, 1914-1918, maintained a strict neutrality which they were prepared to defend by force of arms and were willing to buy aircraft from both Britain and Germany to maintain a strong air arm, but the belligerent nations were not anxious to sell. The Dutch therefore waited for the aircraft to drop in! Their first Sopwith was acquired on April 27th 1916 when Baby floatplane 8140 was stranded off the Dutch coast and towed into internment.

A stretch of Dutch territory near Maastricht, jutting out into German territory, provided the Dutch with a legitimate catchment area, and no less than five 1½ Strutters were netted. The first was on August 25th 1916 after three R.N.A.S. 1½ Strutters had set off to bomb Zeppelin sheds at Cognelee near Namur. One, piloted by Flight Sub-Lieutenant Jamieson was forced down in Dutch territory. Later the very first production 1½ Strutter dropped in; by that time a Pup had already been acquired but the Dutch had to wait a while for a Camel. A list of Sopwith aircraft put into service by the Dutch appears below.

N.B. LA stood for *Luchtvaart Afdeling*, the Dutch Army Air Arm, the change in numbering is due to a 1918 revision of the serialling system.

Dutch Catchment of Sopwith Aircraft

Date	Type	Origin	Dutch service Number
27 Apr. 16	Baby	8140 of R.N.A.S.	T-1 (Navy)
25 Aug. 16	1½ Strutter	9396 A4 of 5 Wing	LA-33
17 Sep. 16	1½ Strutter	9420 A3 of 5 Wing	LA-38 later S-24
1 Mar. 17	Pup	A6164 of R.F.C.	LA-41 later S-212
22 Apr. 17	1½ Strutter	9376 of 5 Wing	LA-42 later S-412
12 May 17	1½ Strutter	N5154 of R.N.A.S.	LA-34 later S-413
7 July 17	1½ Strutter	No. 115 of Escardrille SOP 111	LA-45 later S-701
28 Sep. 18	Camel	C1542 of R.A.F.	(Destroyed landing)
7 Oct. 18	Camel	C1537 of R.A.F.	S-226

Under International Law the crews were interned as having violated Dutch neutrality, but their imprisonment was not arduous and included leave to England on parole. The aircraft would be impounded, but if used an arrangement had to be made with the power concerned. The Dutch, wishing to put the Sopwiths into service, agreed to pay the British Government £1,700 each for the 1½ Strutters; a good price since the British Government had only paid Sopwith £1,750 for the aircraft new.

Another country had Sopwiths drop in—Denmark.

Blackburn - built Sopwith Baby fitted with skis, serving in the Royal Norwegian Navy. Delivered November 1st 1917, this particular Baby crashed May 9th 1919 after 30½ flying hours. The famous explorer Amundsen took Baby F108 to Spitsbergen, the base for his 1923 Arctic expedition.

When 2F.1 Camels made the first carrier strike, three came down in Denmark. Captain W. D. Jackson force-landed in a field west of Guedager near Esbjerg and set his aircraft on fire, Lt. N. E. Williams landed N6823 at Hoje Knolde, Skallingen and Lt. S. Dawson in N6605 went down near the Lymgvig lighthouse. All three pilots were interned, but were back in England before the war ended. The two whole Camels and the remains of the third were taken to Copenhagen and stored in the Royal Danish Navy dockyards. In 1921 they were offered to the Danish Army but due to their poor condition they were not accepted.

Twenty 1½ Strutters had been supplied to Roumania during the war and postwar, Czech, Polish and White Russian Forces all used numbers of 1½ Strutters, Pups, Camels, Dolphins and Snipes. But of all European countries it was Norway that had a Sopwith type in service the longest. The Royal Norwegian Navy had negotiated for Sopwith Baby floatplanes in 1917 and had been supplied with ten crated Blackburn-built models. After erection at Horten they were delivered to stations at Bergen, Horten, Kristiansand and Tromso. After war service on neutrality patrols, scouting for submarines of belligerent nations sheltering in fjords, the floatplanes were retained in use and the last was not scrapped until the end of 1931.

One Eastern nation displayed a particular interest in Sopwith aircraft—Japan. As early as 1916, an example of a 'Schneider Cup floatplane' was requested by the newly-formed Imperial Japanese Naval Air Service, and a representative Baby floatplane (No. 8201) was supplied. It is believed to have been evaluated at Oppama Navy Air Base. Post-war, the Japanese invited a British Air Mission to advise on building up a post-war air fleet, and among their recommendations was the acquisition of six Sopwith Cuckoos which became, in the mid-twenties, the sole torpedo-bombers of the Japanese Navy.

The Japanese were also endeavouring to build up their own aircraft industry and the Mitsubishi Engineering Works opened up an aviation section. Their chief designer was no less than Herbert Smith, late of the Sopwith Aviation Company and designer of the most famous fighter of the war—the Camel. His chief test pilot was W. L. Jordan, D.S.C., D.F.C., a Camel ace with 29 victories. They worked well for the Japanese, with the blessing of the British Government—until events overtook policies.

Apart from Britain and France where Sopwith aircraft were produced, the largest user was the United States of America, both before and just after the Armistice.

Setting off on a training exercise at the 5th Aerial Artillery Observation School of the United States Air Service in France. The aircraft, a French - built 1½ Strutter, was the type used by Americans at La Valdahan, Tours, Coetquidan, Souge, Chatillon, Chateauroux and Cazaux in France, after reception and processing through American Aviation Acceptance Park No. 1 at Orly, where they were overhauled and fitted out to meet American training requirements.

America, making up for lost time, commenced a large-scale purchase of aircraft in Europe for the United States Air Service units serving in their Expeditionary Force.

French-built 1½ Strutters were taken over in their hundreds in 1918 mainly as trainers but some were assigned to squadrons. The 88th Corps Observation Squadron used a few Sopwith 1A2s for training after being assigned to the Front with Salmson 2A2s on May 28th and the 90th and 99th Squadrons, assigned in June, were initially equipped with Sopwith 1B2s.

The bulk of the 1½ Strutters purchased were used at the American Observation Training School at Souge. Both pupils and instructed grew to like the Sopwiths which in general, let them down lightly. Inevitably with training accidents were frequent, but they were rarely fatal. On April 28th 1918 the engine of Sopwith 1A2 No. 640 faltered at about fifty feet as the pilot, 1st Lt. Paul R. Stockton, was taking off. As there were trees ahead, Stockton turned to attempt a landing in a field to the right, but

A Soviet Air Force 1½ Strutter, used for test purposes, fitted with skis for winter flying. This was used by a Soviet Aeronautical Institute which conducted airflow tests, fitting wool tufts over the wings to give a visual indication of the airflow.

the machine side-slipped into the ground and was completely wrecked. The starboard wings having taken the impact, neither the pilot nor his passenger, Corporal Mechanic Robert M. Ammons were hurt. Twelve days earlier landing another 1A2, 1st Lt. Albert B. Potts overshot the field in a fast landing and caught a tree with one wingtip swinging it round into another tree, smashing the wings. Again the pilot and his passenger, 1st Lt. L. R. MacLachlan were unhurt.

With the Sopwith Camel it was different; the Americans took these into action. The 17th Pursuit Squadron forming up in France on June 20th 1918, went into action from July 15th under R.A.F. Command until seven days before the Armistice when it re-equipped with Spad XIIIs and transferred to the American Expeditionary Force. Arriving at the Front on October 1918, the 41st and 185th Pursuit Squadrons both used Camels initially until Spad XIIIs were received in replacement. On the day the war ended, A.E.F. records show 12 Camels on the strength of U.S. front-line units.

By the end of 1918, of the 3434 aircraft received by American Aviation Acceptance Park No. 1 at Orly, 550 were Sopwith types. These comprised 132 Camels from the R.A.F. of which 126 had been issued to field units, and 418 1½ Strutters (314 Type 1A2, 104 Type 1B2) from French units of which all had been issued to training units. Additionally at No. 2 Air Service Production Centre 22 Sopwith 1B2s and 4 1A2s were received but only one had been delivered for service.

The American Expeditionary Force used Camels, as related earlier, and were also interested in obtaining Dolphins and Snipes during the closing stages of the war; and the U.S. Navy purchased examples of Baby floatplanes, Ships Camels and Ships Strutters. Altogether, about 100 Sopwith aircraft of various types were shipped to America after the War.

In South America, the Baby was represented by a few used in Chile in the twenties. By then it was a museum piece —if only one had been preserved. Happily, parts of two had been kept and from these a replica has been constructed in Britain in 1970. This has now been presented to the Fleet Air Arm Museum.

Camel ready for the road. Flying Officer A. Lew and groundcrew of Camel N7367 at Camp Borden, Ontario, 1928. This was one of some seven 2F.1s in Canada.

CHAPTER THIRTY-SEVEN

Transatlantic Venture

Harry Hawker the designer, engineer and pilot who so nearly became first to fly the Atlantic.

Before the war had ended, the £10,000 prize, originally offered in 1914 by the *Daily Mail* for the first non-stop crossing of the Atlantic, was being discussed. When asked about the prospects in July 1918, T. O. M. Sopwith said: 'The Transatlantic flight could be made and the prize won this month, so far as the capacity of the aeroplane of today is concerned. Our Company has been engaged since the beginning of the war with the construction of the Sopwith single-seater fighting machine. Crossing from America to England by air is not the problem it was a few years ago. Undoubtedly the flight is possible. A dozen machines of today could do it. They could do it at once if aeroplane makers and pilots were not all busy with war demands.'

As soon as war demands had finished, T. O. M. Sopwith organised an Atlantic venture with the same enthusiasm as he had arranged the pre-war Schneider racer. The Sopwith B.1 bomber formed the basis of the trans-Atlantic aircraft, re-modelled by W. G. Carter to have a deeper fuselage to accommodate the extra fuel required, with wing area increased commensurate with the extra load, and a 360 h.p. Rolls-Royce Eagle to give the power needed.

There was no question of who should pilot the aircraft; Hawker was the obvious choice and he was already playing an important part in its design. An experienced navigator was needed and Hawker chose Lieutenant Commander K. K. Mackenzie-Grieve, whom he had met aboard H.M.S. *Campania* when concerned with deck-landing trials. Montague Fenn, now that the Sopwith Paris office had closed, was sent to Newfoundland to secure a suitable take-off field.

The aircraft was completed in an astonishingly short time of six weeks and incorporated several unusual features. Of necessity, the fuel tankage was abnormally large and, to reduce head resistance, the undercarriage was designed to be jettisoned. Sopwith was criticised in some quarters for allowing Hawker to take this risk on landing, but it was Hawker himself who persuaded Sopwith to have it this way. He argued that on the old Wrights, pilots had preferred skids to wheels and Sopwith arranged for the bottom longerons to be reinforced with strong wooden runners. Another precaution was to have a boat, upturned and built in as part of the fuselage decking.

The aircraft was carefully tested at Brooklands. An innovation, an airborne directional wireless apparatus, was installed but did not prove successful and was taken out. A normal wireless transmitter, a T.55A set supplied by the Air Ministry, was installed and a wind-driven generator was fitted on the fuselage side to provide the power.

On March 20th Hawker and Cdr. Mackenzie-Grieve left for America accompanied by an engine expert from Rolls-Royce and a cameraman from Jury's Imperial Pictures who had also taken a cine film of the Atlantic in the air on test from the rear cockpit of a Buffalo. The Sopwith Atlantic went with them, in the hold of the S.S. *Digby*. Montague Fenn had taken a ground staff over and had chosen a site at St. Johns, Newfoundland, where he arranged for a log shed to be built.

The crated aircraft arrived at Placentia Bay on March 28th, where it was transferred to the *Portia* and landed at Placentia. It was then sent by train to St. Johns. Transshipment had been due to St. Johns being ice-bound, and the airstrip chosen was snowbound. With a soft surface and heavy snow-falls continuing, an early attempt was out of the question. A decision was made to start as soon as possible, but using a two-blade propeller that had been brought out, not the four-blader used on the initial tests. This was to facilitate take-off at the expense of a slight loss on speed.

Erection of the machine started right away and took a week to complete. After that the aircraft was kept at readiness should conditions be favourable. Other competitors were running them close; in fact the Martinsyde team, F. P. Raynham and Morgan, was staying at the same hotel, Cochrane House in St. Johns. The engine was run daily, the wireless set was tried and controls checked. It was estimated that the aircraft could take-off at two hours notice; time for tanks to be filled, engine warmed, mail to be collected and thermos flasks filled.

On tests, the T.55 set proved unsuitable and a cable was sent for a replacement T.52A set. Meanwhile, the manager of the Marconi station of St. Johns provided a small commercial set with a limited range of 25 miles which was temporarily installed. The T.52A arrived in time for the flight, but not for prior testing.

Early on Sunday 18th May, the Atlantic was made ready and took off at 5.42 p.m. G.M.T. (3.40 Newfoundland summer time). As soon as the coast was crossed,

The Sopwith Atlantic photographed in England on February 21st 1919 before shipment to Newfoundland.

Hawker released the undercarriage trigger and the wheels fell away—he noted that speed increased by 7 m.p.h.

As night fell the weather deteriorated and Hawker was disturbed to find the water temperature higher than it should be. The weather improved but as the water temperature rose steadily past 176°F he decided something must be done. Assuming that the water was not circulating due to some obstruction in the pipes, he stopped the engine and dived from 12,000 to 9,000 feet, both to cool the engine and in an attempt to dislodge any obstruction. This was only temporarily effective, the thermometer reading soon rose again and the dive was repeated. This time, on climbing again the water started boiling. Soon, the centre of the top plane was covered in ice from the steam freezing and the engine was throttled back to try and keep the water below boiling point.

When black clouds loomed ahead, it was out of the question to attempt climbing above them and Hawker went below, stopping the engine again in the dive. But it would not re-start, and the machine dropped further. Hawker prepared for a ditching and the machine was a mere ten feet from the water when Grieve, frantically priming the engine, got it re-started. Handicapped, there was little point in continuing. Hawker headed south towards the shipping lanes. Soon after dawn, through a rising mist, a hull suddenly appeared to loom out of the water; it was the S.S. *Mary* from Denmark.

After circling the ship firing distress signals, they set off a mile or so ahead of the ship's course and ditched the aircraft. Due to the partly-empty fuel tanks it floated well at first. The *Mary* ran out a boat, but it was 1½ hours before they were rescued. As the ship did not have wireless, the world was left to wait and fear the worst.

Muriel Hawker received a message from the Admiralty at ten o'clock the following night that her husband had landed safely in the sea forty miles off the mouth of the Shannon. She went to bed contented but too excited to sleep and her 'phone rang several times with messages of congratulations. Rising early she went to collect the morning papers and opening the *Daily Mail*, read with horror that the sighting report was false and that her husband was lost at sea. At ten o'clock Tom Sopwith came round and, as she herself wrote 'helped me to keep my resolution during the whole of the ensuing week'.

Five days later came a telegram that seemed to set an official seal on the loss of her husband—

'The King, fearing the worst must now be realised regarding the fate of your husband, wishes to express his deep sympathy and that of the Queen in your sudden and tragic loss. His Majesty feels that the nation lost one of its most able and daring pilots to sacrifice his life for the fame and honour of British flying.

STAMFORDHAM'

At the same time, Lord Northcliffe made a generous offer to ensure that Mrs. Hawker would be well provided for. The Sunday papers mirrored the general view that all hope must now be given up—but hope was not given up in the village church at Hook in Surrey, where Muriel

Not a service aircraft, and produced before civil aircraft were registered, the Atlantic did not bear service or civil markings except for rudder striping.

Salvaged from the Atlantic, the wreckage of the Atlantic machine is displayed on the roof of Selfridges, the famous store in Oxford Street, London, which then had the roof set out in gardens. Parts of the machine had been labelled and the oil tank can be seen marked as such above the lady, second from left.

Hawker attended that morning, for prayers were offered for the safety of the two airmen. An hour after leaving church a message was given to Mrs. Hawker that the *Daily Mirror* wanted to contact her with a message that could be entrusted to no-one else. They told her the news of the pick-up.

The Danish ship had signalled by flag to the Butt of Lewis and the Admiralty wirelessed H.M.S. *Woolston* to intercept and the two airmen were taken aboard and conveyed to Admiral Freemantle's flagship, H.M.S. *Revenge*, in Scapa Flow, then by destroyer to Thurso, where sailors rowed them ashore.

Ever the conscientious test pilot, Hawker had already sent in a report to T. O. M. Sopwith. It read as follows: 'My machine stopped owing to the water filter in the feed pipe from the radiator to the watercock being blocked up with refuse, such as solder and the like, shaking loose in the radiator. It was no fault of the motor which ran perfectly from start to finish, even when all the water had boiled away. It had not trouble in landing on the sea. We were picked up by the tramp ship *Mary* after being in the water 1½ hours. We are going to London from Thurso at 2 p.m. on Monday arriving in London between 7 and 8 p.m. on Tuesday".

Meanwhile, Muriel Hawker, escorted by her brother, had rushed round to Fred Sigrist's bungalow on the island at Thames Ditton. Sigrist, who had also heard the news, ran gleefully down the steps to meet the couple and dived fully clothed into the water in his exuberance.

Next day Mrs. Hawker was asked to attend a film of the Atlantic flight at the Majestic Cinema, Clapham. There Tom Sopwith and his wife met her to travel together to Grantham to meet the train in which Harry was coming home from the North, acclaimed as a national Hero.

Another view of the wreckage on the roof of Selfridges showing the giant fuel tanks mounted one above the other. Lying placarded across the front of the lower tank is the long exhaust pipe from the 375 h.p. Rolls-Royce Eagle engine. This was the only Sopwith Biplane to have a Rolls-Royce aero engine.

CHAPTER THIRTY-EIGHT

Post-war Military Sopwiths

Snipe E8213, presented to Canada by Leicester, was one of several contributed by British towns to Canada for a postwar air force in recognition of Canada's magnificent contributions to the air war. This particular aircraft was lent to H. G. Quigley for the 1919 Toronto-New York Air Race, but unfortunately it crashed on August 25th 1919 just prior to the event.

Before 1918 was out, the Pup, 1½ Strutter, Triplane and all land types of Camel had been declared obsolete for British service purposes. The Dolphin, Salamander, Ships Camel and Ship Strutter were retained temporarily and numbers were stored, but the re-equipment of former Camel squadrons by Snipes continued well into 1919. When No. 112 Squadron disbanded in mid-June 1919 their Snipes went over to No. 143 Squadron and when, in turn, that squadron disbanded, they were sent to store. The pattern was similar in other units as the Royal Air Force was reduced to its lowest ever state of 33 Squadrons and eight of these still in the process of reforming.

If the major conflict of the First World War was over, there was no overall peace for the world and Sopwith aircraft were firing their guns in anger for almost as long as they remained in service. In 1919 Sopwith aircraft were being shipped to the White, Black and Baltic seas for operations.

A North Russian Expeditionary Force set out in June 1918 to assist the White Russian cause. Docking at Murmansk they found 1½ Strutters, still crated, from earlier deliveries to the Imperial Russian Forces.

However, numbers of 1½ Strutters from British and French and possibly Russian own production, reached both sides in the Russian Civil War. And at the other end of Russia, in Siberia, the Japanese Army were reported to have 1½ Strutters operating in support of Admiral Kolchak's forces.

The first Camel arrived in Russia from H.M.S. *Nairana* in July 1918 and six more were shipped out in November 1918, to replace the earlier Nieuports. Snipes followed, but were little used as the intense cold adversely affected the B.R.2 engines. At least one Snipe was captured by the Red forces and was later flown by the Soviet fighter ace Georgii Sapozhnikov who was killed giving an aerobatic display at Aleksandrovsk late in 1920.

Meanwhile, in the Baltic, a carrier-base force was in action. H.M.S. *Vindictive*, operating in the Gulf of Finland, sent out five seaplanes, three Camels, two 1½ Strutters and a Grain Griffin to bomb Kronstadt on July 30th 1919. Five direct hits were observed and fires were started.

On the night of August 17th/18th two 1½ Strutters, a Camel and Griffin assisted seaplanes in creating a diversion over Kronstadt, drawing fire from the defences while

Snipe of No. 56 Squadron (believed F2475) patrolling over Constantinople (now Istanbul) to which region the unit was detached from Egypt in September 1922 during the Chanak Incident for policing duties. It will be seen to be fitted with bomb racks for strafing, but this was one of the few crises settled without bloodshed.

coastal motor boats attacked the harbour. The Camel, repeatedly dived and machine-gunned searchlights in this co-ordinated naval/air operation. Later in the year H.M.S. *Furious* arrived from Britain with six seaplanes, two 1½ Strutters and twelve Camels; but welcome as the reinforcements were, the aircraft were found to have been overexposed to the elements and only a few were airworthy.

Camels used for daylight raids, were subject to accurate anti-aircraft fire. One was hit by shrapnel at 15,000 feet and another shot down; others were forced down in the sea. In early December, with the danger of the fleet being frozen in, the force withdrew south to Libau, and were then signalled to return to the U.K. Most of the 1½ Strutters were by then unserviceable and all remaining were scrapped. At the same time the Northern Russian Force was withdrawn and the aircraft were smashed and dropped in the sea to prevent their use by others.

An expedition had also gone to South Russia. It had been decided to help General Denikin in his fight against the Soviet Government and No. 47 Squadron, which had previously served in Salonika, with four Camels from No. 150 Squadron and elements of No. 17 Squadron, was sent to South Russia under Major Raymond Collishaw—the Triplane and Camel ace.

The squadron was divided into three flights of which 'B' was composed of Camels. The first Camel combat occurred on September 30th 1919 when Captain S. M. Kinkead, D.S.O., D.S.C., D.F.C., a former Camel ace with 30 victories, was attacked by an enemy Nieuport. Kinkead promptly shot it down into the Volga.

Operating in co-operation with General Wrangel's Cavalry Corps, the Camels would precede a cavalry charge by attacking the enemy with their machine-guns and bombs. When the force, under a revised governmental policy, was withdrawn early in 1920, the Camels and D.H.9s of the Force were unloaded on to the docks at Novorossisk and crushed by a tank, which was itself then driven into the harbour.

At home, Sir Hugh Trenchard, as Chief of the Air Staff, was planning a solid foundation for the post war air force with the accent, in the formative years, on training. For a home defence fighter, fleet fighter and torpedo bomber, his choice was for Sopwith aircraft and the holdings and stock position, reviewed 1922, was favourable:

	Sqn.	Schools	Store	Total
Snipe	96	39	397	532
Cuckoo II	24	6	15	45
Ships Camel	36	—	43	79
	156	45	455	656

A shortcoming was the lack of an intermediate trainer, thus a pupil went direct from a gentle Avro 504K to a fierce Snipe. However, some forty of the Snipes were converted to two-seaters to bridge the gap to some extent. Nevertheless, to the young pilots of the day it was an awe-inspiring step and one pilot has recorded his views—'The Avro 504K with its very low wing loading, pronounced inherent stability and small engine gave the feel of a gentle, unhurried, ride particularly at low altitude. In

Bonzo, a Snipe at a training unit seen warming up. After starting the engine was run until it was operating smoothly and oil was seen to be circulating in a pulsometer glass in the cockpit. The engine would then be opened up very briefly before final take-off.

normal take-off the angle of climb was so small that the horizon was depressed but very little from its usual place as seen in horizontal flight. Turns and other normal manoeuvres were commenced and completed in a leisurely floating way. To carry out aerobatics a reserve of speed had to be built up by diving, there being little additional power to call upon from the engine by increasing revs with the engine controls. These observations are not in criticism of a very fine aeroplane, but simply an appraisal of the performance to enable a comparison to be made.

'After completing the necessary hours of solo on 504Ks the day came when one was introduced to a dual control Sopwith Snipe. Two cockpits in so short a fuselage? Impossible! But in this stubby machine the impossible was achieved. Following the usual pre-flight palaver, instructor and pupil, the latter in the forward cockpit, harnessed themselves in. The change from the 504Ks safety belt to a Sutton Harness presaged new possibilities but hardly prepared a novice for what was to come.

'The exaggerated nose-up position of the Snipe at rest, due to the very short fuselage and high undercarriage infused a sense of urgency and power—and this, even before the engine was started!

'Taxying out for take-off was a jerky ride, every

irregularity of the ground being transmitted through the tail-skid to the cockpits in the short fuselage. Instructions were given that on take-off full rudder should be applied before opening up the engine, to be followed by rudder in the opposite direction as the tail lifted into flying position. This was to correct the combined effect of deflection of the fin and rudder by the slipstream and the terrific gyroscopic moment imparted by the great mass of the rotating B.R.2. Also the stick had to be pushed right forward to quickly lift the tail into flying position and then eased back. The wheels unstuck after a brief run and as airspeed was built up, assisted by the guiding hand of the instructor on the stick, the aircraft was put into, what seemed at the time, very nearly a vertical climb. After being alarmed at losing sight of the horizon, confidence was restored on finding it somewhere near the leading edge of the lower main plane. This great, roaring engine had taken possession of the situation and was, in a few brief seconds, completely revising the pilot's opinion of flying.

'Although some hours were flown in single-seat Snipes that first dramatic rocketing into the blue is a lasting memory which recalls the amazing and famous climbing quality of the Snipe as no other thing ever did. In most manoeuvres the gyroscopic effect of the large rotary engine could always be felt, in fact, it was so immense that at the top of a loop it was necessary to apply full rudder to prevent the aircraft turning on its back.

'All controls were very sensitive, the aircraft giving immediate response to any movement of the control column and rudder-bar, a feature which made the Snipe highly suitable for aerobatics. It was, due to its sensitive qualities, not an easy aircraft to fly. The low inherent stability in the air and the narrow wheel base and short fuselage in landing and taxying called for unrelaxed attention on the part of the pilot.'

Altogether, Nos. 3, 17, 19, 23, 25, 29, 32, 41, 46, 56 and 111 Squadrons were equipped with Snipes in the United Kingdom, and No. 80 in Egypt, at the same stage in the 'twenties. The two-seat Snipes were allotted one or two to each Snipe squadron and three to five to Nos. 1 and 2 Flying Training Schools and the Central Flying School. Both single and two-seat Snipes served overseas.

Apart from squadron service, several Snipes were used on experimental work. E8137 had a special parachute housing embodied in September 1921 and was used at Martlesham Heath for general parachute testing work. It was joined by E7534 which earlier, in March 1921, had been flown on comparative trials with the Westland Wagtail and B.A.T. Bantam.

Among the more unusual experiments was yet another method of launching Camels—from airships. The idea had originated late in the war as an anti-aircraft measure, with the purpose of having aircraft that could be lifted to a patrol area by airship and released at height when a raid warning was received either by wireless or searchlight signal. Two Ships Camels of No. 212 Squadron at Yarmouth, N6622 and N6814, had been flown to Pulham and fitted with quick-release hooks. After a dummy drop by a Camel with locked controls and ballast, Lt. R. E. Keys, D.F.C. made the first live drop. Following release he dived gently, started the engine, climbed to circle his parent ship and landed at Pulham. After the Armistice, the experiment had lapsed for two years.

In the post-war experiments the R33 was used.

Take-off from a carrier in a French-built, British-modified, Ship Strutter postwar. At the right a stream of smoke blowing back with the forward motion of the ship, indicates to the pilot the amount of wind drift to compensate for by use of the rudder.

Whereas the R23 had a single aircraft position, the R33 had two and this time Clerget-engined F.1 Camels were used, B2312 and H7363, fitted with self-sealing tanks.

During the Iraqi insurrection of the early twenties, the only Snipe squadron in India, No. 1 at Bangalore, was moved to Hinaidi in Iraq.

Flying in the desert brought new hazards. Apart from the effect of the climate on the wooden airframes, there was the sanddust which caused excessive wear on the B.R.2 engines.

After Sheik Mahmud, a fanatical and ambitious Kurd with a large following, had swept through parts of Kurdistan the 2nd Levy Cavalry Regiment was sent out on a punitive expedition. The extent of the Sheik's following had not been appreciated and the force was beset on all sides by armed bands showing allegiance to the Sheik. A situation which might have resulted in a massacre was changed by the arrival of a flight of Snipes, which dispersed the tribesmen with machine-gun fire. From then onwards the Snipes flew column protection patrols.

The Snipes were withdrawn from overseas in 1926 and in 1928, together with the B.R.2 engine, were declared obsolete for Royal Air Force purposes. By that time the R.A.F. was getting Hawker fighters.

CIVIL SOPWITHS

Left column, top to bottom: Pup G-EAVX with a Le Rhone engine under a Monosoupape cowling, a Snapper prototype numbered 17 for the 1919 Aerial Derby, the first production Gnu and Harry Hawker's aerobatic Scooter. Right column, top to bottom: Two of three Snipes entered in the 1920 Aerial Derby of which the top one (G-EAUW) came in fifth place, the prototype Gnu and the Rainbow fitted with a Dragonfly engine for use by Harry Hawker as a racer.

CHAPTER THIRTY-NINE

Civil Sopwiths

Captain H.R.H. The Prince of Wales (now Duke of Windsor) with T. O. M. Sopwith (left) about to fly in the prototype Dove with Major Barker V.C. (right). Barker still has his arm in a sling as a result of his epic combat in a Snipe related earlier. The Prince's brother Albert, who later became King George VI, was at this time reported to have flown in a two-seat Camel at Cranwell.

At the end of the war the four Sopwith plants, the original skating rink, the works in Canbury Park Road, and the premises at Ham and at Brooklands, employed a total of 3,500 persons of which over a thousand were women. After the Armistice many of the older workers and most of the women, who had been more concerned with 'doing their bit' than earning a living, left. Even so, there was little enough work to keep the remaining staff occupied in spite of Sopwith's personal efforts to branch into other fields.

Within a month of the Armistice, Sopwith had obtained a licence to build the A.B.C. motorcycle. This entailed considerable expenditure in re-organising and re-tooling. Early in 1919, while using 1 Albemarle Street, temporarily, premises were taken over at 67, South Molton Street, London, W.1., to bestow the prestige of a London office and showroom, as well as serving as a London depot. Although some structural alterations were necessary, the work was pushed ahead and completed in four weeks.

A formal petition was made to the Board of Trade to confirm the altered objects of the company—to make motor cycles and motor vehicles, act as coach and carriage builders, and to make all kinds of furniture. A. R. Fenn, formerly of the Paris Office, was made sales manager, while the executive posts remained as before.

If the aviation side did not flourish, it was certainly not through lack of enterprise. As related, a Sopwith aircraft had already attempted an Atlantic crossing and even before that a civil Sopwith aircraft was on the market. Appropriately named the Dove, this sporting biplane was virtually a two-seat Pup.

The Dove made its debut at Hounslow in May 1919 in the hands of Major Barker, whose gallant fight, related earlier, was rewarded with a Victoria Cross. He was still convalescent, when he asked T. O. M. Sopwith if he could fly the aircraft in Hawker's absence in Newfoundland. Sopwith, delighted to have an expert opinion, gave Major Barker full rein. The young Prince of Wales (now Duke of Windsor), visiting Hounslow at the time, was greatly impressed by Barker's flying and asked to be taken for a flight. Apparently the King, on hearing of his son's flight, in a prototype aircraft, with a pilot who had his arm in a sling, was not unnaturally concerned. He asked his son to refrain from flying—a request he did not rescind for some ten years.

Hawker was back in time to present the next Sopwith civil aircraft, the Gnu. This was one of the first passenger cabin biplanes, having an open cockpit for the pilot and an enclosed cabin for two passengers. Hawker flew the prototype K-101 from Brooklands to Hendon on May 29th for the reception given to American crews of the Navy-Curtiss NC-1, 2 and 4 flying boats that had crossed the Atlantic via the Azores.

Meanwhile, his Atlantic biplane, abandoned in the Atlantic, had turned up! The S.S. *Lake Charlotteville*, bound from Montreal to Danzig, sighted the tail of the aircraft sticking vertically out of the water. Not knowing that the crew had been rescued, a search was first made, then the wreckage was winched aboard and lashed on deck. It was taken to Falmouth where mail, in a waterproof casing, was extracted and posted. Later the wreckage was taken to London. The undercarriage, jettisoned at the Newfoundland coast, was also found and taken to a museum at St. Johns—the first article of Sopwith manufacture to become a museum piece.

For the re-constituted Aerial Derby in 1919, Hawker had an inspiration. The latest Sopwith prototype fighters on the stocks were three R.M.1 Snappers of which only the first, due to lapsed requirements, was complete. This became the Sopwith entry for the Derby but, as the Dragonfly engine was still on the Secret List, the Air Ministry banned its participation.

Everyone expected a Sopwith entry for the first post-war Schneider Trophy competition. T. O. M. Sopwith built a special racing floatplane, to take the 450 h.p. Cosmos Jupiter engine. Like Sopwith, Hawker had a propensity for boats and cars and took with him to Southampton his racing motor-boat *Kangaroo II*, and his powerful Sunbeam racing car—both were to be needed.

Staying with his wife in Southampton, he used the motor-boat as a ferry to the floatplane which was being asembled, after transit from Kingston, at Hythe. The maiden flight, on September 10th, nearly ended in disaster. Shortly after opening up the engine to take off, the machine lurched and nosed over in the water. When the machine was beached, it was found that the floats had been fitted too far back. It was quickly loaded on the Sopwith lorry for return to the shops at Kingston, but the lorry's engine failed. This was where Hawker's powerful Sunbeam came in—he towed the lorry to the works.

Within two days the Schneider floatplane was back. The Sopwith experimental shop had worked at wartime tempo. Its chief rivals for representing Britain were a Supermarine flying boat and a Fairey floatplane, both powered by 450 h.p. Napier Lions. In an eliminating trial the floatplane reached a satisfactory 180 m.p.h., but again the floats were damaged on landing. Hawker had the floats removed and conveyed them back to Hythe in his motor-boat, built for two, which as well as carrying two mechanics, had his wife and her sister aboard: on a turn the craft capsized depositing all five in the water. Australian soldiers on the shore set out to the rescue in a rowing boat which had the bung out—and they too floundered! Fortunately no lives were lost.

The Schneider Trophy was almost as much a fiasco. At the appointed hour a sea fog came up and blotted out the turning points. The Sopwith entry had yet again damaged its floats landing at Bournemouth and only just managed to get off the water. On landing at Hythe it was pulled quickly ashore to prevent it sinking and its chance was lost. To cap it all, Hawker and his wife had arranged to join Sopwith in Scotland after the race and were prevented by a railway strike.

The design of the Atlantic biplane had not been wasted. The Australian Government offered in March 1919 a prize of £10,000 to the first Australian airman who succeeded in flying back to Australia in 30 days, in an aircraft of British or Commonwealth manufacture, starting from Hounslow (landplanes) or Calshot (seaplanes). Since this was essentially for returning servicemen, Hawker

The Prince of Wales and Major Barker in the Dove's cockpit. With an arm in a sling, flying a prototype, Barker stunted with the heir to the Throne.

did not fly the aircraft, but acted in an advisory capacity on the design. Again, a Sopwith was first off the mark.

Aptly named the Wallaby, the aircraft closely resembled the Atlantic biplane, with the same type of engine, but was rigged with three-bay wings. It set out on October 21st piloted by Captain G. C. Matthews, a master mariner, who did his own navigation and who had flown Camels in the war with No. 4 A.F.C. Squadron. His companion, Sergeant T. D. Kay, ex-No. 3 A.F.C. Squadron, was chosen for his skill as a mechanic. They were dogged with bad luck. Delayed by bad weather at Vienna, then arrested as Bolsheviks in Yugoslavia, they damaged the machine landing in Iran.

Anticipating the arrival of a Sopwith in Australia, T. O. M. Sopwith gave his name to an associate company, the Larken-Sopwith Aeroplane Company (of Australia) Limited which had offices at 18 Gurner Street, St. Kilda,

Back in the passenger-carrying business. From an auction conducted by Grahame-White, Miss Daisy King of Leeds, seen entering the Gnu, paid 60 guineas for this first postwar passenger flight in a Sopwith aircraft piloted by Hawker. T. O. M. Sopwith can be seen be-between the wings at extreme right.

Melbourne. Two Gnus and four Doves were shipped out early in 1920 but, as in England, there were too many ex-service aircraft on the market to attract sales for new aircraft.

In 1920 a number of ex-R.A.F. Snipes and Pups were registered as civil aircraft and three of the former were entered in the Aerial Derby of that year—competing against the Sopwith entry—the Rainbow. This was the Schneider Trophy machine's airframe, fitted with a 320 h.p. A.B.C. Dragonfly engine and a wheeled undercarriage. Unfortunately, Hawker was disqualified for an incorrect finish and two of the three Snipes made forced landings.

The fortunes of the firm were at their lowest. Hawker himself, ever on the go, turned to racing on land and sea. Sigrist told a story of him coming into the factory one day about this time and picking up the latest product, a kitchen utensil and saying, 'Well, Fred, What about this? Saucepans! Where do I come in? I never thought I'd be on a job that Mrs. Beeton could do better than I'.

In spite of Sopwith's attempts to keep the factory alive, the slump of 1920 proved too much. It was an-nounced that the works would close from September 3rd for a fortnight. On the 10th the employees, most of whom realised the way things were, received a notice: 'We much regret we find it impossible to re-open the works as the difficulties caused by restricted credit prevent the company from finding sufficient working capital to carry on the business and it will therefore be wound up. G. H. Mitchell, Works Manager.'

The Company went into voluntary liquidation and paid its creditors twenty shillings in the pound. As an independent, unbiased, view of the reasons for winding up the comment of *Flight* in their September 16th 1920 edition is quoted verbatim:

'The closing down of the Sopwith works at Kingston conveys a lesson for all. In these works were constructed during the war some of the finest fighting aeroplanes possessed by any of the belligerent Powers. When the Armistice came to put an end to hostilities, the directors with their accustomed foresight, decided to employ the works in the manufacture of motorcycles, for which the plant was eminently adapted. They secured a magnificent design

Watching other competitors at the 1919 Schneider Trophy contest are (left to right) T. O. M. Sopwith, C.B.E., Harry Hawker who piloted the Sopwith entry, Harold Perrin the Royal Aero Club Secretary, and Lieutenant-Colonel F. K. McLean who later became Sir Francis. The 1919 course was off Bournemouth.

Harry Hawker had many different cars in his short life, including an Austro-Daimler which he 'hotted up' for breakneck trips between his house at Hook and his work at Brooklands. He is seen here at the wheel of his famous 150 h.p. Sunbeam racer on the Brooklands racing track.

in the A.B.C., and orders poured in upon them for the new machines. In the interval, like too many other firms in like case, they have had to pay out on account of wages and other outgoings antecedent to production, and at last their resources of credit became exhausted. In the circumstances there were but two courses of possible adoption. The one was to hang on and eke out resources until the coming of better times; the other to close down altogether at once. They have taken the second, and although we very naturally deplore the passing of a firm so famous in the annals of the War, it is difficult to see what else could be done. We are quite satisfied that the works' management has done its best to get the production stage, and that the efforts which have been made have not been successful is not to be laid at their door. But the payment of high wages for a low standard of production cannot be maintained for ever, and one day the inevitable is bound to happen.

'According to an official of the Company, the ultimate decision was taken as the result of the slump in the motor trade. Valuable orders from all parts of the overseas Dominions and Scandinavia have been cancelled, and this state of affairs, coupled with the competition from America, has made it impossible for a lucrative trade to be carried on. So far as we are competent to judge, there is a very wide market for motor vehicles if manufacturers can deliver the goods. It is too much, however, to expect people who have ordered a year or more ago to wait indefinitely for the completion of their orders, and they quite naturally turn elsewhere for fulfilment of their requirements.

'We trust most sincerely that some way out of an apparent impasse may yet be found and that the firm of Sopwith, with all its traditions in the world of aviation, will not disappear.'

The same was happening with other companies. Martinsyde, like Sopwith Aviation, had turned to motor production but they, too, were forced to cease trading. Several of the contractors for Sopwith aircraft had greater difficulties. Glendower Aircraft, who had been expanding throughout 1918 to get the Salamander into production, had only delivered a small number when orders were cancelled and the company was wound up. Kingsbury Aviation, nearby, only completed thirty of their Snipes before they went into liquidation. The Portholme Aerodrome Ltd., who had relied solely on Sopwith contracts, took over a little work from Ruston Proctor before they, too, went out of business. The old established firms were better placed, it was literally a case of returning to ploughshares for firms like Ruston Proctor who had a pre-war reputation for building agricultural machinery.

Harry Hawker, taking the Grasshopper round the Brooklands motor track, passes under the bridge that led to the Sopwith sheds to the right of this picture.

CHAPTER FORTY

Forward as Hawker

First of the H. G. Hawker Engineering Company's aircraft was the Duiker—a kind of antelope — which was the name given to the last of the Sopwith-designed aircraft.

In the face of enormous taxes due to the Exchequer under a wartime Ministry of Munitions control of profits, and civil aviation still in its birth pangs, T. O. M. Sopwith had advised his board to wind up the company while still solvent. Both R. O. Cary, who had become Joint Managing Director, and H. P. Musgrave, the Company Secretary, voluntarily relinquished their posts. In place of the old company, and in the same works, a new organisation was brought into being. It was thought better to drop the old name to avoid any confusion with the Sopwith Company in liquidation and Harry Hawker's name was given to the new organisation.

H. G. Hawker Engineering Company Limited was registered as a private company on November 15th 1920 with a capital of £20,000 in £1 shares. Company objects were to acquire from F. I. Bennett all patent rights, etc., relating to the manufacture of motor cycles and dealers in cycles of all kinds of internal combustion and steam engines, motor cars and aircraft. The first directors were F. I. Bennett (also Company Secretary) of 19, Cadogan Road, Surbiton, engineer; H. G. Hawker of 'Ennerdale' Hook Road, Surbiton, aeroplane pilot; T. O. M. Sopwith of Horsley Towers, Surrey, engineer; F. Sigrist of Torrington House, Wolsey Road, East Molesey, engineer and V. W. Eyre of Honeyhanger, Hindhead, Surrey, also engineer.

While the Hawker story is a subject on its own, such highlights as carry forward the major events in Sir Thomas Sopwith's life, and events concerning the buildings where once Sopwith aircraft were built are briefly touched upon. Some of the first orders given to the new company, concerned their former products. Spares were required for maintaining naval Camels in store and a contract was given for re-conditioning Snipes, chosen as standard home defence fighters in the new peacetime Royal Air Force. This work was conducted under the eagle eye of Fred Sigrist who became managing director. Hawker at this time had also formed a Company in his homeland, Australia, with an agency for D.F.P. cars, and he was also showing a particular interest in A.C. cars. At the same time, the new Hawker Company continued to produce the A.B.C. motorcycles and both Hawker and Sopwith entered them in races on occasions. A racing two-stroke motor-cycle was in the design stage when Harry Hawker turned again to aircraft.

For the Aerial Derby of 1921, Hawker had elected to fly the Nieuport Goshawk which was powered by a 320 h.p. A.B.C. Dragonfly 1A engine. A one-off single-seat racing biplane, the aircraft had been designed by H. P. Folland and built at Cricklewood in 1920 by the Nieuport & General Aircraft Company. Four days before the Aerial Derby, on July 12th, Hawker took up the Goshawk on test at Hendon. Precisely what happened is not known. It is thought that he suffered a haemorrhage while in the air. Over Burnt Oak the aircraft was seen to lurch and then it appeared as if the pilot was attempting to land, but the machine crashed and took fire. Hawker, unfortunately, was instantly killed.

It was not only a disaster for the new Company, it was national disaster. A message from the King read, 'The nation had lost one of its most distinguished airmen, who by his skill and daring has contributed much to the success of British aviation' and Lloyd George wrote. 'The nation is the poorer for the loss of one who always displayed such splendid courage and determination. To such pioneers we owed our supremacy of the air during the war'. There were many other messages of sympathy.

The new Company slowly recovered from this loss. Captain B. Thomson had been taken on as an aircraft designer and the Duiker, the first Hawker aircraft, appeared in 1923. The following year, W. G. Carter who had been on design work with the Sopwith Company, took over as Chief Designer. Other work had a more direct association with their former work—a few Ship's Camels in store were re-conditioned for the Admiralty and a number of Snipes were converted to two-seaters as advanced trainers.

The 1919 Sopwith Schneider, re-built as the Rainbow, was further rebuilt by the H.G. Hawker Engineering Company with a Bristol Jupiter engine. As such it was the only aircraft type that might be called a Sopwith/Hawker. Flying it, Flight Lieutenant W. H. Longton came second in the 1923 Aerial Derby.

Sopwith was always shrewd in his choice of men. Among those joining the firm in the twenties was the pilot Fred Raynham who had attempted to fly the Atlantic in a Martinsyde about the same time as Harry Hawker. The services of a promising young draughtsman, who had also been with the Martinsyde Company, was obtained shortly afterwards—his name was Sydney Camm.

The first Hawker aircraft to win a production contract was the Woodcock of which a total of 64 was built. They entered squadron service in the mid-twenties, replac-

Fred Sigrist, who died December 10th 1956, seen after receiving the M.B.E. After serving with Sopwith Aviation throughout the whole of its existence, he became a Founder Member of H. G. Hawker Engineering, and was Joint Managing Director of Gloster Aircraft, among other directorships.

ing Snipes in Nos. 3 and 17 Squadrons. One, registered G-EBMA, was retained by the Company as a demonstration aircraft and Sopwith entered this in the King's Cup Air Race of 1925. Flown by Flight Lieutenant P. W. S. Bulman, who became the company's chief test pilot, it was unfortunately forced down at Luton by bad weather on the opening stages of the race.

The first of many export orders for aircraft came from Denmark who ordered a development of the Woodcock as the Danecock; others were produced under licence in Denmark. Sopwith, watching the trend towards metal construction, set Camm and Sigrist to work on designing in metal. Together, they devised the system of bolted duralumin tubes that became a Hawker constructional characteristic.

While the Canbury Road Works were changing internally from the Sopwith days, with metal presses, millers and lathes replacing the earlier wood-working machinery, a new Hawker torpedo-bomber of wooden construction was awarded a production order for the R.A.F. But following orders specified wood/metal construction and the final batch was required to be all-metal. Orders passed the hundred mark and it was much like the old days, with lorries taking fuselages, built at Kingston, to Brooklands for assembly and flight-testing. At this time Sopwith, as well as directing work at Kingston, was chairman of the Society of British Aircraft Constructors; an appointment he held from 1925 to 1927.

It was Sopwith who suggested that the Horsley might well be modified to meet an Air Staff proposal to fly a service aircraft non-stop from England to India. An attempt was made, but the Horsley was forced down in the Persian Gulf—nevertheless it achieved a world non-stop distance record of 3,420 miles. In the 1927 King's Cup Race, Sopwith's personal entry was the Horsley J8606 which Bulman brought in to sixth place.

Being so occupied with organisation, design and manufacture of aeroplanes, Sopwith had not piloted an aircraft for some fifteen years. One morning at Brooklands, just as Major George Bulman, the firm's chief test pilot, was about to fly the new Tomtit trainer to Martlesham Heath, he said on impulse that he would fly the machine. While putting on flying overalls, Sopwith evidently had second thoughts, for he said, 'George *you* had better take her off'. It was just as well for Sopwith confessed that he was surprised at the rudder bar movement in the right bank made after take-off, which he followed with his hands and feet lightly on the controls of the dual-control fitted aircraft. Once in the air, his old confidence returned and he took over. He made a successful, but not altogether perfect, landing at Martlesham Heath; a station to which he was no stranger.

With the Hart and its variants, the future of the company was assured. Orders were received in such large quantities that the former Ham works, then leased to Leyland Motors, would have proved a great asset.

During these years of experience Sopwith's first wife died and early in 1932 he married Phyllis Brodie, the present Lady Sopwith. On November 9th 1932 a son was born—and the following year a new Company was formed.

The outstanding success of the Hawker Hart series, with large export orders, led to Sopwith seeking additional capital and forming a public company, which adopted in 1933 the name Hawker Aircraft Limited. The necessary floorspace to facilitate new orders was obtained next year by the outright purchase of the Gloster Aircraft Company and the production of certain Hawker aircraft types was transferred to the Gloster works at Hucclecote, Gloucestershire.

In 1935 T. O. M. Sopwith announced the formation of a Trust, to acquire the shares of the Armstrong Siddeley Development Company, and a holding company which was named the Hawker Siddeley Aircraft Company Limited. Within this group were Armstrong Siddeley Motors, Armstrong Whitworth Aircraft, A. V. Roe (Avro) and Air Service Training. Of the four directors, three— T. O. M. Sopwith, Fred Sigrist and F. S. Spriggs—were all from the former Sopwith Aviation Company.

In spite of his aeronautical interests, Sopwith retained his love of boats and remained a keen yachtsman. Up to that time, for almost the whole of his adult life, he had been keenly following Sir Thomas Lipton's successive, but unfortunately unsuccessful, challenges for the *America's* Cup. Now, he was in a position to take up the challenge himself, with his beautiful steel Marconi-rigged sloop *Endeavour* designed by Charles Nicholson. Sailed by Sopwith under the colours of the Royal Yacht Squadron, his series of races in 1934 proved some of the longest and most

Maple Leaf IV *when T. O. M. Sopwith was at the helm of this boat in 1912.*

gruelling in the history of the Cup—which dates back to 1851. It was close run—a grand endeavour—but it failed. However, by winning the first two races, of the four out of seven necessary for a win, Sopwith came nearer to gaining the *America's* Cup than any other challenger before or since. Sopwith tried again with similar results in 1937 using his *Endeavour II*, the last of the J Class yachts to complete before the change to the smaller boats of today.

At Hawker Aircraft, further floorspace was still needed as the R.A.F. Expansion Scheme got underway. A new factory and airfield was constructed at Langley and was soon producing Hurricanes together with the earlier plants at Brooklands and Kingston. Again T. O. M. Sopwith was in control of a vast complex, but whereas the Sopwith Company had employed some 3,000 persons at the end of the 1914-1918 War, the Hawker element alone of the new Group employed some 4,000 persons at the beginning of the 1939-1945 War.

Fortunately both the Brooklands and Langley plants escaped serious damage in the war, but the old Sopwith Canbury Road Works received one hit that damaged a shop in which the prototype Typhoon was being assembled; part of the building was severely damaged but the aircraft only superficially. By the number of bombs that fell near, it is possible that it was a specific German target.

T. O. M. Sopwith's 'Hawker empire' grew during the war to an organisation employing 12,500 people. The Hurricane remained in production until 1944, but by that time the Typhoon was already in production and as early as 1941, a jet engine as a means of aircraft propulsion was being studied in the Canbury Park Road Works—and it was the Gloster element of the group that produced Britain's first jet. Post-war a new airfield was acquired—Dunsfold—as test facilities at Langley were becoming increasingly difficult due to growing traffic at London Airport which was not much over five miles distant.

One other change, must have much pleased Sopwith. The 20-year lease held by Leyland Motors on the Richmond Road Works expired and they were quickly re-occupied. And so in another factory Hawker aircraft were being built, where once Sopwith aircraft were on the lines.

In the fifties, under Sopwith's chairmanship of the Hawker Siddeley Group, the Hunter was introduced and once again he was at the head of an organisation producing his country's premier fighting aircraft—and, one in this case which found favour in many countries worldwide and is still flying in service at home and abroad.

The year 1953 was a memorable one for Sopwith; he

T. O. M. Sopwith with the record-breaking Horsley—a machine that he suggested should be named Kingston, but that name had already been used for an earlier flying boat. It was eventually named Horsley—the name of Sopwith's residence of the time, Horsley Towers, a Surrey mansion house.

received his Knighthood in the Queen's Coronation Honours List for his services to aviation. As if to endorse this service, it was in that year that a Hawker Hunter achieved a new World Absolute Speed Record at 727.6 m.p.h. piloted by Neville Duke.

For another ten years Sir Thomas remained at the helm as Chairman of the Hawker Siddeley Group. During that time, on November 19th 1960, the Hawker P.1127 made its first untethered hovering flight and developed, as the Harrier, the Group have the finest VTOL strike aircraft in the world. In 1963, as related earlier in the book, Sir Thomas relinquished his chairmanship and accepted the Presidency of the Group, an office he currently holds. He has not forgotten his earlier organisation. An active Sopwith 'Old Boys' Association, of which Victor Derrington is secretary, meets at intervals. On these occasions, before dining, the gong—a Gnome Monosoupape engine cylinder is ceremonially sounded. It is with regret that it was learned, just as the book was going to press, that Jack Pollard had died.

In 1967, in recognition of his long association and enthusiastic interest in sail and powerboating, Sir Thomas was invited to open the 13th International Boat Show at Earls Court, London. Already, by this time, his son Tommy had made his name as a keen competitor and organiser of powerboat racing and had won the International Power Boat Race in 1961. He is also an experienced aeroplane and helicopter pilot, and in 1968—a year in which he again won the International Power Boat Race—he joined the boards of C.S.E. Aviation Limited and C.S.E. International Limited.

For the 1970 International Offshore Power Boat Race Tommy Sopwith was at the wheel of *Miss Enfield II*, owned by Mr Goulandis.

Starting from Cowes, where his father had launched the first Sopwith seaplane, he led throughout and returned the winner. As Sir Thomas himself said when recently asked about the reaction of people meeting one of the famous air pioneers: 'Around here', he replied 'I'm known as the father of Tommy Sopwith'!

Miss Enfield II *with T. E. Sopwith, driving his father, Sir Thomas Sopwith in 1968.*

Top, Sir Thomas and Lady Sopwith visited by their son T. E. Sopwith, who flew 'Sister May', at right, to Compton Manor in his Bell Jet Ranger helicopter, G-AVZG, seen in the background. This photograph was taken by The Earl of Lichfield on October 12th 1970. Below left, Sir Thomas and Lady Sopwith at Camper & Nicholson's boatyard, Gosport, in the days of Endeavour 1934; and at right T. E. Sopwith in his youth.

FAMILY ALBUMS

Top, Sir Thomas and Lady Sopwith photographed on January 21st 1969 in the grounds of their home, Compton Manor. Bottom left, T. E. Sopwith, in childhood days, shows his propensity at an early age; and at right Sir Thomas and his sister who first flew with him in America in 1910, and who, 60 years later, in October 1970, was flown over to Compton Manor by his son T. E. Sopwith in his Jet Ranger helicopter.

SOPWITHS FORTY-FOLD—PRODUCTION DETAILED

Since only ten percent of the most famous of all Sopwith aircraft, the Camel, was built by Sopwith Aviation and similarly of such famous types as the Pup or 1½ Strutter only five percent were built by the parent firm, the story of the production side of Sopwith aircraft cannot be confined to the Sopwith Aviation Company.

In this section a brief is given on all the plants and depots that produced or received contracts for Sopwith type aircraft, with a listing of the contracts awarded. Dates of first flights or deliveries of individual aircraft are included, when known, in the remarks column as a guide to the actual time of production. A summary table of total production of all Sopwith aircraft types is given at the end of this Part.

THE AIR NAVIGATION Co. Ltd.

At their Bleriot SPAD works at Addlestone, Surrey, busily engaged in producing S.E.5As in quantity, an order was received by the Air Navigation Company to produce Salamanders, but was cancelled before any were delivered.

Contracts for Sopwith Types

Type	Quan.	Serial Nos.	Remarks
Salamander	150	F7801-7950	Cancelled

Total built: Nil

AMIOT-S.E.C.M.

M. Felix Amiot had only just joined partnership with the Société d'Emboutissage et du Constructions Mécaniques when contracts were received at their premises, Nos. 171-183 Boulevard du Havre (now Boulevard Charles-de-Gaulle) at Colombes, Seine, France, for the construction of Sopwith 1½ Strutters for the French Government. The Sopwith contracts replaced Moranes in the assembly shops, which were in turn superseded by Breguet 14s. It is presumed that around 600 1½ Strutters were built.

Example of a French-built 1½ Strutter Type 1A2, shown serving as a trainer with the American Expeditionary Force.

ARROL-JOHNSTON Ltd.

A Scottish engineering concern that built cars including their engines, engaged in various sub-contract work at their Dumfries works in Scotland including, in the final months of the 1914-1918 War, the erection of complete aircraft. Initially the firm were asked to erect ten Ships Camels as a sub-contract from Beardmore and the first, N7140, is

A characteristic of all Beardmore-built aircraft was that the elevators as well as the rudder were striped as shown on this Pup.

recorded back at Beardmore's works on October 17th 1918 ready for delivery. The last of the ten was delivered in January 1919.

Shortly before the Armistice the firm received a direct contract (AS24907) for 40 Ships Camels and of these only twenty can be traced as completed.

Contracts for Sopwith Types

Type	Quan.	Serial Nos.	Remarks
Ships Camel	10	N7140-7149	Ex-Beardmore order
Ships Camel	40	N7350-7389	20 built

Total built: 30 Ships Camels

BARCLAY, CURLE & Co. Ltd.

This shipbuilding company of Whiteinch, Glasgow, had built B.E.2cs and Fairey Campanias when, on November 1st 1918, a contract was awarded for Snipes—and cancelled soon afterwards.

Contracts for Sopwith Types

Type	Quan.	Serial Nos.	Remarks
Snipe	75	J3917-3991	Contract 35a/3447/C4052

Total built: Nil

WILLIAM BEARDMORE & Co. Ltd.

Wm. Beardmore & Co., founded in 1835 and now a subsidiary of Thomas Firth & John Brown Ltd., had acquired building rights for the German D.F.W. biplane in 1913. After war was declared the company established an Aviation Department on receipt of contracts for B.E.2cs. These were built in the cabinet shops and factory sheds together with Wight Seaplanes and Nieuport 2-seaters.

New shops were erected at the Dalmuir works, on the banks of the Clyde near Glasgow, until some 98,000 sq. ft. of factory floor was devoted to aircraft construction. Nearby was an aerodrome and the firm's airship shed and factory at Inchinnon.

Their first order for Sopwith aircraft was an Admiralty Contract for fifty Pups, CP117318/16, before the Sopwith firm had received a production order. The first ten Pups, known as Type 9901, had 80 h.p. Clerget engines initially fitted and the remainder 80 h.p. Le Rhônes. This batch had a particularly active life, some going to R.N.A.S. Squadrons on the Western Front, others to the Eastern Mediterranean, and a few to ships of the Fleet. In all, sixteen of the fifty were fitted with airbags for shipboard use. By the end of February 1918 21 standard models and 9 with airbags were

Blackburn-built aircraft could be identified by their trade-mark seen on the fin of this Baby floatplane which in 1918 was serving with No. 455 Flight of No. 229 Squadron at Yarmouth.

still in use, and several were even in commission at the end of the war, viz. Nos. 9931, 9932 and 9940 at the Scapa, Smoogroo and Turnhouse fleet bases respectively, 9942 under repair at Mudros island and two actually at sea—9944 and 9949 on H.M.S. *Vindictive* and H.M.S. *Argus* respectively.

The last of the batch, No. 9950, was converted by Beardmore to a new standard for shipborne use. Known as the Beardmore W.B.III a hundred were built to this new standard with unstaggered wings of reduced dihedral designed to fold to facilitate stowage. Although of basic Sopwith design, like the Fairey adaptation of the Baby, they are not regarded as Sopwith type aircraft.

However, a further order for basic Pups with skid undercarriages was placed as Admiralty Type 9901A followed by an order for shipboard Camels. Their first Camel, N6750, flew in February 1918 and was delivered to Rosyth from the works in March; 140 had been completed by the time of the Armistice. One survives today in the Imperial War Museum.

Contracts for Sopwith Types

Type	Quan.	Serial Nos.	Remarks
Pup	50	9901-9950	Admiralty Type 9901
Pup	30	N6430-6459	Admiralty Type 9901A
Pup	150	(D4211-4360)	Cancelled, Nos. re-allotted
Camel	100	N6750-6849	Type 2F1. N6812 survives
Camel	50	N7100-7149	Type 2F1. Last 10 to Arrol Johnson for erection
Camel	30	N7650-7679	Type 2F1. Cancelled

Total built: 80 Pups, 140 Ships Camels

BESSONEAU

The Société Anomonye Industrie Bessoncau with offices at 29, Rue du Louvre, Paris and works at 21, Rue Louis Gain, Angers, were among the French contractors awarded orders by the French Government for the production of 1½ Strutters.

THE BLACKBURN AEROPLANE & MOTOR Co. Ltd.

In 1916 Blackburn's took over production of Sopwith Baby type floatplanes from Sopwith and continued development work with the type including modifications to accommodate the 130 h.p. Clerget. They were built in the former skating rink Olympia Works in Roundhay Road, Leeds, and were taken by road to Brough, for test-flying on the Humber by the Blackburn test pilot R. W. Kenworthy.

Due to the inability of other contractors to produce the Cuckoo quickly, large contracts were placed for this torpedo-bomber. These were built at Sherburn-in-Elmet some twelve miles from Leeds from April 1918. A Leeds Aircraft Acceptance Park was established at Sherburn in 1918 and was renamed Sherburn A.A.P. but in May 1919, when work ceased on the Park, it had not been completed. Arrangements were made with the firm for the Cuckoos to be tested and delivered by service pilots from Sherburn.

Contracts for Sopwith Types

Type	Quan.	Serial Nos.	Remarks
Baby	1	N300	110 h.p. Clerget. Contract C.P.16706/16
Baby	30	N1010-1039	110 h.p. Clerget. Contract C.P.118060/6 N1030-1039 had experimental wing section
Baby	40	N1060-1069	110 h.p. Clerget Contract anti-sub model. Contract C.P.133305/16
Baby	40	N1410-1449	110/130 h.p. Clerget. Anti-Zepp version. Contract AS679
Baby	75	N2060-2134	130 h.p. Clerget. Anti-sub model. Contract A.S.10059
Cuckoo	30	N6901-6929	Contract A.S.10375/18 ex-Pegler
Cuckoo	50	N6950-6999	Contract A.S.3298/18/3
Cuckoo	50	N7150-7199	Contract A.S.23688
Cuckoo	100	N7980-8079	Contract A.S.32641. Not all delivered

Total built: 186 Babies, 162 Cuckoos

Most manufacturers had a characteristic way of interpreting official and Sopwith instructions for finishing aircraft, that betrayed the contracting firm—with Boulton and Paul it was the way the serial number was marked and outlined on the rudder which, being part of the national markings, was not affected by service, functional or on individual's fanciful markings.

BOULTON & PAUL Ltd.

The second largest contractor of Sopwith aircraft, Boulton and Paul Ltd. formed in 1873 under that title, had a lineage that can be traced in Norwich back to 1797. A hundred years ago they were the largest producers of wire-netting in the world and this led to their production of chicken runs and aviaries, which changed to portable bungalows, vehicle sheds and hangars by the early part of the 1914-18 War, and continued throughout the War.

In 1915 the firm were given orders for F.E.2b and F.E.2d aircraft, of which they eventually built over 500. The first had been delivered to Farnborough by Howard Pixton, then a Captain in the R.F.C. and formerly the Sopwith test pilot. Orders for Camels followed in 1917 and the initial order for 100 was completed by the end of the year. By that time repeat orders had followed, supplemented in 1918 by orders for Snipes.

New buildings were erected and older buildings were taken over as the orders increased. The main offices remained at the Riverside Works. At Rose Lane Works, where propellers were made, four floors were used for component work. Large new erecting sheds were put up on nearby Mousehold Heath from where the aircraft were passed over to No. 3 (Norwich) Aircraft Acceptance Park.

Contracts for Sopwith Types

Type	Quan.	Serial Nos.	Remarks
Camel	100	B5151-5250	Ordered 1917
Camel	200	B9131-9330	Completed by March 1918
Camel	100	C1601-1700	Completed May 1918
Camel	100	C3281-3380	Ordered April 1918
Camel	300	D6401-6700	Completed by June 1918
Camel	150	D9381-9530	Completed by October 1918
Snipe	400	E6137-6536	Ordered mid-1918
Camel	250	F1301-1550	Completed by end of 1918
Camel	75	F1883-1957	Completed by October 1918
Camel	200	F6301-6500	Completed by early 1917
Camel	200	F9496-9695	Cancelled. Nos. re-allotted.
Camel	100	H2646-2745	75 of order delivered
Snipe	100	J451-550	15 confirmed delivered

Total built: 1550 Camels, 415 Snipes

Built and exhibited in London—a Hooper-built Dolphin in the 1918 Lord Mayor's Show

BRITISH CAUDRON Co. Ltd.

In April 1915 W. H. Ewen Aviation Co. Ltd. was re-formed as the British Caudron company and built limited numbers of Caudron G.III and G.IV aircraft to Admiralty orders. The works and head office at Broadway, Cricklewood, were extended 1915/1916 and a new factory and airfield was opened at Alloa in Scotland.

Sub-contracts for B.E.2cs in the Scottish works and D.H.5s in the English works were replaced in 1917 by contracts for Camels. It would seem that the initial order for 100 was divided between the two works. It is reported that in mid-1917 the Camel went into production at the London works before D.H.5 production had been completed and the first was delivered to Hendon for test. However, of the initial batch, C6978 was tested at Alloa on June 4th 1918. Orders followed for Snipes from both factories but the

Built as a single-seater by Clayton and Shuttleworth, this Camel has been converted subsequently to a two-seater.

war ended before deliveries were effected.

Contracts for Sopwith Types

Type	Quan.	Serial Nos.	Remarks
Camel	100	C6701-6800	Built
Camel	50	H3996-4045	Ordered 9.8.18. Cancelled
Snipe	30	J651-680	Ordered 26.10.18. Cancelled
Snipe	150	J2392-2541	Ordered 1.11.18. Cancelled Numbers re-allotted

Total built: 100 Camels

CLAYTON & SHUTTLEWORTH Ltd.

The Stamp End Works, in Lincoln, of this engineering firm covered 100 acres and the employees numbered nearly five thousand. Their products—tractors, steam wagons and agricultural machinery, were exported throughout the world and while building aircraft, 300 steam wagons were also supplied to the War Department.

Surprisingly, their introduction to aircraft came with an order for Triplanes. The first built, N5350, was delivered on December 2nd 1916. After completion they were carted uphill and tested at the Robey flying ground on Bracebridge Heath where No. 4 (Lincoln) Aircraft Acceptance Park was soon afterwards established and to which the subsequent Camels built in the works were delivered. The first Camel order for one hundred is believed to have been completed at the end of 1917 from record of B5731 with a 130 h.p. Clerget first flying on December 12th 1917 at No. 4 A.A.P. Of this batch B5747, transferred to the Belgian Flying Corps, is preserved today in Brussels.

While several Camels of the first batch went to the Belgians, several of the second batch went to the Greeks. The firm, taking the tip from the parent firm who marked Sopwith Aviation Co. clearly on the fin, adopted the practice of marking their name at the rear of the fuselage aft of the serial number.

The firm soon gained a reputation for speedy and robust construction. Before 1917 was out the second batch had been started upon, confirmed by record of B7189 with a 150 h.p. B.R.1. leaving Lincoln for Dover en route to France on December 28th. This aircraft served in 'Naval Eight' and was one of 9 other Clayton-built Camels destroyed by No. 208 Squadron personnel at La Gorgue to prevent them falling into enemy hands during the German spring offensive of 1918.

Camel B7202 by this manufacturer was much prized by the ace Captain A. W. Carter. Delivered first to No. 9 (Naval) Squadron it was transferred to No. 10 (Naval) on February 4th 1918. Fifteen days later, Carter achieved his twelfth victory on this Camel, shooting down a German biplane in flames to crash into the Ypres-Combines canal on our side of the lines. Carter was still using this Camel when the squadron moved from Teteghem to Treiszennes on April 1st. He had used it for ground strafing, carrying 20lb bombs as well as for offensive patrols. Several times the fuselage was riddled with bullets and repaired; the top starboard plane had to be replaced and on one occasion an elevator was shot away. After shooting down two balloons on April 12th for his 14th and 15th victories, the tailplane had to be replaced. But the machine was never the same again and was discarded by Carter three days later, for a different Clayton-built Camel. Captain R. A. Little, another ace of the naval squadrons, flew Camels from Clayton's—B7220, B7275 and D 3416.

The firm received orders in 1918 to build Handley Page 0/400s which restricted the orders for Camels and because of this, the firm did not follow on with Snipes. Their final deliveries of naval Camels were completed in 1919 after contracts had been cut.

Contracts for Sopwith Types

Type	Quan.	Serial Nos.	Remarks
Triplane	106	A9813-9918	Allocation cancelled
Triplane	40	N5350-5389	Built late 1916/early 1917
Triplane	6	N533-538	Twin-Vickers version
Camel	100	B5651-5750	Delivered mainly to naval sqns.
Camel	100	B7181-7280	
Camel	100	D3326-3425	
Camel	100	D9581-9680	Built April-May 1918
Camel	50	E4374-4423	Built June-July 1918
Camel	50	F3096-3145	Built Aug.-Sep. 1918
Camel	100	F4974-5073	Approximately 50 built
Ships Camel	50	N8180-8229	Up to N8204 delivered

Totals built: 46 Triplanes, 550 Camels, 25 Ships Camels

THE COVENTRY ORDNANCE WORKS Ltd.

The Coventry Ordnance Works had taken over the Warwick Wright concern, including Howard T. Wright, in 1912 and it was their entry in the War Office trials that T. O. M. Sopwith had flown that year.

Soon after war was declared, the Works were given contracts for government designed aircraft, in addition to their ordnance work, and they built in succession numbers of R.E.7s, B.E.12s and R.E.8s, until 1918 when they were switched over to Snipe production.

It was a radical change, from Royal Aircraft Factory standardised aircraft to the new Sopwith Snipe with continuing modifications from the issue of drawings to the works in mid-1918 to the end of the year. Production did not get underway until late in 1918 and it is believed that none of their aircraft saw active service during the war, but many served in squadrons in the early 'twenties.

A Snipe prototype (B9967) being tested with a Dragonfly engine April 27th 1918. T. O. M. Sopwith can be seen standing second from left.

Contracts for Sopwith Types

Type	Quan.	Serial Nos.	Remarks
Snipe	150	E6537-6686	Built
Snipe	150	F9846-9995	Cancelled

Total built: 150 Snipes

Wm. CUBITT & Co.

An order for Sopwith Snipes (C7901-8200) was given to Cubitt's at a time when Messrs. Holland, Hannen, Cubitt and Sopwith had been asked by the Ministry of Munitions to manage a National Aircraft Factory. Sopwith opted out of the scheme and the order was cancelled, the serial numbers being re-alloted. A new order was placed and Sopwith built the Snipes in the new Sopwith works at Ham.

Contracts for Sopwith Types

Type	Quan.	Serial Nos.	Remarks
Snipe	300	C7901-8200	Cancelled. Nos. re-allotted

Total built: Nil

DUKS FACTORY

The Duks Factory at St. Petersburg, Russia, was reported to have built a number of 1½ Strutters during 1917. This information cannot be confirmed.

THE DARRACQ MOTOR ENGINEERING Co. Ltd.

Darracq & Co. (1905) Ltd. with motor works in S.W. London at Townmead Road, Fulham, changed their title to The Darracq Motor Engineering Co. Ltd. in October 1916, shortly after contracts for airframes had been awarded. The company built F.E.8s and D.H.5s in turn before receiving contracts for Sopwith 5F1s (Dolphins) which they built in greater numbers than any other aircraft type. In fact the whole of British Dolphin production was centred South-West of London at the Darracq, Hooper and Sopwith works.

The first Darracq-built fuselage was inspected by the A.I.D. at the end of 1917. French-built Hispanos were

Girls at the Whitehead factory engaged on fabric work. The fabric was normally Irish for operational aircraft and Egyptian cotton for trainers, which was doped with cellulose acetate for tautening.

intended as initial installations, but a Wolseley-built Hispano was delivered in error for this first airframe. Production was well underway by the spring of 1918 and all of the first order for 200 was completed before the Armistice.

Contracts for Sopwith Types

Type	Quan.	Serial Nos.	Remarks
Dolphin	200	C8001-8200	Ordered 13.7.17
Dolphin	100	F7034-7133	Ordered 8.6.18
Dolphin	100	J151-250	Ordered 28.9.18. Cancelled

Total built: 300 Dolphins

Darracq evidently followed the Sopwith way in marking the serial on the fuselage.

FAIREY AVIATION Co. Ltd.

This famous company became firmly established as a result of orders for Sopwith aircraft. It was registered in July 1915 with a capital of £35,000 to "manufacture, prepare, let or hire and deal in aeroplanes, waterplanes, and aerial conveyances and aircraft of all kinds. To acquire ground, build and maintain hangars, garages, sheds, aerodromes and accommodation for aerial conveyances . . .".

The works at Clayton Road, Hayes, Middlesex, with a base at Hamble, Hampshire, had built a single aircraft to their own design and were completing a batch of a dozen Short 827 seaplanes when Contract 87A499 for 100 Sopwith 2-seaters (1½ Strutters) was placed on July 6th 1916. First deliveries of the Sopwiths were made before the year was out and by early 1917 these were in use on the Western Front.

Like Blackburn, Fairey were given a contact to improve the Sopwith Baby. For a sample aircraft, Baby No. 8134 was already in the works October 1916 on a repair contract. The wings were completely redesigned incorporating the Fairey Patent Camber Gear, a form of trailing edge flap to increase lift. With type floats and a redesigned tail unit, only the fuselage retained the Sopwith characteristics. Known as the Fairey Hamble Baby, fifty were built to Contract AS4765 as follows:

Quan.	Power Unit	Serial Nos.
10	110 h.p. Clerget	N1320-1329
10	130 h.p. Clerget	N1330-1339
30	130 h.p. Clerget	N1450-1479

The Hamble Baby was declared obsolete shortly before the Armistice when a total of a dozen, Fairey and Parnall production, remained in service.

Contracts for Sopwith Types

Type	Quan.	Serial Nos.	Remarks
1½ Strutter	100	A954-1053	2-seat version
Ships Camel	100	N7200-7299	Cancelled

Total built: 100 1½ Strutters

FAIRFIELD SHIPBUILDING & ENGINEERING Co. Ltd.

When the Sopwith Torpedo-plane was ordered into production in late 1917, Fairfield was one of two firms new to aircraft construction that were awarded contracts; in order that existing aircraft production arrangements would not be disrupted. Difficulties arose from lack of experience in airframe construction, at a time when skilled labour was extremely hard to obtain, and complicated by frequent design changes. The Sopwith Drawing Office handed over 347 drawings to Fairfield's at Govan, Glasgow in November 1917, but many were later withdrawn and re-issued to embody changes.

Fairfield's contract AS27863/17 was for 100 Sopwith T.1s (C7901-8000 later changed to N7000-7099) and the first was completed on August 6th 1918. Deliveries were by road to No. 6 (Renfrew) Aircraft Acceptance Park where the aircraft were erected and sent to service, the first seven (N7000-7006) going to No. 208 Training Depot Station, East Fortune. By the time of the Armistice, twenty had left the factory and were deployed at East Fortune and Fleet Bases. Only one became shipborne, N7011 which was taken aboard H.M.S. *Argus* in November 1918 and was returned to store the following month.

After the Armistice the contract was cut to fifty. By the end of January 1919 up to N7035 had reached Renfrew and plans were made to deliver the remainder (N7036-7049) direct to store.

Contracts for Sopwith Types

Type	Quan.	Serial Nos.	Remarks
Cuckoo	100	N7000-7099	Order cut to N7000-7049

Total built: 50 Cuckoos

GLENDOWER AIRCRAFT Co. Ltd.

Coming late into the aircraft business, with a loan by he Western Australian Insurance Company, the newly-formed Glendower company bought freehold land and premises at Kew, North Sheen and Mortlake, and established offices at Harrington Road, South Kensington.

D.H.4s were first produced, followed by a contract for Salamanders which was placed too late in the war to be completed. The firm went into liquidation soon after the war and production details cannot now be ascertained, but an eye-witness report of some twenty completed Salamander airframes confirms that production of this type did get underway.

Contracts for Sopwith Types

Type	Quan.	Serial Nos.	Remarks
Salamander	100	J5892-5991	Ordered 1.11.18

Total built: 20+ Salamanders

Constructional shot of a Snipe tail unit.

THE GLOUCESTERSHIRE AIRCRAFT Co. Ltd.

Formed at Cheltenham for aircraft construction, starting with components, the company was building Bristol Fighters in 1918 when an order was received for Snipes too late to be effected.

Contracts for Sopwith Types

Type	Quan.	Serial Nos.	Remarks
Snipe	300	J3042-3341	Ordered Nov. 18

Total built: Nil

Nieuport (foreground) and Sopwith-built (background) Snipes built during the war shown serving in the early 'twenties with the Royal Air Force

THE GRAHAME-WHITE AVIATION Co. Ltd.

Established at Hendon and producing aircraft prewar, the wartime contracts allotted to Grahame-White Aviation were mostly for trainers. Only in the final month of the war did the company receive orders for truly operational aircraft.

Contracts for Sopwith Types

Type	Quan.	Serial Nos.	Remarks
Snipe	500	J2542-3041	Ordered 1.11.18. Cancelled

Total built: Nil

HANRIOT

The Société Anonyme des Appareil's d'Aviation Hanriot with works at Avenue des Moulineaux, Billancourt, Seine, France was given contracts for the building of Sopwith 1½ Strutters to French Government contracts in 1916-1917. It is not known how many were built but the various French manufacturers making these Sopwiths averaged six hundred each.

HOOPER & Co. Ltd.

This firm of coach-builders with London offices at St. James's Street, London, S.W.1., and a factory at Chelsea, received their initiation into aircraft work in 1916 with an order for 1½ Strutters, and commenced deliveries late the same year. A further order for the bomber version of the 1½ Strutter kept this type in production to the summer of 1917. These were mainly delivered to store.

Camel and Dolphin orders followed including 100 Camels (H734-833) fitted for night fighting and Ships Camels.

Contracts for Sopwith Types

Type	Quan.	Serial Nos.	Remarks
1½ Strutter	50	A1511-1560	Ordered 24.6.16
1½ Strutter	100	A6901-7000	Mainly bomber version
Camel	50	B5401-5450	Ordered 1917
Camel	50	C1551-1600	Ordered 26.6.17
Dolphin	200	D5201-5400	Ordered 28.6.17
Camel	100	F2083-2182	F2101 first flew 25.6.18
Camel	100	H734-833	Ordered 1.8.18
Camel	70	H7343-7412	Ordered 25.9.18
Dolphin	150	J1-150	Ordered 27.9.18
Ships Camel	50	N8130-8179	Deliveries post-war

(Final three orders may not have been completed, deliveries confirmed up to H7363, J16 and N8159)

Total built: 150 1½ Strutters, 321 Camels, 30 Ships Camels, 216 Dolphins.

KINGSBURY AVIATION Co. Ltd.

Messrs. Berningham with premises in Albany Street, N.W. London, known as 'Friswell's', sponsored an aircraft manufacturing business and acquired some 150 acres at Kingsbury, Surrey, to construct an aerodrome. Following the lead of Whitehall, a hangar was built—reported the largest in the country—with premises suitable for turning into a club-house in time of peace. A London office was opened at 175, Piccadilly.

A small sub-contract for the installation of Le Rhône engines in 1½ Strutters was placed in 1917 and to utilise the spacious accommodation an order for 150 D.H.6s followed. When these were completed in mid-1918 a small contract for Snipes was arranged. According to records Contract 35a/3785/C.4439 was placed in February 1919, presumably a paper confirmation to endorse earlier agreements and to help the labour situation. There is evidence that the Snipes were built, but delivered direct to long-term storage.

Contracts for Sopwith Types

Type	Quan.	Serial Nos.	Remarks
Snipe	30	J6493-6522	Ordered 12.2.19

Total built: 30 Snipes

V.A. LEBEDEV

It was reported that the V. A. Lebedev factory at Novoi Derevnii, near St. Petersburg in Russia, completed 140 1½ Strutters during 1917. This information cannot be confirmed.

Another example of a French-built 1½ Strutter, in this case one captured by Austro-Hungarian troops.

LOIRÉ ET OLIVIER

The Establishment Loiré et Olivier, 46-48 Boulevard de la Revolte, Levallois-Perret, was among the French Government contractors that averaged production of some 600 Sopwith 1½ Strutters.

MANN, EGERTON & Co. Ltd.

Motor engineers and specialists in coachwork, with head offices and works in Prince of Wales Road, Norwich, Mann Egerton were co-opted by the Admiralty early in the war to build Short 184 floatplanes and later Short Bombers. Their next orders were for Sopwith 1½ Strutters to meet R.N.A.S. requirements and these were built at the turn of the years 1916/17. Probably the majority of the 75 built by the firm were sent out East, to R.N.A.S. units in the Aegean and several were offset to the Belgian, Greek and

Russian Governments, but a few served in warships.

Contracts for Sopwith Types

Type	Quan.	Serial Nos.	Remarks
1½ Strutter	20	N5200-5219	Single-seat bombers
1½ Strutter	30	N5220-5249	Two-seat fighters
1½ Strutter	25	N5630-5654	Two-seat fighters

Total built: 75 1½ Strutters

MARCH, JONES & CRIBB Ltd.

This Leeds engineering company received contracts to produce Camels after building a hundred D.H.5s as their introduction to aeronautical work. Snipe orders logically followed on from Camel production but as orders for them were placed just over a month before the Armistice, those built were delivered to store.

Contracts for Sopwith Types

Type	Quan.	Serial Nos.	Remarks
Camel	100	C8301-8400	Contract AS29339
Camel	75	F5174-5248	Delivered 1918/1919
Snipe	100	J301-400	Ordered 7.10.18. Built
Snipe	50	J681-730	Ordered 26.10.18. Cancelled

Total built: 175 Camels, 100 Snipes

A Robey-built Sopwith Gunbus.

MORGAN & Co. Ltd.

This motor company with London offices at 10 Bond Street, started aircraft production in February 1917 at their Leighton Buzzard, Bedfordshire, works after receiving Contract 87A983 for Sopwith 2-seaters. Although this 1½ Strutter contract used the term Sopwith 2-seaters, it was changed to include a large proportion of the single-seat bomber version. Since the Expeditionary Force had asked for reserves of these bombers, although they did not use them in any quantity, they were held in store, and it is possible that the contract was cut to 100. Contracts for D.H.6s and Avro 504J/Ks followed the Sopwith order.

At the height of Sopwith production, June 1917, during which the firm completed a dozen Sopwith bombers, 378 employees were on the payroll.

Contracts for Sopwith Types

Type	Quan.	Serial Nos.	Remarks
1½ Strutter	200	A5950-6149	Order believed cut to 100

Total built: 100 (approximately) 1½ Strutters

D. NAPIER & SON Ltd.

Famous as manufacturers of cars and aero engines, Napier's at Acton, N.W. London, had large scale contracts for R.E.8s. In 1918 the firm received an order for 150 Snipes. Some delay in getting these aircraft into service was engendered by the Badin petrol system on the first 76 being modified and deliveries were effected in 1919.

Contracts for Sopwith Types

Type	Quan.	Serial Nos.	Remarks
Snipe	150	E6787-6930	Ordered March 1918

Total built: 150 Snipes

The first 1½ Strutter built at Leighton Buzzard by Morgan's, a single-seat bomber.

NATIONAL AIRCRAFT FACTORY No. 1

Situated at Waddon near Croydon and managed by Messrs. Holland, Hannen & Cubitt after T. O. M. Sopwith had declined the Ministry of Munitions offer of management, the factory employed 1,595 men and 1,638 women at peak producing D.H.9 airframes and spares, erecting Handley Page 0/400s and manufacturing gun interrupter gear sets.

There were considerable labour difficulties, including strikes at the works and the large order for Sopwith aircraft did not get underway before being cancelled. Postwar, the buildings were used as a salvage and storage depot and housed several hundred Snipes.

Contracts for Sopwith Types

Type	Quan.	Serial Nos.	Remarks
Salamander	400	J6092-6491	Ordered 1.11.18

Total built: Nil

NATIONAL AIRCRAFT FACTORY No. 3

Situated at Aintree near Liverpool and managed by Cunard, No. 3 National Aircraft Factory employed 1,580 men and 1,054 women. Their main order was for Bristol Fighters which they commenced building in March 1918 but only 126 airframes were eventually delivered, due, at Waddon, to labour unrest. Orders were placed for large numbers of Sopwith aircraft in the last month of the war but none were completed.

Contracts for Sopwith Types

Type	Quan.	Serial Nos.	Remarks
Snipe	500	J4092-4591	Ordered 1.11.18

Total built: Nil

THE NIEUPORT & GENERAL AIRCRAFT Co. Ltd.

Formed early in 1914 to produce Nieuport designs in Britain, the Nieuport & General Company had offices at 45, Great Marlborough Street, London, W, and a factory fronting on Langton and Temple Roads at Cricklewood, N.W. London. Their contracts in the early war years were mainly for components, but after an order for fifty Nieuport single-seat scouts contracts for Camels followed in 1917.

The first Camel contract was placed on August 20th 1917 and while the first delivery cannot be traced, it is known that the ninth machine (C9) which served at the School of Special Flying, Gosport, first flew on February 4th 1918. Presumably 150 had been completed by mid-1918 since C158 first flew at No. 2 (Hendon) Aircraft Acceptance Park, to where deliveries were normally made, on June 3rd 1918.

Orders for Snipes were received in the spring of 1918,

followed by further but limited orders for Camels.

Contracts for Sopwith Types

Type	Quan.	Serial Nos.	Remarks
Camel	200	C1-200	Contract No. AS14412
Snipe	100	E6937-7036	Delivered postwar
Camel	50	F3196-3245	Delivered July-Aug. 1918
Camel	50	F3918-3967	Delivered Sept.-Oct. 1918
Camel	100	F8496-8595	Delivered Nov.-Dec. 1918

Total built: 400 Camels, 100 Snipes.

OAKLEY Ltd.

Oakley Ltd., with London offices at 83, Regency Street, Westminster, controlled the Ilford Aeroplane Works which had previously done light foundry work. Aeroplane work started in 1916 with B.E.2c Curtiss engine installation work and repairs.

Their first, and last, contract for complete aircraft was for Sopwith Triplanes, 25 being ordered under Admiralty Contract CP216457/17. Work was started on these in the spring of 1917 but was brought to a complete halt in the summer due to instructions to install two Vickers machine guns, instead of the usual one, in these machines. By then the Triplane was obsolescent and it was decided to cancel the contract but accept delivery of the three completed. The last to leave the works, N5912 on October 19th 1917, is the example that survives today.

Further areonautical work by the firm was restricted to component and repair work. At least one Sopwith type was repaired in the Ilford works, 1½ Strutter No. 9717 in June 1917.

Contracts for Sopwith Types

Type	Quan.	Serial Nos.	Remarks
Triplane	25	N5910-5934	N5910-5912 only built

Total built: 3 Triplanes

The only surviving Triplane, N5912, one of the only three built by Oakley, shown at an exhibition between the wars.

PALLADIUM AUTOCARS Ltd.

This London coachbuilding firm, engaged on building D.H.4s, received Contract 35a/3622c4236 to build Salamanders, which was rescinded shortly afterwards, due to the Armistice.

Contracts for Sopwith Types

Type	Quan.	Serial Nos.	Remarks
Salamander	100	J5992-6091	Placed 1.11.18. Cancelled

Total built: Nil

PARNALL & Sons Ltd.

Parnall's, with works at Mivart Street, Eastville, Bristol, built Hamble Babies under licence from Fairey (recorded

A two-seat fighter and single-seat bomber versions of the 1½ Strutter built by Westland.

earlier in sequence) but, in the process, came nearer to the original Sopwith design by retaining on their Baby floatplanes the characteristic Sopwith type fin of the Sopwith Baby. By a decision made by the Air Board in November 1917 the remaining Babies on order were produced with a land-type wheeled chassis and these were known as Hamble Baby Converts. Production was as follows:

Type	Quan.	Serial Nos.	Remarks
Hamble Baby	30	N1190-1219	Delivered from mid-1917
Hamble Baby	100	N1960-2059	75 produced as Converts

N.B. This production not included on the Sopwith production summary since the aircraft are normally classified as Fairey aircraft, although of basic Sopwith design.

PEGLER & Co. Ltd.

Like Fairfield, new to aircraft construction, Pegler were awarded a contract for 50 Sopwith Torpedoplanes (D3276-3325 later changed to N6900-6949) under Contract A.S.35976/17. Due to inexperience in airframe manufacture and changing design requirements this Doncaster company were unable to effect deliveries and Blackburn's were asked to take over part of their contract, which was reduced to twenty (N6930-6949).

The first (N6930) was completed in September 1918 and delivered, as were all subsequent, to No. 9 (Newcastle) Aircraft Acceptance Park. Forwarded by air to East Fortune in October, it was the only Pegler-built aircraft in service before the Armistice. Four had been delivered to Newcastle by the end of 1918, but N6931 had crashed making a forced landing at Ossett en route on November 19th. Striking a pole, a wing was torn off N6931 but the pilot, Lt. H. L. Taylor, was unhurt. The remainder of the

Apart from Sopwiths themselves, Ruston Proctor were the largest of the contractors, and pictures show an example of each of the Sopwith types they produced. The 1½ Strutter (top) is an early model with Nieuport type gun ring, the Camel is without armament and the Snipe shown is the first one built by the firm.

twenty were delivered in 1919 to direct to store.

Contracts for Sopwith Types

Type	Quan.	Serial Nos.	Remarks
Cuckoo	20	N6930-6949	Larger order reduced

Total built: 20 Cuckoos

PORTHOLME AERODROME Ltd.

Founded by Mr. James Radley, racing driver and pioneer pilot, the Portholme company boasted the 'largest meadow in England' with 390 acres outside Huntingdon and offices in St. John's Street in the town. Unfortunately, when the Ouse rose in Winter, much of the area was under water but fortunately not the works in which coachwork for cars was made pre-war and armoured cars and recovery vehicles in the early war years.

With the extensive ground available for flying, it was natural that Radley should seek orders for aircraft but in the first two years of war he had received orders only for a few Wight floatplanes. In 1917 the first Sopwith contract for fifty Camels was placed and this order was repeated several times throughout the war. Although the grounds were large, the workshops were relatively small and the grounds were used by the Royal Flying Corps for training; the station became No. 59 Training Depot Station in 1918.

The Portholme contracts included test flying, and deliveries of Camels were mainly to No. 3 (Norwich) Aircraft Acceptance Park. Apart from the aircraft listed below, the airframes of 44 Snipes erected by Ruston Proctor were completed by Portholme.

Contracts for Sopwith Types

Type	Quan.	Serial Nos.	Remarks
Camel	50	B4601-4650	Ordered 2.6.17
Camel	50	B7131-7180	Built early 1918
Camel	50	D9531-9580	D9538 flew 29.5.18
Camel	50	E5129-5178	Built in mid-1918
Snipe	100	E8307-8406	Ordered 20.3.18
Camel	50	F1958-2007	Delivered by Sep. 1918
Camel	50	F8646-8695	Delivered Oct.-Dec. 1918
Snipe	100	H8663-8762	Ordered 26.9.18. Cancelled

Total built: 300 Camels, 100 Snipes

R.E.P.

The organisation of Robert Esnault Pelterie, the French aviation pioneer, was geared to Sopwith 1½ Strutter production in 1917. Construction was centred at Lyons where the firm had premises at 47A Boulogne-sur-Seine, 37 Rue de Abundances and 47 Chemin de Croix Marlon A St. Alban.

ROBEY & Co.

First of the Sopwith contractors for complete airframes, this old established Lincoln engineering firm received Admiralty order CP51744/15 early in 1915 for thirty Sopwith Type 806 Gunbuses. Harry Hawker went to Lincoln to test the first, No. 3833, which is believed to have made its initial flight from Bracebridge Heath. Unfortunately the Sopwith Gunbus was not successful and the type was little used. Robey's were committed to Short 184 floatplane production for the rest of the war.

Contracts for Sopwith Types

Type	Quan.	Serial Nos.	Remarks
Gunbus	30	3833-3862	Last 13 delivered as spare parts

Total built: 17 Gunbuses

RUSTON PROCTOR Ltd.

The largest sub-contractor of Sopwith aircraft, Ruston Proctor of Lincoln had established their name in engineering, particularly for agricultural machinery. Their experience in working in both wood and metal led to orders for both aero engines and airframes as well as many miscellaneous items including portable pigeon lofts and over half a million horse shoes.

Following the building of 200 B.E.2c/ds orders were placed for Sopwith aircraft in 1916 and thereafter the firm's airframe construction was restricted to Sopwith types starting with 1½ Strutters. Their first was delivered to Farnborough for type testing in July 1916 and with further orders deliveries averaged 43 a month during the first half of 1917. Initial deliveries had the early Nieuport type gun rings, but the bulk had Scarff gun rings.

The last 1½ Strutter of the first batch of fifty, No. 7811, survived the war after much flying in 1917 with 'B' Squadron of the Central Flying School and conversion for naval use in 1918. Few of the others lasted through 1917; A2406, for example, ready for collection at Lincoln in January 1917 was issued new from No. 2 Air Depot on the last day of March and was shot down at Farbus by von Richthofen on April 7th as his 38th victory—five days earlier he had dispatched A2401, another of the firm's 1½ Strutters.

The first Camel, B2301, finished in June 1917, flew over Lincoln and Cambridge on the 14th on delivery from Lincoln to Hounslow. These early Camels were fitted with a variety of engines. A number of French-built Clergets were delivered to the works but B2311 (the 11th built) had one of the firm's own 130 h.p. Clergets (RP No. 1498/WD11647) and no doubt many others had both airframe and engine constructed at the works. Several of the first batch were used for a variety of tests including B2312 (the 12th built) which in November 1917 had a 110 h.p. Le Rhone built by W. H. Allen & Sons Ltd. of Bedford, later a 130 h.p. Clerget and in 1920 was still being used for experiments which included self-sealing fuel tanks. Many of this early batch were flown by famous pilots; it was B2313 (the 13th built) that W. G. Barker, who eventually became seventh in the British list of aces, was flying on October 26th 1917 when he shot down two Albatros Scouts after a long patrol and landed out of fuel near Arras.

Camel production rose at the works and from November 1917 over a hundred a month were being delivered. Up to February 1918 the 130 h.p. Clerget had, in general, been the power unit installed, but from then onwards many had the 140 h.p. Clerget. From No. 4 (Lincoln) Aircraft Acceptance Park, to where the output was delivered, Camels with 140 h.p. Clergets were allotted to squadrons and those with the 130 h.p. Clergets to training stations. A peak of 128 Camels produced in a month was reached in May 1918; thereafter production tailed off as preparations were made to turn over to Snipe production.

Two Camels did not reach service. One was E1593 with a 140 h.p. Clerget. Flown by Lt. Reid, M.C. on August 10th 1918 for its initial flight at Lincoln, the lower plane crumpled in a dive. Miraculously the pilot maintained control and brought it in to a crash landing at which it turned over but without injuring the pilot.

During September 1918 the firm amalgamated with Richard Hornsby & Sons of Gainsborough and J. & H. Andrews & Company of Stockport to form Ruston & Hornsby Ltd.

Due to engine shortages many late production Camels were sent to store at Ascot without engines in October 1918. That same month the first Snipes were delivered and from December 1918 only Snipes were in production. Post-war, in January 1919, the firm reached their peak Snipe production with 84 delivered that month. During 1919, eleven Snipes on the firm's contracts were delivered to Portholme for completion, and production of aircraft finally tailed off in October 1919.

Contracts for Sopwith Types

Type	Quan.	Serial Nos.	Remarks
1½ Strutter	50	7762-7811	Ordered in 1916.
1½ Strutter	50	A2381-2430	Deliveries from Nov. 16
1½ Strutter	200	A8141-8340	Built summer 1916
1½ Strutter	50	B2551-2600	Completed Aug. 16
Camel	250	B2301-2550	Ordered 22.5.17
Camel	100	B5551-5650	B5595 flew 13.12.17
Camel	200	B7281-7480	B7368 flew 18.1.18
Camel	100	C8201-8300	C8291 flew 23.3.18
Camel	200	D1776-1975	Deliveries from 8.3.18
Camel	150	D8101-8250	Built summer 1918
Camel	200	E1401-1600	E1532 flew 19.7.18
Camel	200	E7137-7336	E7137 flew 2.7.18
Snipe	500	E7337-7836	Ordered 20.3.18
Camel	75	F2008-2082	Built Sep. 18
Camel	100	F3968-4067	Delivered Oct/Nov. 18
Snipe	300	H351-650	Ordered 25.7.18

(N.B. Snipes. H376-378, 382, 386-403, 419-450 partially completed only. E7798, 7802-7832, 7834 and H408-418 delivered to Portholme for completion, H451-556 cancelled.)

Total built: 350 1½ Strutter, 1573 Camels, 498 Snipes

FREDERICK SAGE & Co. Ltd.

A large shop-fitting firm of Peterborough, the Sage Company produced several of their own aircraft designs as well as Short 184 floatplanes and Avro 504K trainers. They received orders for Camels too late in the war to be effected.

Contracts for Sopwith Types

Type	Quan.	Serial Nos.	Remarks
Ships Camel	130	N7850-7979	Cancelled

Total built: Nil

SARAZIN FRERES

The French firm formed by the Sarazin brothers at 81, Rue Arago, Puteaux, was among the contractors given orders for Sopwith 1½ Strutters.

SOPWITH AVIATION COMPANY

The details of the parent firm directing all the sub-contracting, totalling forty different plants building complete airframes of Sopwith aircraft types, is contained throughout the narrative and the type-by-type review in this book. To summarise, building commenced in sheds at Brooklands in 1911 which were expanded and used later as sheds for flight testing and modifications. The skating rink works at Kingston acquired in 1912 which became the experimental sheds where prototypes were built and the Canbury Park Road works the following

Standard Motors built more Pups than any other firm, including Sopwith Aviation. These two late production aircraft, straight from the Coventry factories, bear presentation names.

Up to early 1918 Sopwith stencilled their name and address clearly on the fins of their aircraft, which were left clear doped for the purpose, as illustrated by this B1 and Camel.

year were expanded considerably during the war.

When the Ham works were opened early in 1918, the size and number of contracts greatly increased as the list below shows. The early aircraft of various types are recorded in the Type-by-Type Review.

Contracts from 1915

Type	Quan.	Serial Nos.	Remarks
Schneider	12	1436-1447	Contract CP31834/15 placed Nov. 14
Schneider	24	1556-1579	Contract CP38624/15
1½ Strutter	1	3686	Prototype. Became B394 of R.F.C.
Pup	1	3691	Prototype. Contract CP109458/16
Schneider	100	3707-3806	Contract CP45562/15
Baby	100	8118-8217	8123-8186 Clerget engines, remainder Gnôme Monos
1½ Strutter	50	9376-9425	Type 9400. Contract CP104237/16
Pup	2	9496-9497	Type 9901. 80 h.p. Clerget
1½ Strutter	100	9651-9750	65 Type 9400, 35 Type 9700
1½ Strutter	1	9891	80 h.p. Gnôme. Trainer project
1½ Strutter	6	9892-9897	Type 9400. Contract CP113240/16
Pup	3	9898-9900	Type 9901. Contract CP120390/16
Baby	1	N4	130 h.p. Clerget. Prototype
Camel	1	N5	Prototype. Contract AS26088
Cuckoo	1	N74	Prototype T1
Triplane	1	N500	Prototype. 110 h.p. Clerget
Pup	1	N503	Type 9901
Triplane	1	N504	Prototype. 130 h.p. Clerget
Triplane	2	N509-510	Hispano-engined prototypes
Camel	2	N517-518	F1 Prototypes
Triplane	4	N524, 541-543	For French Government
Circuit Seaplane	20	N1340-1359	Cancelled
1½ Strutter	40	N5080-5119	21 Type 9700, 19 Type 9400
1½ Strutter	10	N5170-5179	Type 9400. Contract CP118231/16
Pup	20	N5180-5199	Type 9901. Contract CP119901/16
Triplane	75	N5420-5494	Contract CP125849/16
1½ Strutter	50	N5500-5549	Type 9700. 12 cancelled.
Pup	50	N6160-6209	Type 9901 built early 1917
Triplane	20	N6290-6309	Contract CP138323/16
Camel	50	N6330-6379	Contract CP102581/17
Pup	70	N6460-6479	Type 9901. 20 only built
Ships Camel	50	N6600-6649	Camel 2F1. Contract CP103733/17
B1	1	B1496	Experimental Contract AS15242
Camel	200	B3751-3950	Delivered from June 1917
Camel	250	B6201-6450	Delivered from August 1917
Snipe	6	B9962-9967	Prototypes. Contract AS31668
Dolphin	500	C3777-4276	Ordered 29 Jun 17
Snail	6	C4284-4289	Prototypes. Contract AS37484
Dolphin	200	D3576-3775	Contract AS35977 29 Nov. 17. D3773 left works 15.6.18
Dolphin	200	E4424-4623	E4449 first flew 20.6.18
Dolphin	500	E4629-5128	140 confirmed built
Salamander	6	E5429-5434	Ordered 20 Mar 18. 3 cancelled
Snipe	300	E7987-8286	E8176 delivered 9.10.18
Dolphin	1	E9997	Contract AS38905
Snipe	200	F2333-2532	Ordered 25 Apr 18
Salamander	500	F6501-7000	160 believed delivered
Snipe	30	F7001-7030	Several completed as Dragons
Snapper	3	F7031-7033	Completed post-war
Camel	3	F7144-7146	Cancelled
Cobham	3	H671-673	Completed post-war
Hippo	2	H4420-4421	X11 and X18
Bulldog	2	H4422-4423	X3 and X4
Snipe	200	H4865-5063	Ordered 9 Aug 18. 32 only confirmed delivered
Buffalo	2	H5892-5893	Experimental
Snipe	3	H9964-9966	Escort fighter order. Cancelled
Dragon	300	J3717-3916	Originally ordered as Snipes. Around 200 believed built.

Total built: 1043+ Dolphins, 560 Snipes, 503 Camels, 50 Ships Camels, 200 approx. Dragons, 163 Salamanders, 246 1½ Strutters, 137 Schneiders, 100 Babies, and miscellaneous early aeroplanes and seaplanes, prototypes and civil aircraft as tabled at the end of this section

THE STANDARD MOTOR Co. Ltd.

This famous Coventry motor works was weaned on aircraft by the allocation of a contract for Royal Aircraft Factory B.E.12s in 1915. The following year contracts were placed for R.E.8s and concurrent with the production of these, the firm turned out Sopwith Pups.

The initial Pup contract 87A461 for 50 in 1916, which was later extended, was awarded to the Sopwith works who sub-let the contract to Standard. From mid-1917 the firm had instructions to concentrate more on R.E.8s and then in 1918 on Bristol Fighters, with Pup production a secondary consideration as it was then out-dated at the Front. Nevertheless, contracts were not cancelled, for the Pup proved an excellent fighter trainer with none of the vices of the Camel, and deliveries from late 1917 onwards were to training units via No. 1 (Coventry) Aircraft Acceptance Park. Some were shipped out East and, for example, B6053 first flew on May 15th 1918 at Aboukir. Pups were still on the lines at Standard's when the war ended.

Contracts for Sopwith Types

Type	Quan.	Serial Nos.	Remarks
Pup	50	A626-675	A663 delivered 5.2.17
Pup	50	A7301-7350	A7302 collected 8.2.17
Pup	150	B1701-1850	B1751 serving 1.7.17
Pup	250	B5901-6150	B5978 serving 1.2.18
Pup	350	C201-550	Built in 1918

Total built: 850 Pups

VICKERS LTD. (AVIATION DEPARTMENT)

A well-known munitions firm, with an aviation department controlling facilities at Crayford, Dartford and Weybridge, Vickers received Contract 87A495 on June 15th 1916 for 100 Sopwith 2-seaters, later increased by 50. Delivered from late 1916 the order was completed mid-1917 at the Crayford factory.

Contracts for Sopwith Types

Type	Quan.	Serial Nos.
1½ Strutter	150	A1054-1153 & A8744-8793

Total built: 150 1½ Strutters

WELLS AVIATION Co. Ltd.

Starting as R. F. Wells & Company under the management of R. F. Wells, "sculptor and designer of artistic pottery", using the Coldrum Pottery of Chelsea as premises, a new firm entered the aviation industry. As agents for the American Benoist flying boat they had built an example and obtained a sub-contract from Vickers to build F.B.14 fuselages.

Their great chance came in 1916 when Contract 87A999 for 100 1½ Strutters was placed. The original works at Elystan Street were expanded and houses opposite at Whitehead's Grove, with gardens adjoining the works, were acquired for new offices. Empty shops near the works entrance were incorporated and a building in Jubilee Place, used for the painting and storage of scenery for the Alhambra, was made into a component assembly bay.

With assistance from Sopwith's the firm started to produce 1½ Strutters, concurrently with building an aeroplane and seaplane station at Cobnor near Chichester, and in February 1917 a School of Flying. Next month the firm, by then employing 2,000, were in financial difficulties and went into voluntary liquidation in May. That month, Mr. Sydney Pickles, who was to test many Sopwith aircraft from contractors, resigned from the firm. As Wells and Company Limited, work on Sopwith aircraft continued. The 1½ Strutter production line was apparently abandoned, but a few machines were completed later in 1917 when the type was already outdated. Later the premises were used for aircraft component manufacture including Camel parts.

Contracts for Sopwith Types

Type	Quan.	Serial Nos.	Remarks
1½ Strutter	100	A5238-5337	Limited completion of 100 ordered.

Total built: 20 + 1½ Strutters

WESTLAND AIRCRAFT WORKS

The Westland Aircraft Works were formed at Yeovil in 1915 by Petters Ltd., makers of marine engines and agricultural machinery. After initial orders for Short Seaplanes, contracts were placed by the Admiralty for 1½ Strutters. The new works were controlled by R. A. Bruce, M.Sc., who had been released by the Royal Navy from his post as Admiralty overseer at the Sopwith Aviation Company.

Two Admiralty contracts were placed for 1½ Strutters, CP119334/16 for 50 and CP129281/16 for 25. Most of the first contract were delivered crated by rail and thirty of these were given to the French. In the Spring of 1917

A Vickers-built 1½ Strutter.

the firm purchased Northover Fields from the Yeovil and District Hospital Board and, after levelling and draining, used it as an airfield from which the last of the 1½ Strutters was flight delivered. The firm then concentrated on the production of D.H.4s.

Contracts for Sopwith Types

Type	Quan.	Serial Nos.	Remarks
1½ Strutter	50	N5120-5169	Type 9700
1½ Strutter	25	N5600-5624	First 5 Type 9700. Remainder Type 9400

Total built: 75 1½ Strutters (55 single-seat bombers and 20 two-seat fighters)

WHITEHEAD AIRCRAFT Ltd.

J. A. Whitehead, before emigrating to America, had worked for the Aircraft Manufacturing Company at Hendon and after the outbreak of war returned to his homeland to put his experience to use. After buying the old Richmond Drill Hall as premises he received a large contract to build Maurice Farman trainers after an initial small batch of B.E.2bs.

It was a contract for Sopwith Pups that led to an enormous expansion of his plant by new premises at Townsend Road, Richmond, near the drill hall, fronting on Manor Park Lane. These works were opened on July 10th 1916 by Sir Charles Wakefield. An airfield with erecting sheds was opened at Feltham, which post-war became the Hanworth airfield. Soon in financial difficulties, Whitehead, on the strength of his Sopwith contracts, acquired capital in the city to promote Whitehead Aircraft (1917) Ltd. with himself as managing director.

Such was Whitehead's publicity tactics, with full-page adverts in Flight and Aeroplane showing a Pup-like aircraft, that post-war one well-known annual published a picture of a line-up of Pups captioned as Whitehead Fighting Scouts! And this has had repercussions in far more recent books.

The firm's first Pup was delivered at the beginning of 1917 after close co-operation with the Sopwith works

One of the few Wells-built 1½ Strutters.

Two Sopwith aircraft built up from salvage, the Pup at No. 2 A.R.D. and the Camel at a Depot in France.

only a mile or two away. At the time the Sopwith contract was received J. Ward was works manager with A. J. Macphail, A. Stebbing, W. J. Loosemore, A. E. Hagley, H. Crump, J. Kemp, J. Garner, E. Boyle, F. Wright, E. C. Baker, W. Chandler, O. Allard, M. Byl and W. Cory heads of departments.

It was not only on aircraft work that there was an association with Sopwith's—on August 17th 1917 the Sopwith Aviation tug-of-war team beat the Whitehead team at a Sports Carnival on Feltham airfield. Further Pup contracts followed in 1917 but there were several hold-ups that year and at one time thirteen Pups were in the works awaiting Vickers guns. Early in 1918 Monosoupape-engined Pups were being delivered for which the firm produced their own welded engine cowling with stiffening brackets

Final deliveries to the Pup orders were made as spares as the firm turned over to D.H.9 and D.H.9A production.

Contracts for Sopwith Types

Type	Quan.	Serial Nos.	Remarks
Pup	100	A6150-6249	Deliveries from early 1917
Pup	100	B2151-2250	
Pup	150	B5251-5400	Built during 1917
Pup	100	B7481-7580	
Pup	100	C1451-1550	Ordered 20.8.17
Pup	70	C3707-3776	Ordered 27.7.17
Pup	200	D4011-4210	Ordered late 1917

Total built: 820 Pups

WOLSELEY MOTORS Ltd.

The Adderley Park, Birmingham, works of this famous motor company were concentrating on S.E.5A production when, in mid-1918, an order was placed for Salamanders to be built concurrently with a continuing order for an equal number of S.E.5As. Since the Salamander contract meant re-tooling and delays were occasioned by modifications, the order was cancelled at the time of the Armistice before any had been delivered.

Contracts for Sopwith Types

Type	Quan.	Serial Nos.	Remarks
Salamander	150	F7601-7750	Ordered 11.7.18

Total built: Nil

RECONSTRUCTED SOPWITHS

Numbers of aircraft during the war years were built up from the salvaged remains of others, using an amount of newly built spares to make up parts as necessary. Aircraft so re-constructed were given a new military identity and for the sake of completeness they are included here, but being rebuilt airframes these are *not* included in the production totals.

No. 1 (Southern) Aircraft Repair Depot, South Farnborough, Hants, from late 1916 to the end of 1918 erected from scrap some seven hundred aircraft including the following Sopwith types: 1½ Strutters B7903, B7914, B7915, B7916. Ship Strutter B744; 1½ Strutter modified as single-seater for Home Defence B762; 1½ Strutter B799; Pup B804; Camels B778, B7743, B7745, B7756, B7860, B7869, B7896, B8025, B8155; Dolphins B7849, B7855, B7927, B7928, B8189.

No. 2 (Northern) Aircraft Repair Depot, Coal Aston, Sheffield, from early 1917, rebuilt aircraft including Sopwith types: Ship Strutter B4044 and Pups B4128/B4136.

No 3 (Western) Aircraft Repair Depot, Yale, Bristol. This A.R.D. was awarded a special contract in April 1918 to build up twenty Camels (E9964-9983) from salvage and spares as one of the measures to meet the serious situation at the Front. A few months later the material used was questioned as several crashed in quick succession. E9973, which first flew at the Depot on May 23rd 1918, crashed on a height test at Saughall the following July 31st. E9975 and E9976, both at No. 42 Training Depot Station, crashed on the same day—August 8th 1918; E9975 crashed at Stanwell after pulling out from a dive at a floating target on Staines Reservoir and E9976 broke up over Hounslow.

A further twenty Camels (F2189-2208) were ordered on May 16th 1918 followed by an order for forty (F4177-4216). Again there were accidents which called for investigation. F4177, which first flew on August 22nd, 1918, crashed 12 days after, by which time F4178 had also crashed.

Aircraft Depots in France during 1918 built up Camels F5914, F5918, F5919, F5925, F5926, F5927, F5932, F5938, F5939, F5941, F5943, F5946, F5951, F5958, F5959, F5967, F5968, F5972, F5981, F5990, F5991, F5993, F6022, F6024, F6030, F6032, F6033, F6034, F6037, F6063, F6064, F6084, F6089, F6102, F6110, F6117, F6123, F6132, F6135, F6138, F6149, F6152, F6169, F6176, F6180, F6183, F6185, F6191, F6192, F6194, F6201, F6210, F6211, F6221, F6223, F6240, F6245, F6249, F6250, F6254, F6257, F6258, F6259, H6847, H6860, H6997, H7003, H7007, H7012, H7092, H7098, H7160, H7239, H7272 and H7281 and Dolphins F5916, F6961, F6144, F6145 and H7245.

Formation or Unit Built

Sopwith types were reconstructed at several stations in the United Kingdom from salvage and spares. Known examples, by their military serial and place of buliding, are as follows:

Production scenes at Sopwith factories, Dolphin fuselages before and after covering late in 1917.

No. 6 Wing, Dover: Pups B8784-8786, B9455, B9931, C3500-3503; a 1½ Strutter single-seater and Camels B8830, F9623-9624.

No. 7 Wing, Norwich: Pups B8801, B8829, C8653-8654.

No. 26 Wing, Thetford: Pups B8795, B8821, C9993.

No. 30 Wing, Turnhouse and Montrose: Pup C4295.

No. 3 Training Depot Station, Lopcombe Corner: Camels F9628-9630.

No. 7 Training Depot Station, Feltwell: Pups C9990-1.

No. 29 Training Depot Station, Beaulieu: Camel F9634.

No. 43 Training Depot Stations, Chattis Hill: Camels F9631-9632, H8200-8202, H8258-8259 of which the last two were built as 2-seaters.

No. 207 Training Depot Station, Chingford: Camel F9548.

Central Flying School, Upavon: Camels F9695, H8264, H8291-8292.

No. 112 Squadron, Bekesbourne: Camel F4175.

No. 42 Training Squadron, Wye: Camels F9637, H8253.

No. 63 Training Squadron, Joyce Green: Pup B9440.

Aircraft Repair Depot, Salisbury: Pup F4220.

First in the field of aeronautical products at Kingston were T. W. K. Clarke & Company, who sold model aircraft before Sopwith established his works there, and who later set up as propeller makers supplying the Sopwith works.

SOPWITH AIRCRAFT PRODUCTION SUMMARY

Type and Manufacture	Misc. Seaplanes	Misc. Aeroplanes	807 Seaplane	Gunbus	Schneider	Baby	1½ Strutter	Pup	Triplane	Camel F1	Ships Camel	Cuckoo	Dolphin	Snipe	Dragon	Salamander	Totals
Arrol Johnson	—	—	—	—	—	—	—	—	—	30	—	—	—	—	—	—	30
Beardmore	—	—	—	—	—	—	—	80	—	—	140	—	—	—	—	—	220
Blackburn	—	—	—	—	—	186	—	—	—	—	—	162	—	—	—	—	348
Boulton & Paul	—	—	—	—	—	—	—	—	—	1550	—	—	—	415	—	—	1965
British Caudron	—	—	—	—	—	—	—	100	—	—	—	—	—	—	—	—	100
Clayton & Shuttleworth	—	—	—	—	—	—	—	—	46	575	—	—	—	—	—	—	621
Coventry Ordnance Works	—	—	—	—	—	—	—	—	—	—	—	—	—	150	—	—	150
Darracq	—	—	—	—	—	—	—	—	—	—	—	—	300	—	—	—	300
Fairfield	—	—	—	—	—	—	—	—	—	—	—	50	—	—	—	—	50
Fairey	—	—	—	—	—	—	100	—	—	—	—	—	—	—	—	—	100
French manufacturers	—	—	—	—	—	—	4200	—	—	—	—	—	—	—	—	—	4200
Glendower	—	—	—	—	—	—	—	—	—	—	—	—	—	—	—	20+	20+
Hooper	—	—	—	—	—	—	150	—	—	321	30	—	216	—	—	—	717
Kingsbury	—	—	—	—	—	—	—	—	—	—	—	—	—	30	—	—	30
Morgan	—	—	—	—	—	—	100	—	—	—	—	—	—	—	—	—	100
Mann Egerton	—	—	—	—	—	—	75	—	—	—	—	—	—	—	—	—	75
Marsh, Jones & Cribb	—	—	—	—	—	—	—	—	—	175	—	—	—	100	—	—	275
Napier	—	—	—	—	—	—	—	—	—	—	—	—	—	150	—	—	150
Nieuport & General	—	—	—	—	—	—	—	—	—	400	—	—	—	100	—	—	500
Oakley	—	—	—	—	—	—	—	—	3	—	—	—	—	—	—	—	3
Pegler	—	—	—	—	—	—	—	—	—	—	—	20	—	—	—	—	20
Portholme	—	—	—	—	—	—	—	—	—	300	—	—	—	100	—	—	400
Robey	—	—	17	—	—	—	—	—	—	—	—	—	—	—	—	—	17
Ruston Proctor	—	—	—	—	—	—	350	—	—	1573	—	—	—	498	—	—	2421
Sopwith Aviation	30	70	12	12	137	100	246	97	103	503	50	1	1043+	560	200	163	3327
Standard	—	—	—	—	—	—	—	850	—	—	—	—	—	—	—	—	850
Vickers	—	—	—	—	—	—	150	—	—	—	—	—	—	—	—	—	150
Wells	—	—	—	—	—	—	20+	—	—	—	—	—	—	—	—	—	20+
Westland	—	—	—	—	—	—	75	—	—	—	—	—	—	—	—	—	75
Whitehead	—	—	—	—	—	—	—	820	—	—	—	—	—	—	—	—	820
Totals	30	70	12	29	137	286	5466+	1847	152	5497	250	233	1559	2103	200	183+	18054+

In addition Sopwith Aviation built prototypes as follows: 6 Snails; 3 each Snarks, Snappers, Cobhams; 2 each Buffalos, Bulldogs, Rhinos, Hippos; 1 each B.1, Antelope, Atlantic, Wallaby, Grasshopper, Schneider, and civil aircraft production of 13 Gnus and 10 Doves = 52 aircraft, making a Grand Total of .... 18106+

N.B. Aircraft rebuilt from salvage not included. Aircraft in miscellaneous columns detailed in Type-by-Type Review.

FROM BEGINNINGS AT BROOKLANDS

The hangars at Brooklands were the original buildings of the Sopwith establishment. First used to house T. O. M. Sopwith's aircraft when he took up flying, then as the Sopwith School sheds, they became workshops for the early aircraft built by Sopwith. After an ex-skating rink had been acquired in 1912 and the buildings shown in 1914, both in Kingston, the Brooklands sheds were used for housing Sopwith prototypes under test. The building, but now without the wall, is currently an art school.

THE SOPWITH AVIATION COMPANY LIMITED, KINGSTON-UPON-THAMES, SURREY, ENGLAND

EMPLOYEES OF CANBURY PARK ROAD WORKS, 1914

These are some of the original employees, numbering some 200 in 1914. Many later served with the Colours and their places were taken by others, including women as the firm rose to its 1918 peak of 3,000 employees. Not all of the employees in this photograph can be identified and the publishers and author would be interested to know of any others that can be identified: meanwhile, apologies to those not named.

Employees identified are Messrs. L. A. (Jack) Pollard (3), Joe Fairweather (5), Dick Snelling (15), Parkinson (17), Jim Mortlock (18), Ernie Fryer (24), Joe Childs (25), Ridley (26), Chambers (29), Joe Wheeler (30), Jim Baker (31), Ernie

Richard Garrett (45), Joe Robson (52), Dick Venn (53), Len Cherry (54), Lew Budd (55), Tom Dumbrell (63), Shirley (65), Wally Harris (66), Fred Sedgley (67), T. Watters (74), Ashurst (75), B. Stonard (80), Ketteringham (88), Frank Odell (89), Brothers (90), Jack Goy (94), Joe Pratt (96), Apps Snr (100), Edwards (103), Markwick (106), Mitchell (107), E. Yardley (117), H. Cobb (118), Young Jnr (119), Hayter (121), W. Tyrell (122), J. Dale (124), F. Tadd (127), H. J. (Bert) Hinkler (129), H. P. Musgrave (Company Secretary 131), R. O. Cary (General Manager 132), Fred Sigrist (Works Manager 133), Jack Whitehorn (134), H. Wilbur (135), Bob Shaw (137), Madgwick (138), Baigent (139), Materface Jnr (143), Joe Baysting (145), Lazell Apps Jnr (160), Arthur Bush (162), Bob Potter (163), Wally Headen (165), Pain (167), Young Snr (173), Beckwith (174), Jennings (175), and Tillbrook (sailor) 178.

SOPWITH AVIATION COMPANY CANB[URY]
PORTRAYED ON A RECENT AERIAL PHOT[OGRAPH]

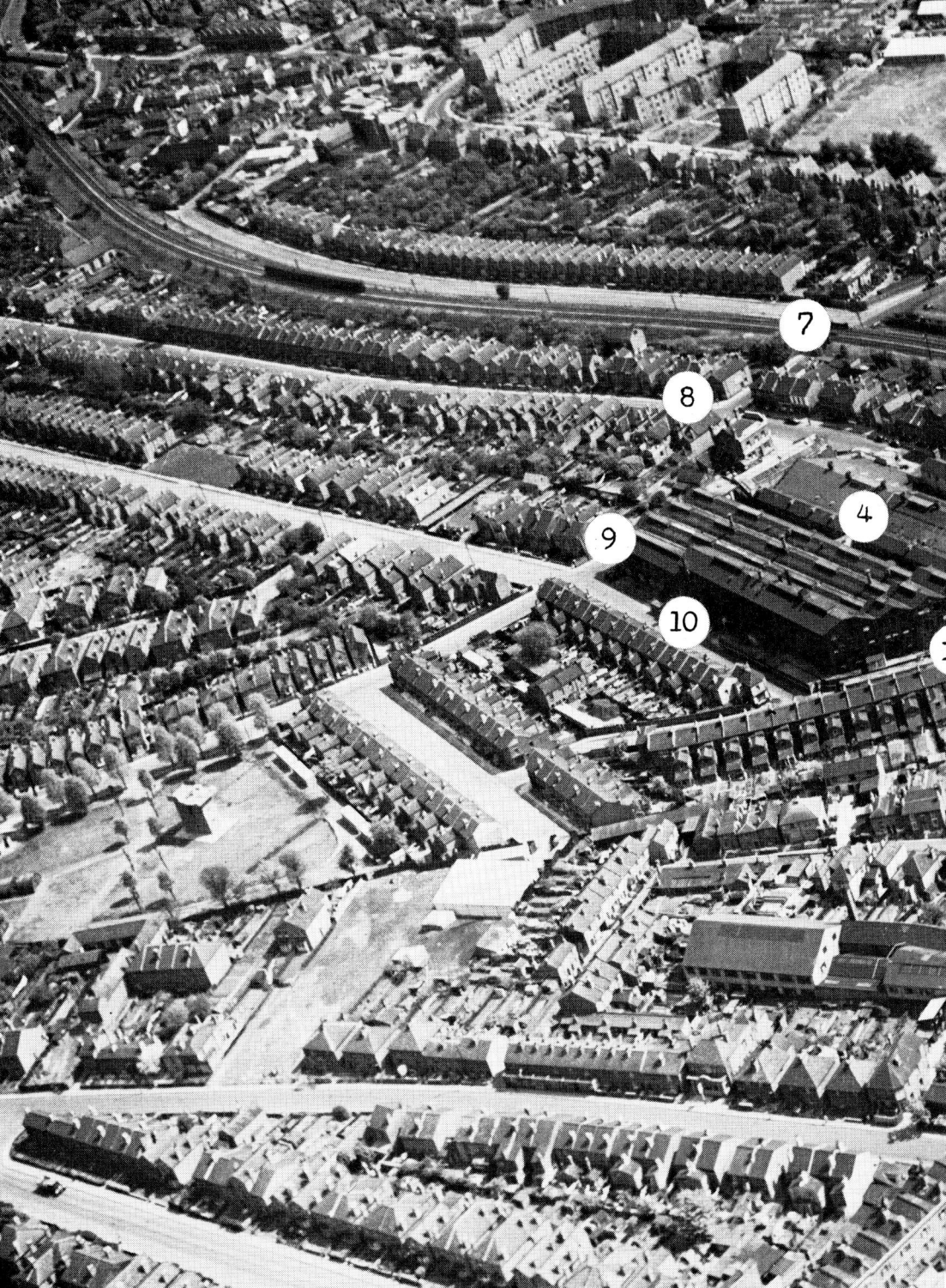

1 Ex-skating rink original works, acquired in 1912. Became Experimental Shop 1915-1919.

2 Original Drawing Office built on roof here.

3 Offices and Works acquired early in 1914. T. O. M. Sopwith's office was at the corner of this building.

4 Wartime expansion of works acquired in 1914, making this the Main Works 1915-1920.

5 Further wartime expansion on opposite side of Canbury Park Road.

6 Richmond Road leading to Ham Works, acquired in late 1917, about a mile away, downwards and to the left.

The original works acquired in 1912 were a former skating rink, illustrated in its origin[al] subsequently. When buildings were obtained further down Canbury Park Road in 19[14] and Works. Progressive expansion of the Main Works was limited by the 'Elm stree[t'] three groups of buildings, all in use at peak

**ROAD WORKS LOCATIONS, 1912–1920,
F THE KINGSTON-UPON-THAMES AREA**

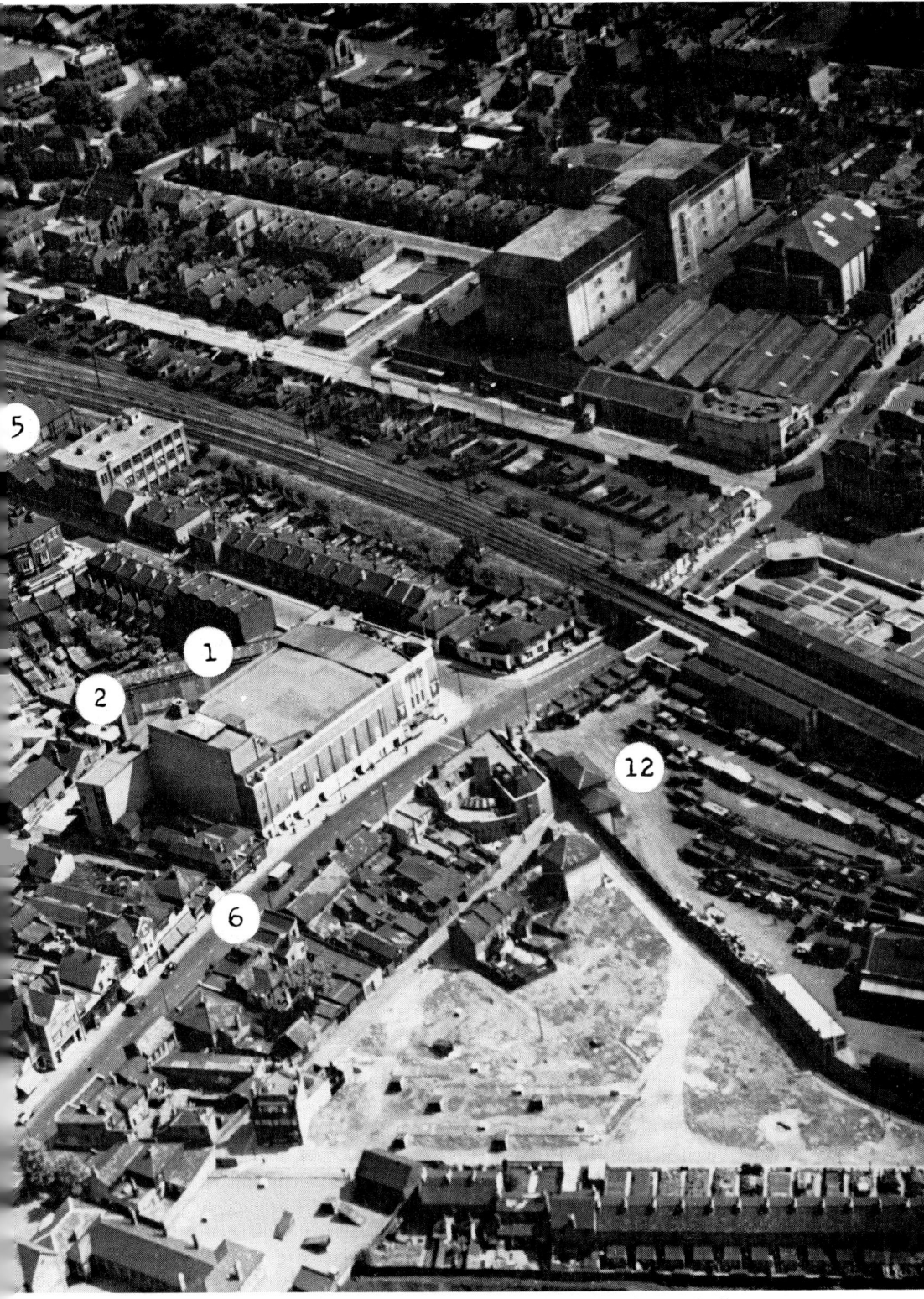

7
Railway to Waterloo Station, London, travelling left (roughly northeast).

8
Canbury Park Road on which the three separate building blocks fronted, and the road of the official address of the Company.

9
Elm Road bounding the Main Works.

10
Elm Grove bounding the Main Works.

11
Elm Crescent bounding the Main Works.

12
Kingston Station Yard which was conveniently only a matter of a few hundred yards to all parts of the Works.

6. The building still stands and can be seen here behind the large cinema built
ing rink became the Experimental Shop and the new buildings the Main Offices
urther expansion buildings were obtained on the other side of the road; but all
1918, fronted on Canbury Park Road.

FROM PLANKS TO PLANES

The Ham works at Richmond showing the Saw Mill as it was in 1919 and the main Assembly Bay on September 11th 1918 with Snipes and Salamanders under construction. The Salamanders can be distinguished by their streamlined headrests.

Sopwith Wright

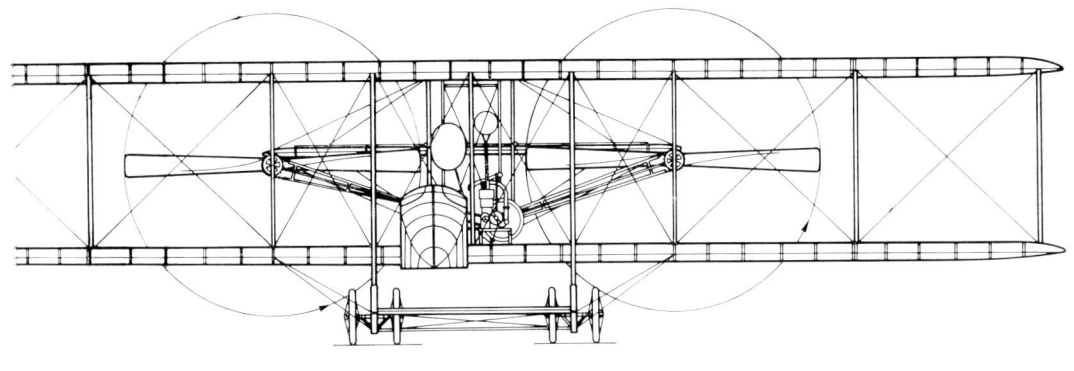
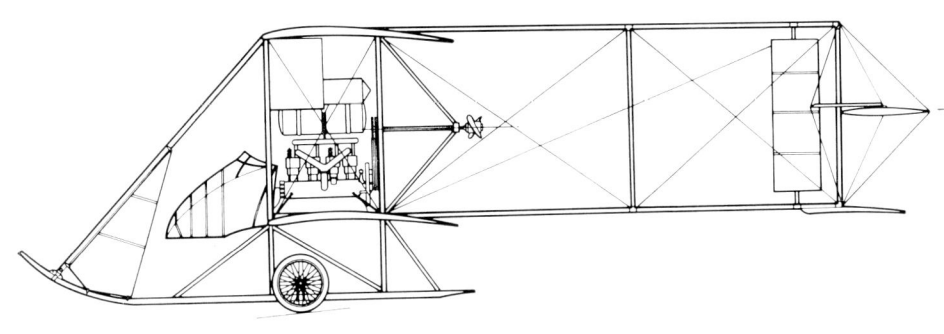
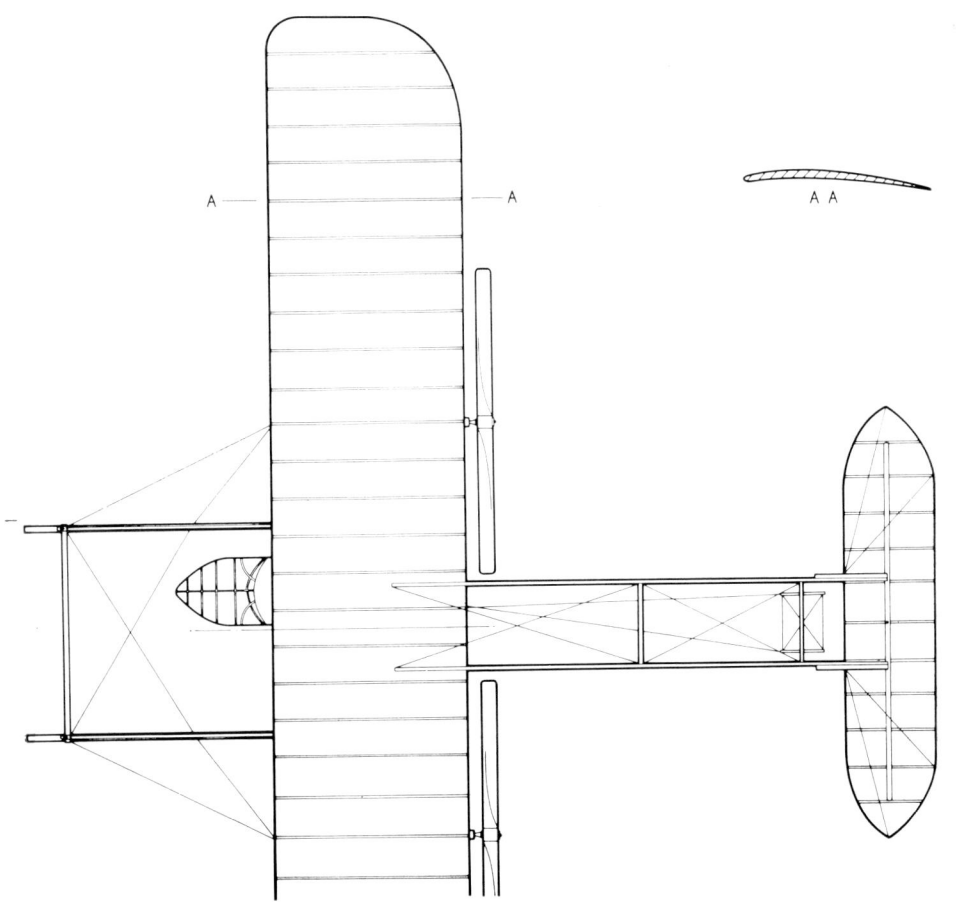

Batboat 1a

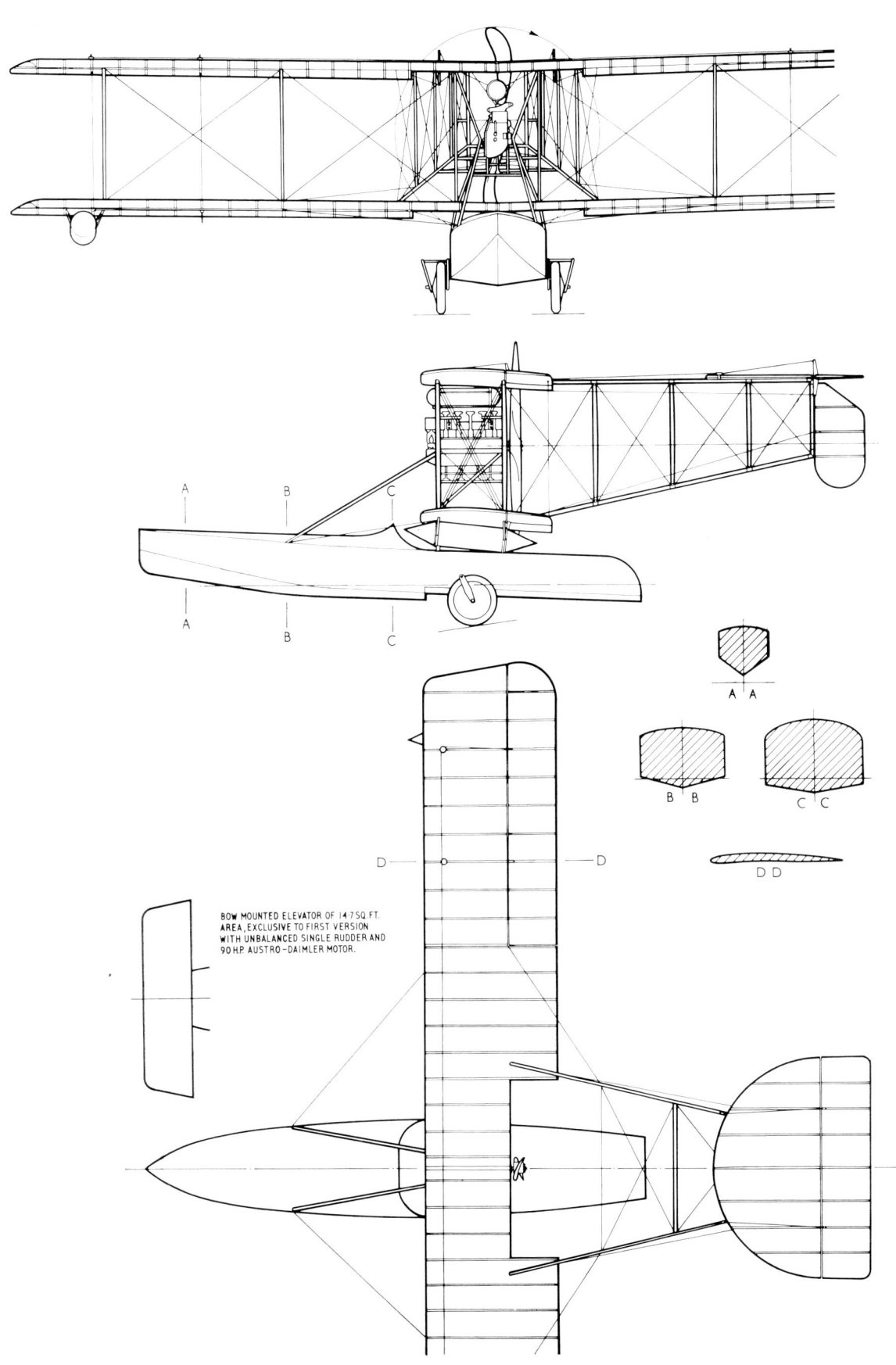

BOW MOUNTED ELEVATOR OF 14·7 SQ. FT. AREA, EXCLUSIVE TO FIRST VERSION WITH UNBALANCED SINGLE RUDDER AND 90 H.P. AUSTRO-DAIMLER MOTOR.

Gun-Bus (land version)

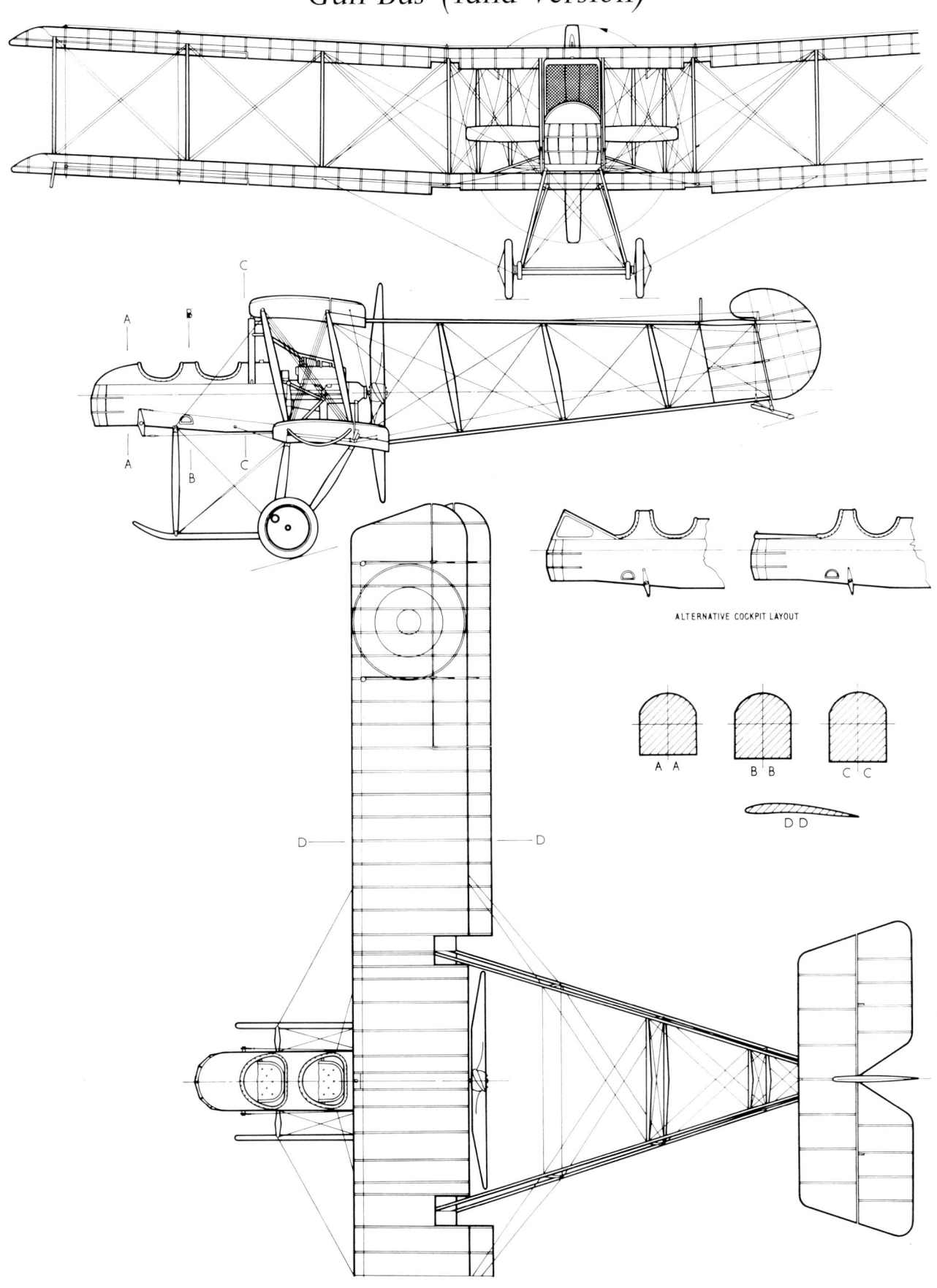

ALTERNATIVE COCKPIT LAYOUT

Churchill—Sociable

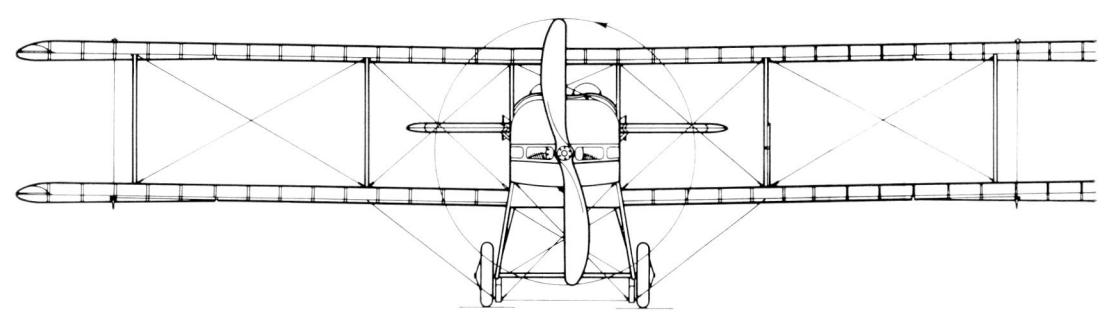
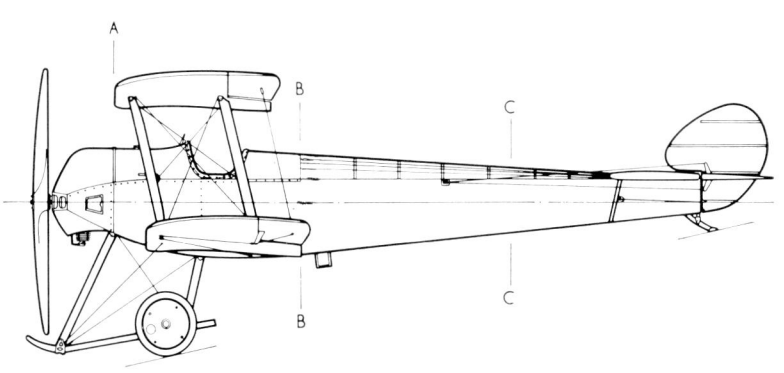
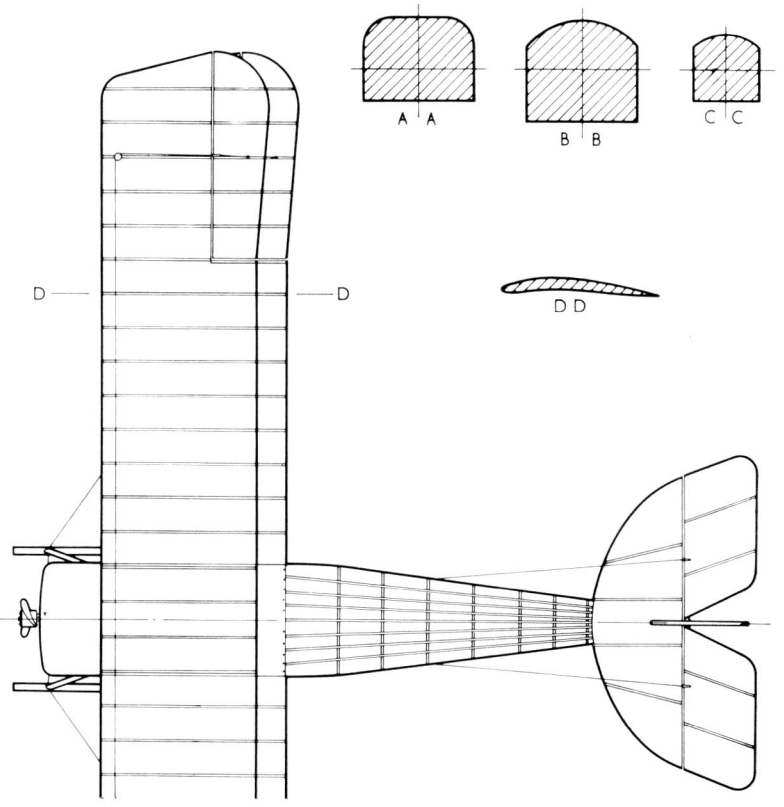

Tabloid (RNAS 1914)

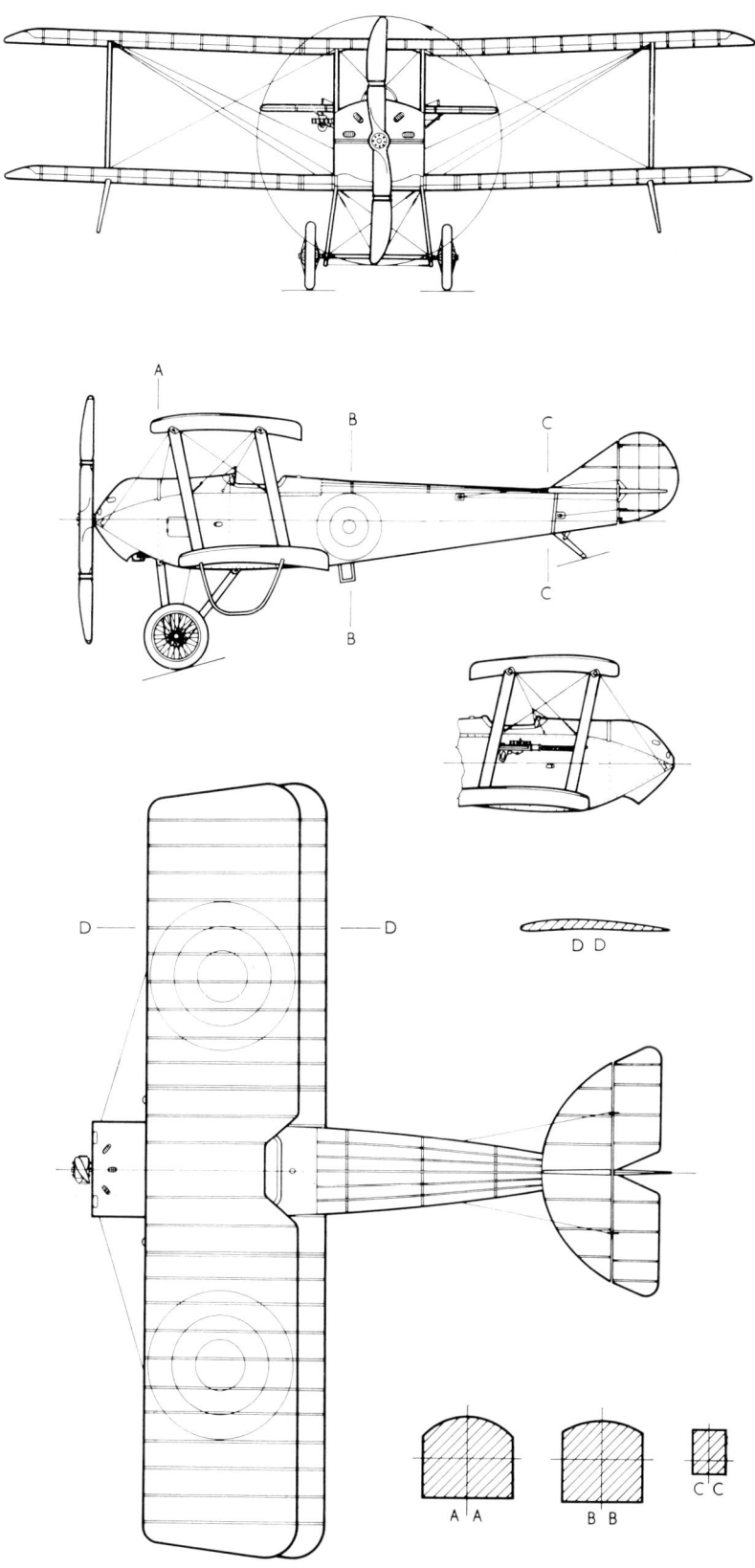

Schneider

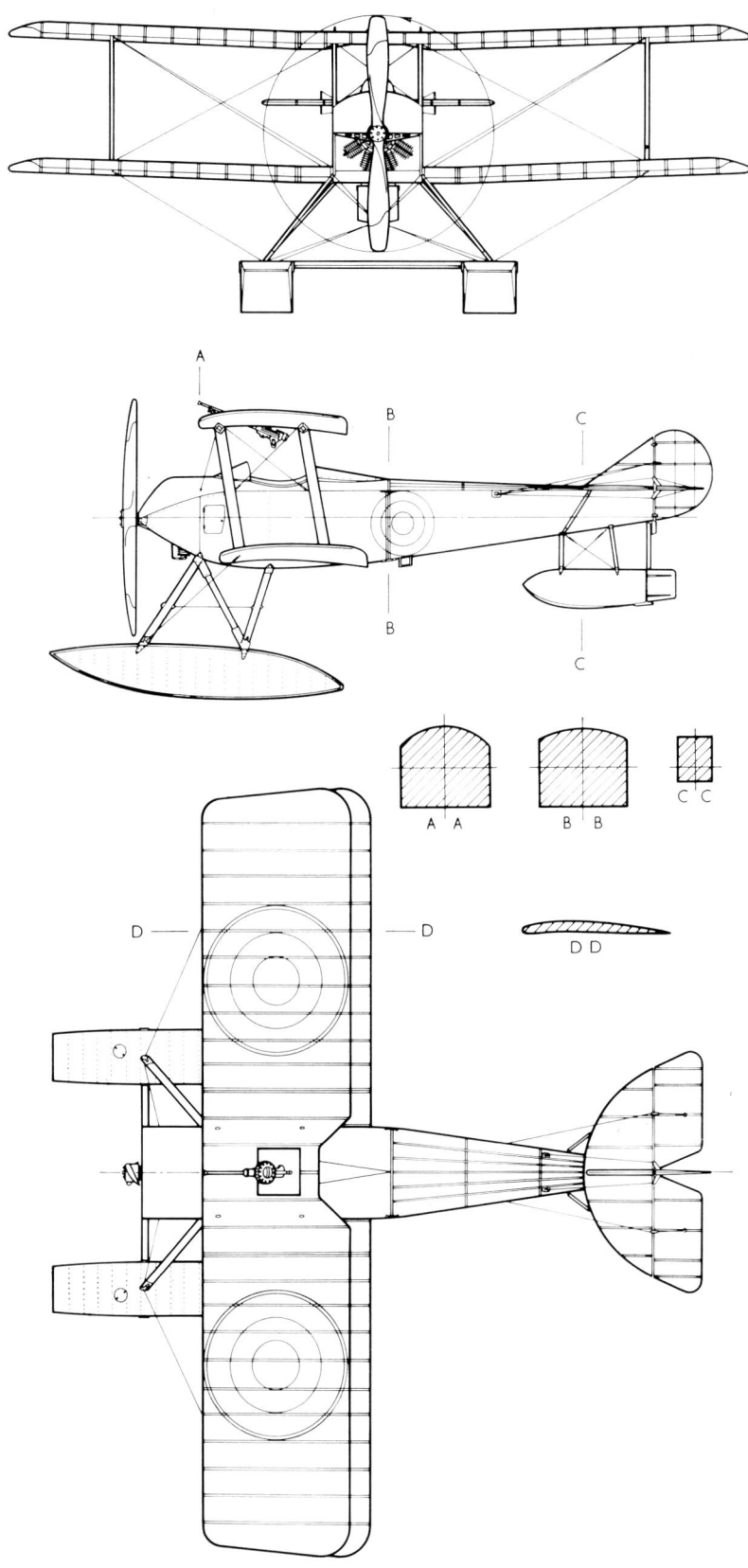

Three-Seat Biplane

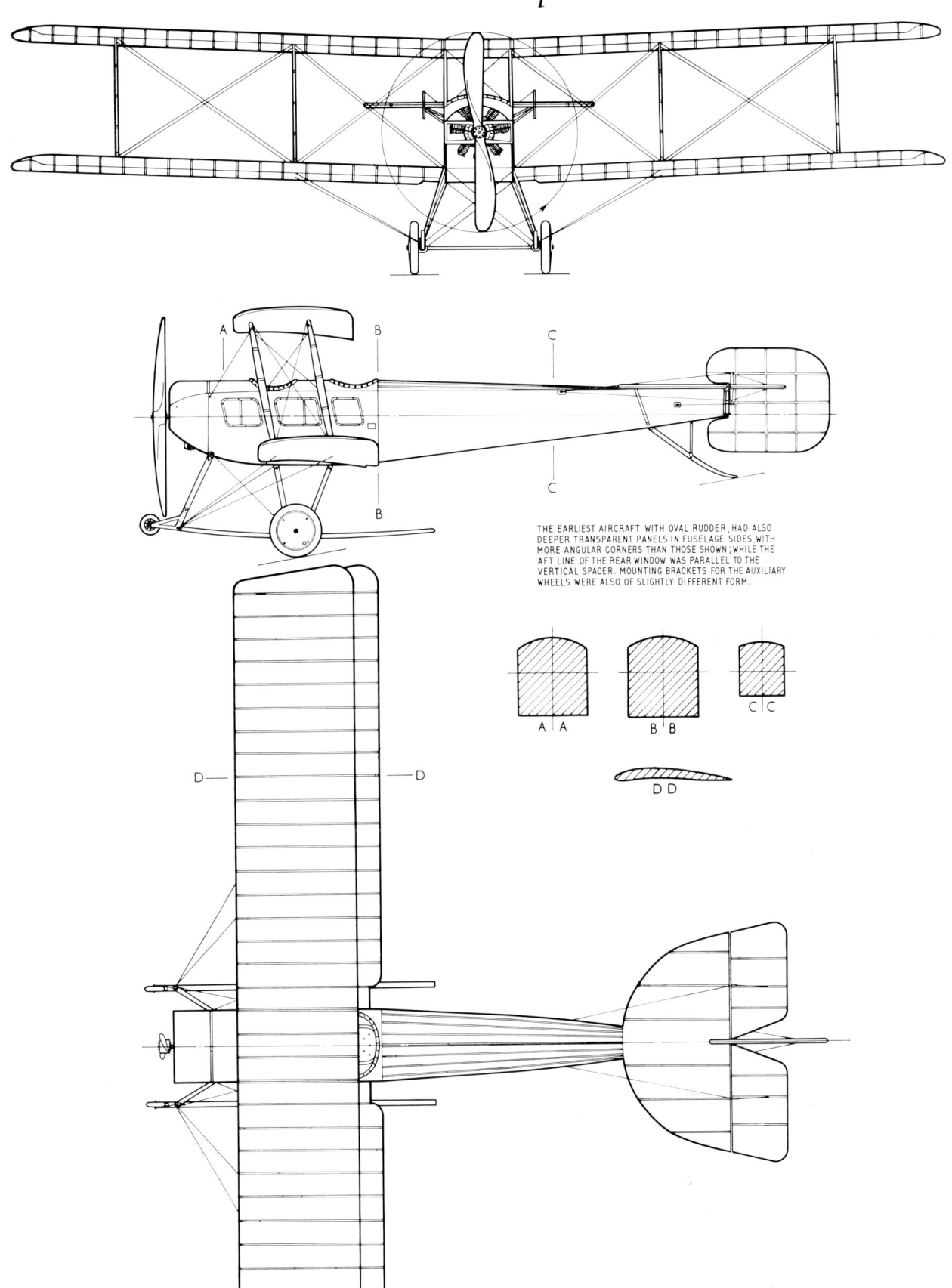

THE EARLIEST AIRCRAFT WITH OVAL RUDDER, HAD ALSO DEEPER TRANSPARENT PANELS IN FUSELAGE SIDES, WITH MORE ANGULAR CORNERS THAN THOSE SHOWN; WHILE THE AFT LINE OF THE REAR WINDOW WAS PARALLEL TO THE VERTICAL SPACER. MOUNTING BRACKETS FOR THE AUXILIARY WHEELS WERE ALSO OF SLIGHTLY DIFFERENT FORM.

Folder Seaplane (Type 807)

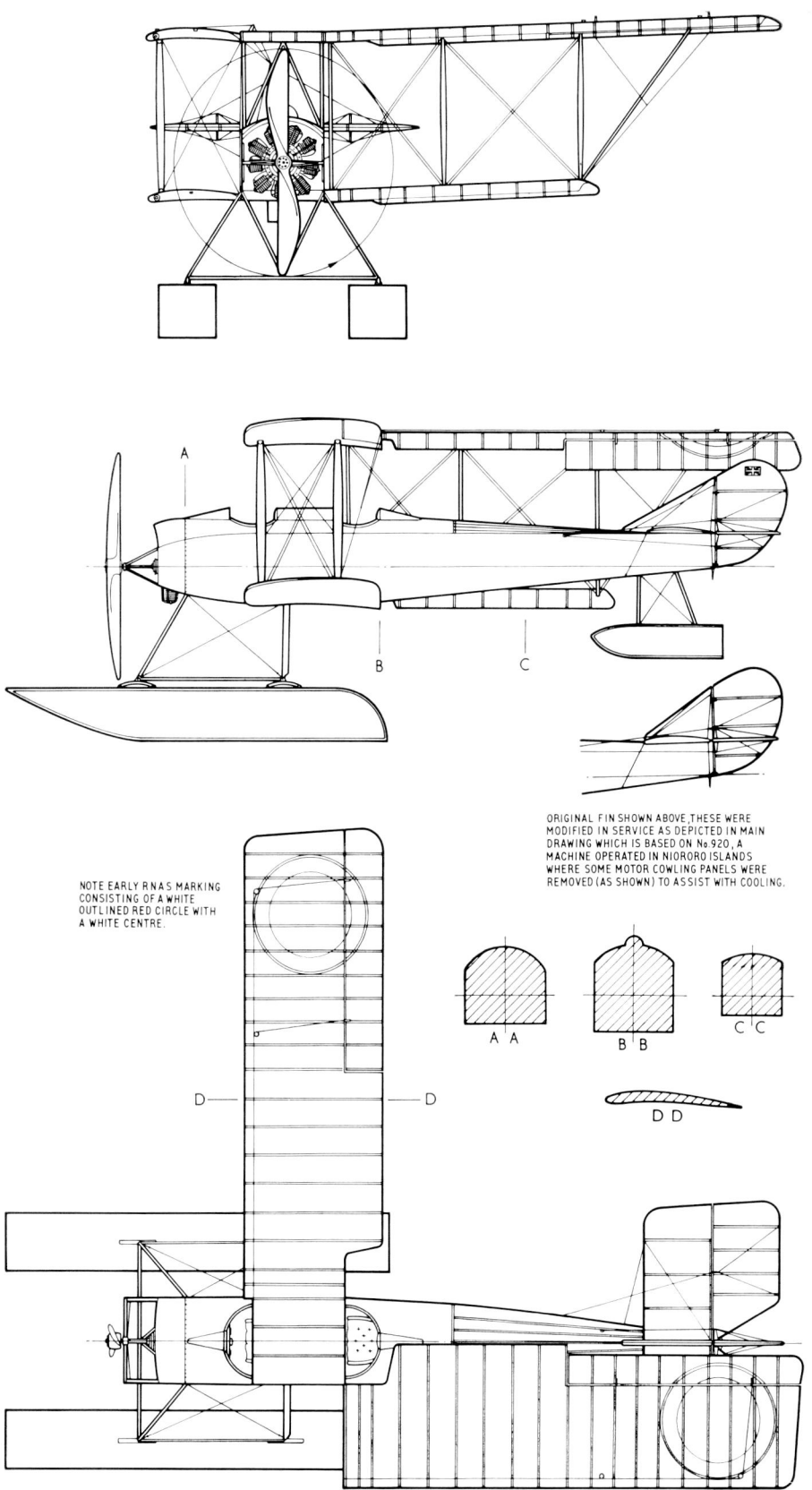

Baby (110 Clerget)

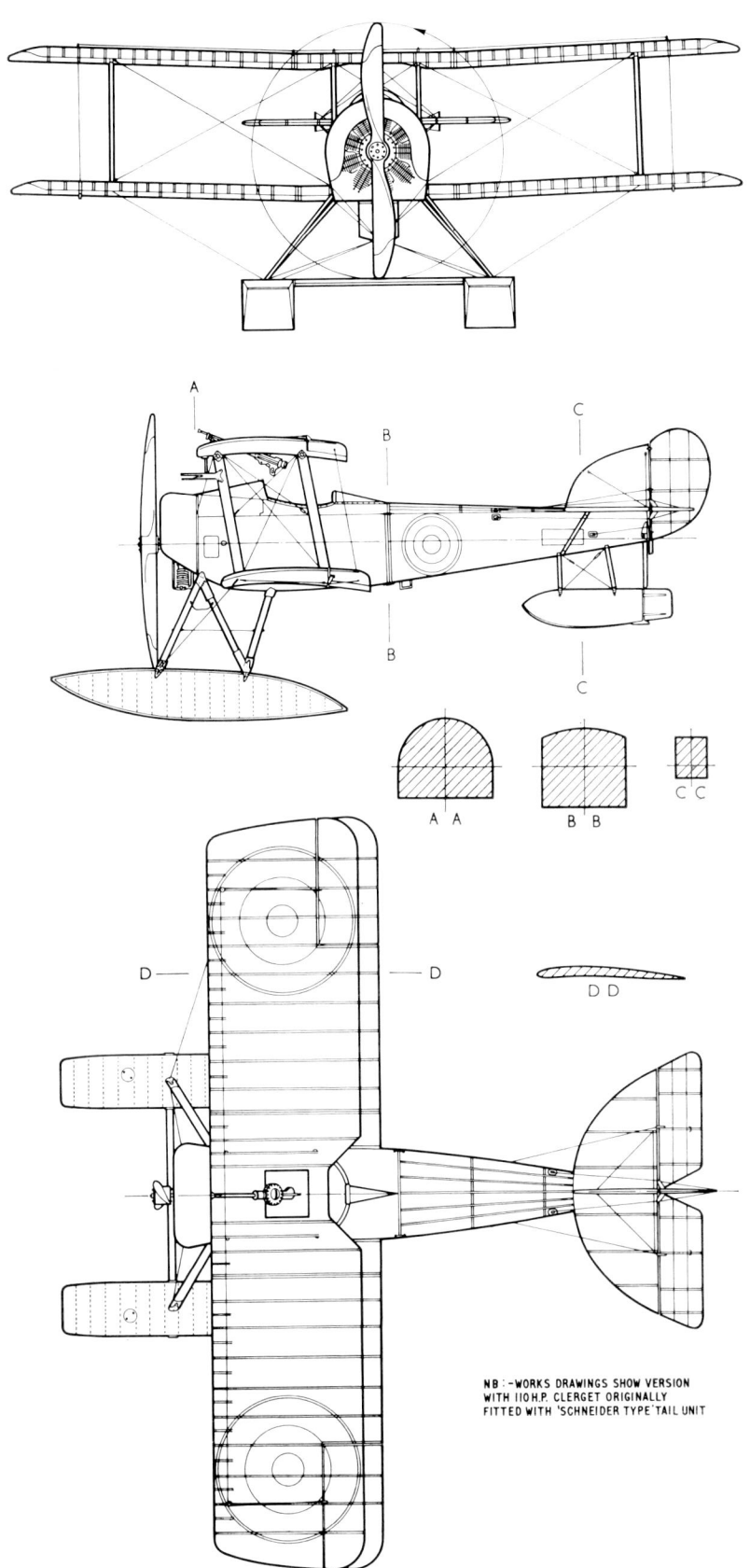

NB:- WORKS DRAWINGS SHOW VERSION WITH 110 H.P. CLERGET ORIGINALLY FITTED WITH 'SCHNEIDER TYPE' TAIL UNIT

1½ Strutter (Ship Strutter)

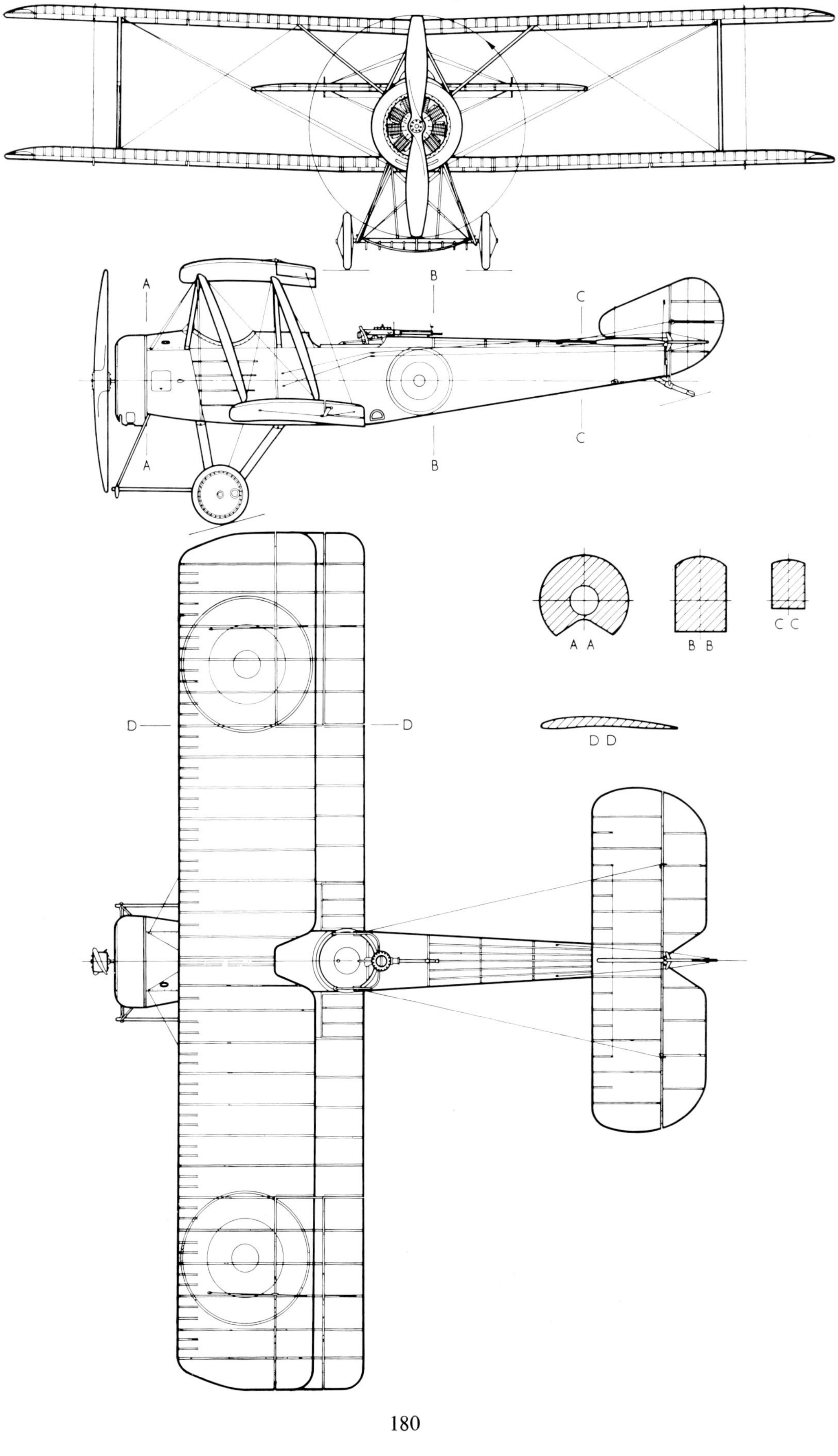

Pup (Ships Pup)

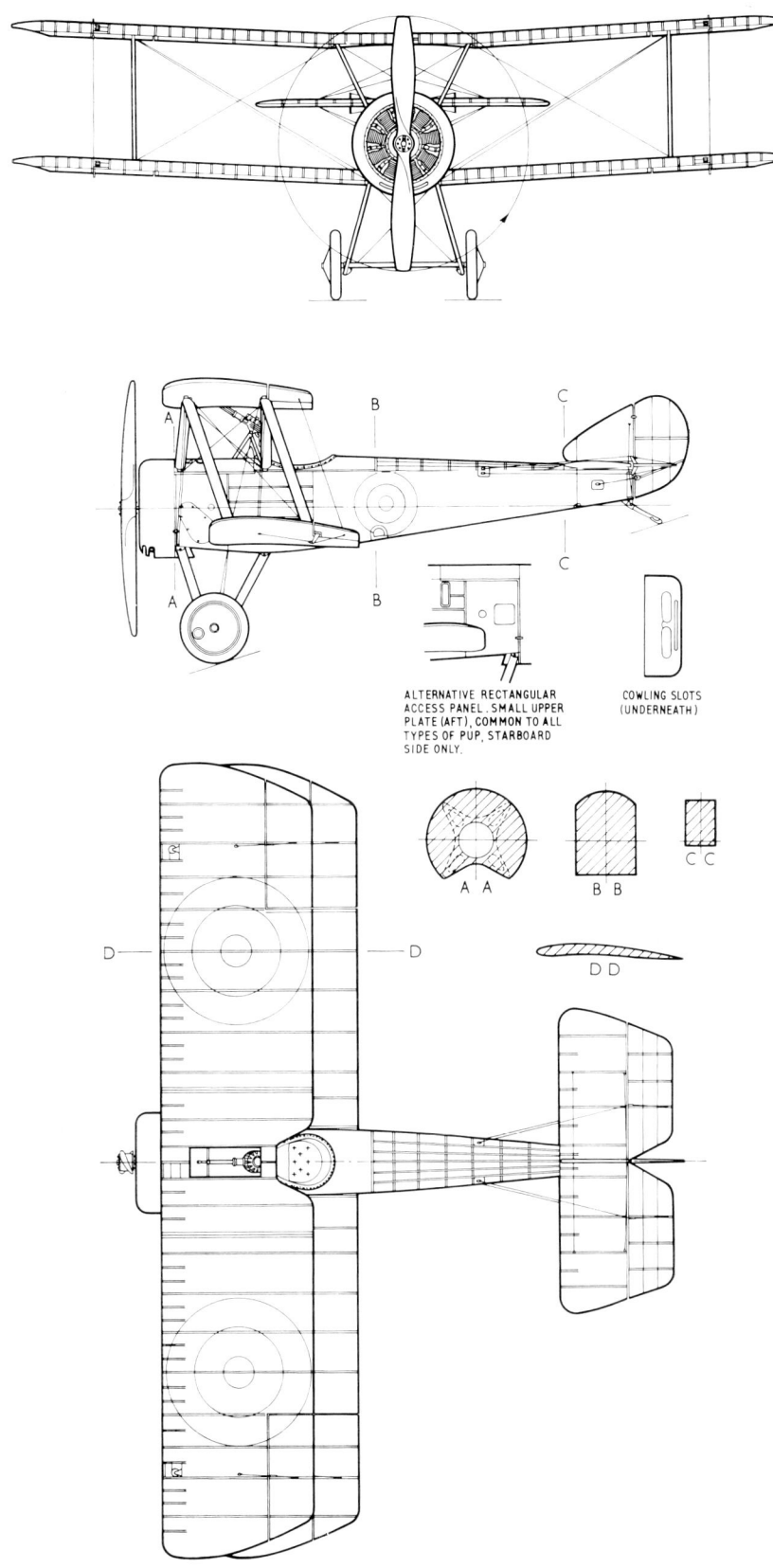

Triplane

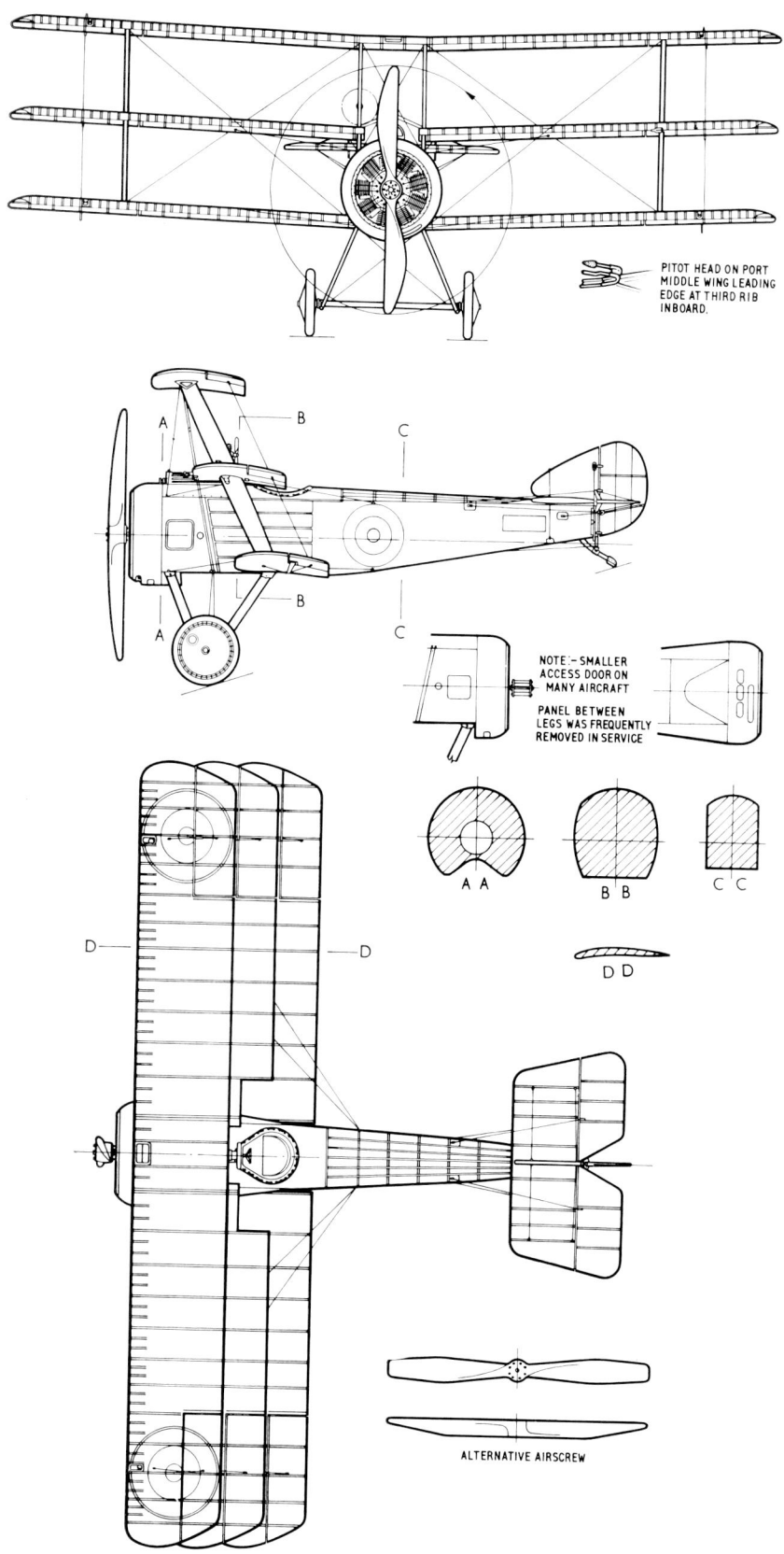

Triplane (150 h.p. Hispano)

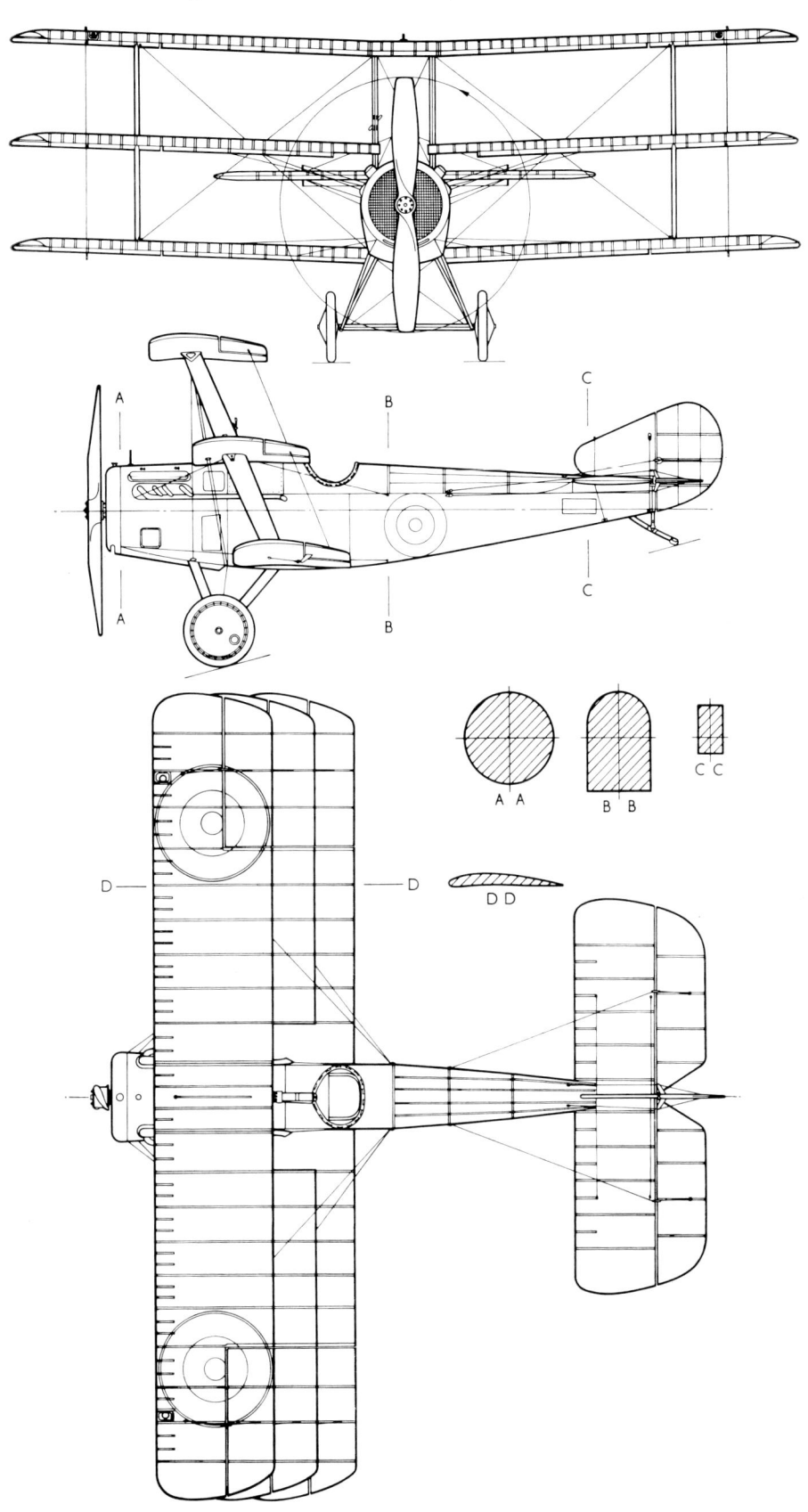

Camel F.1

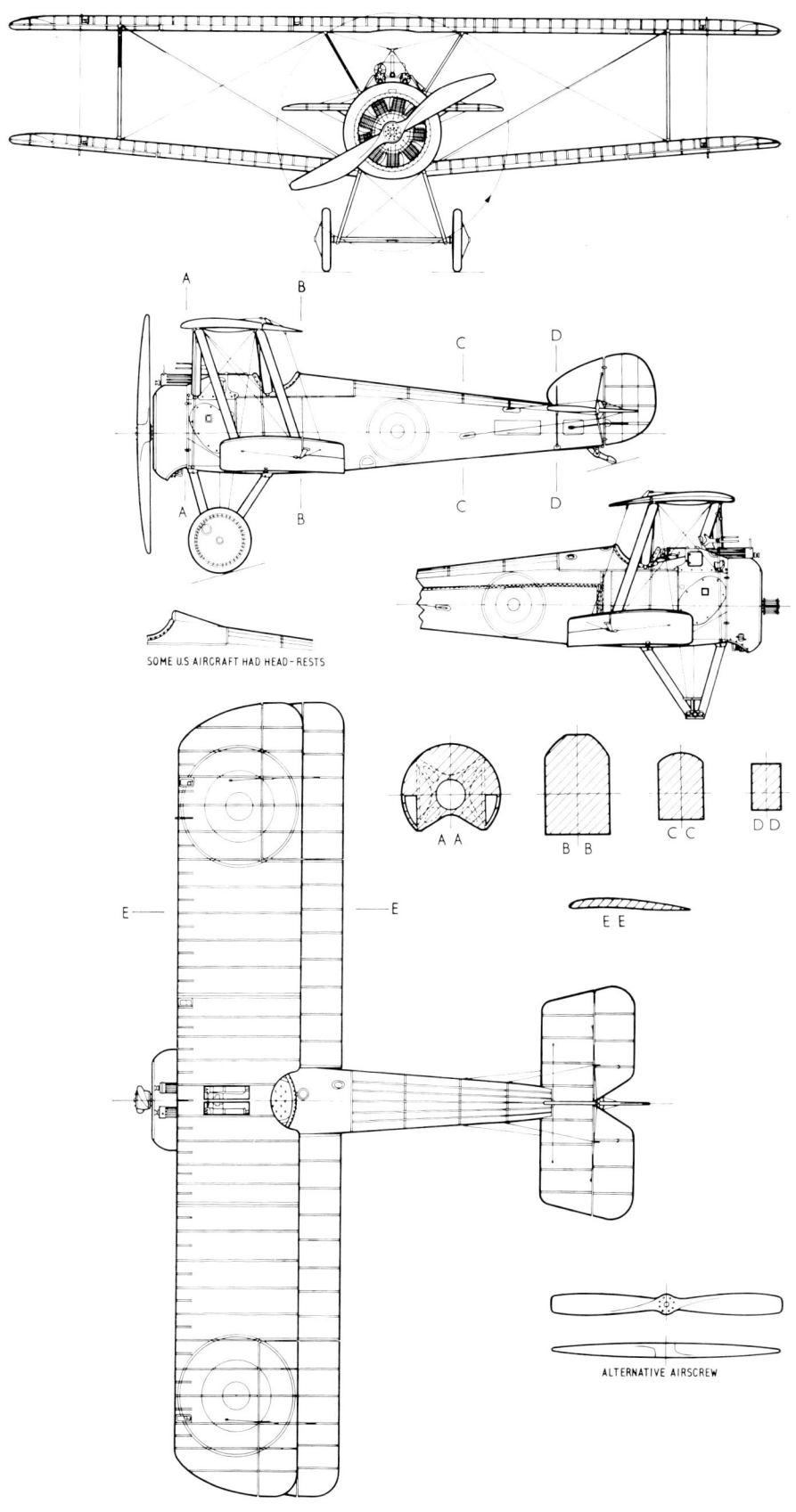

SOME U.S AIRCRAFT HAD HEAD-RESTS

ALTERNATIVE AIRSCREW

CAMEL CARRIERS

Experiments were conducted with the air-lifting of aircraft by airships and releasing them at height, having reserved their fuel for longer patrol work or for release only when enemy raiders were reported. The initial experiments were carried out with Camels, first dropped with ballast and locked controls, and then manned. These pictures show hook-up, carriage and release from an airship, details of which are given in chapter 38.

Camel 2F.1

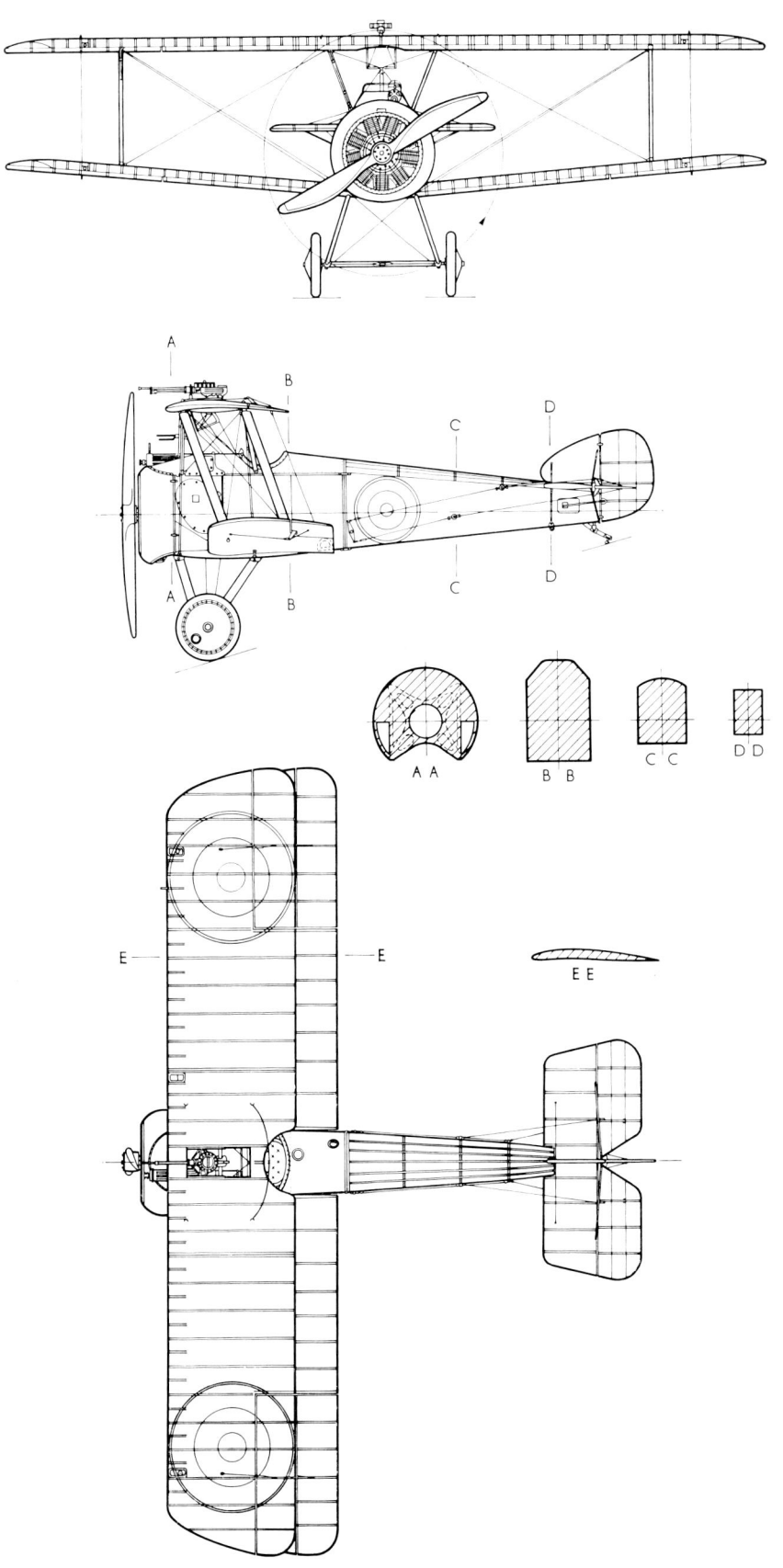

Camel Seaplane

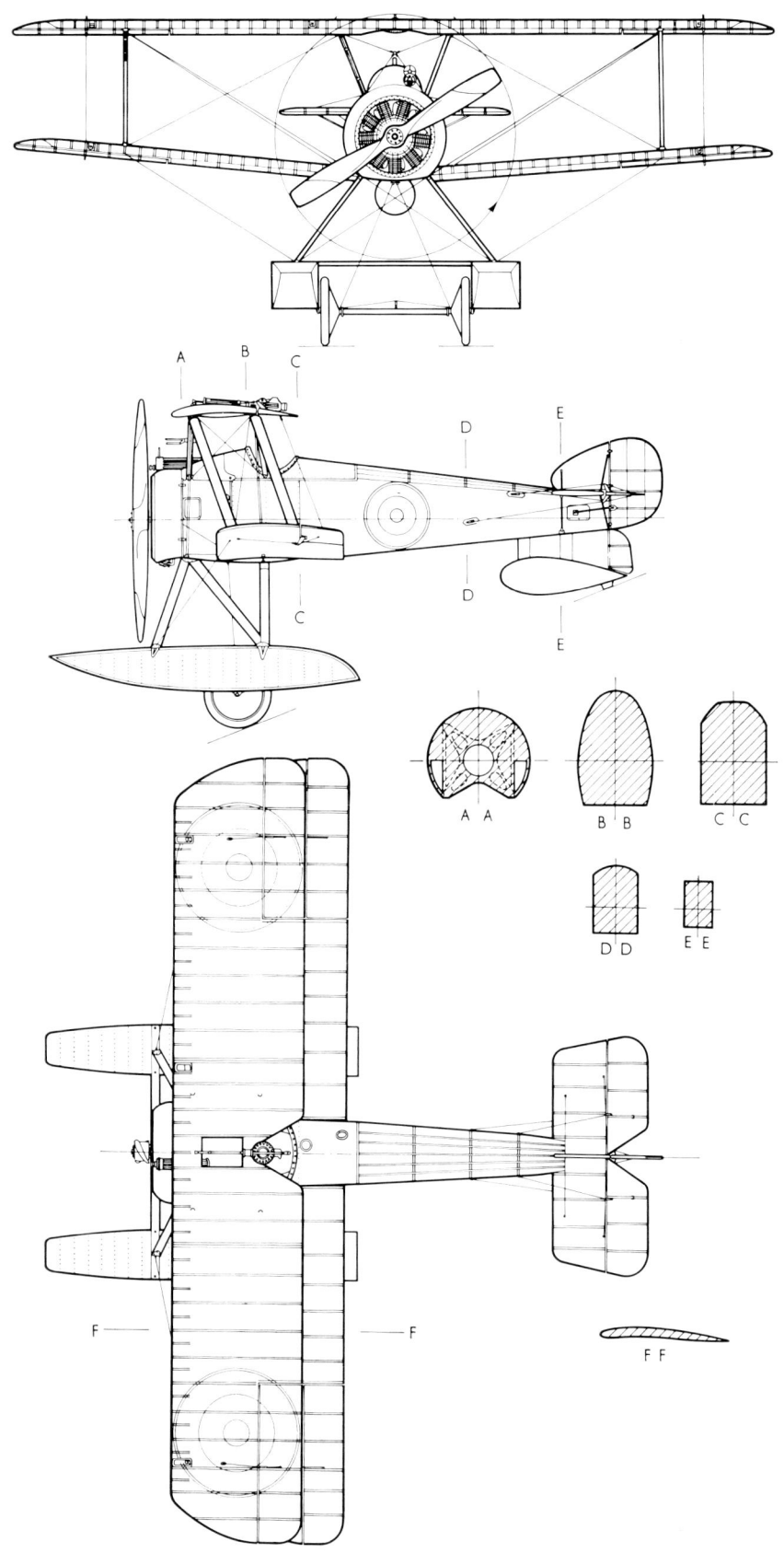

Cuckoo

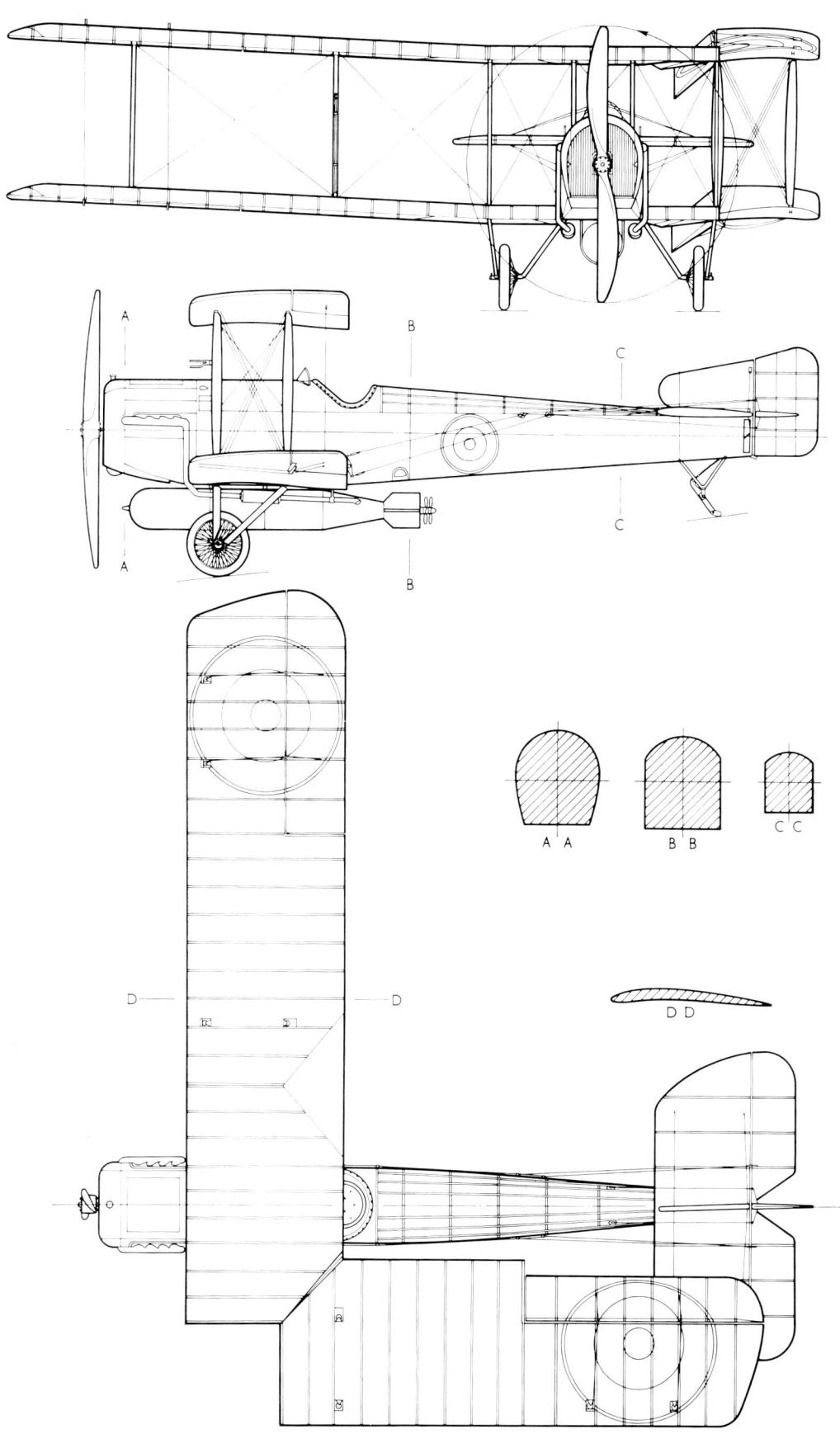

B1

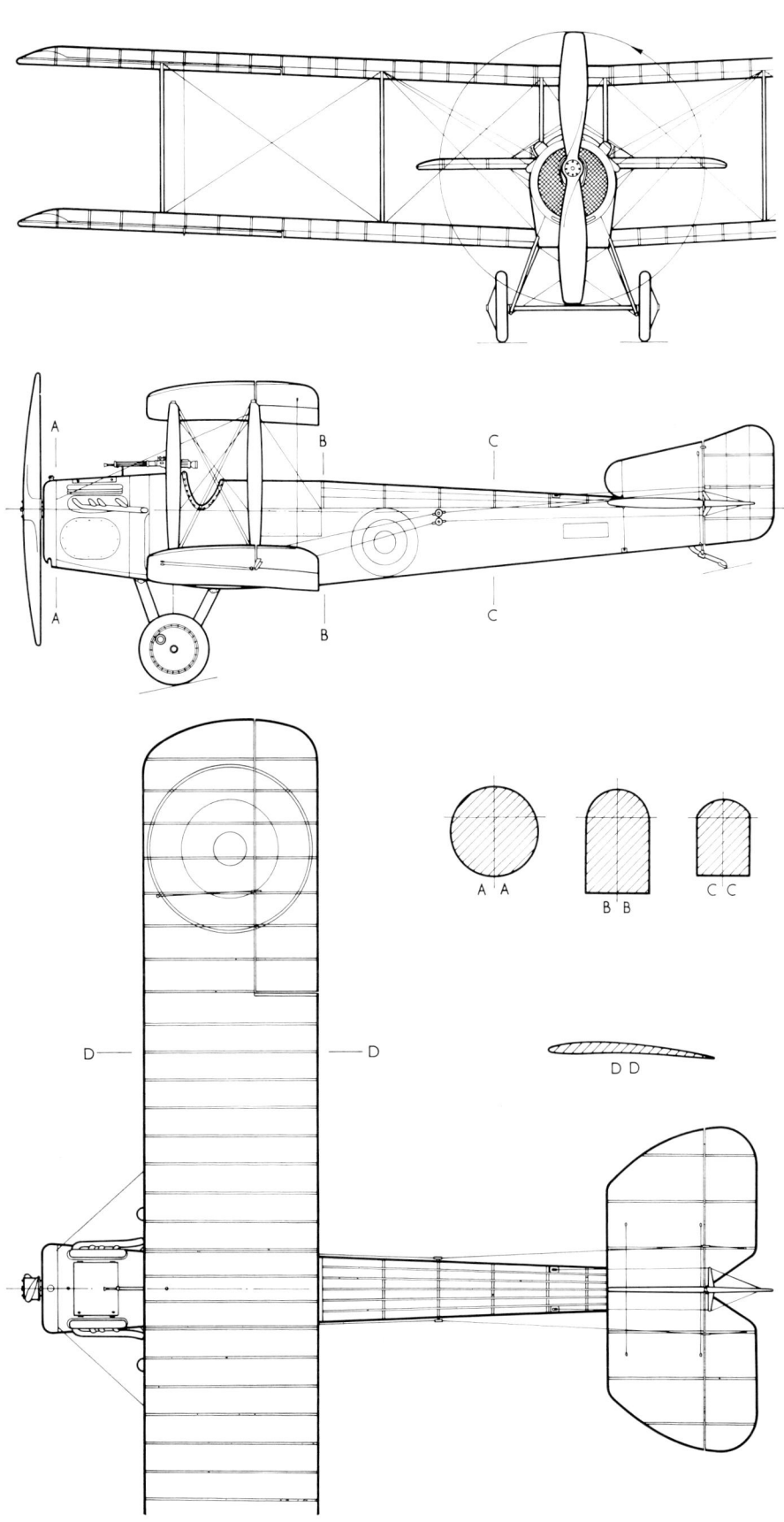

Dolphin

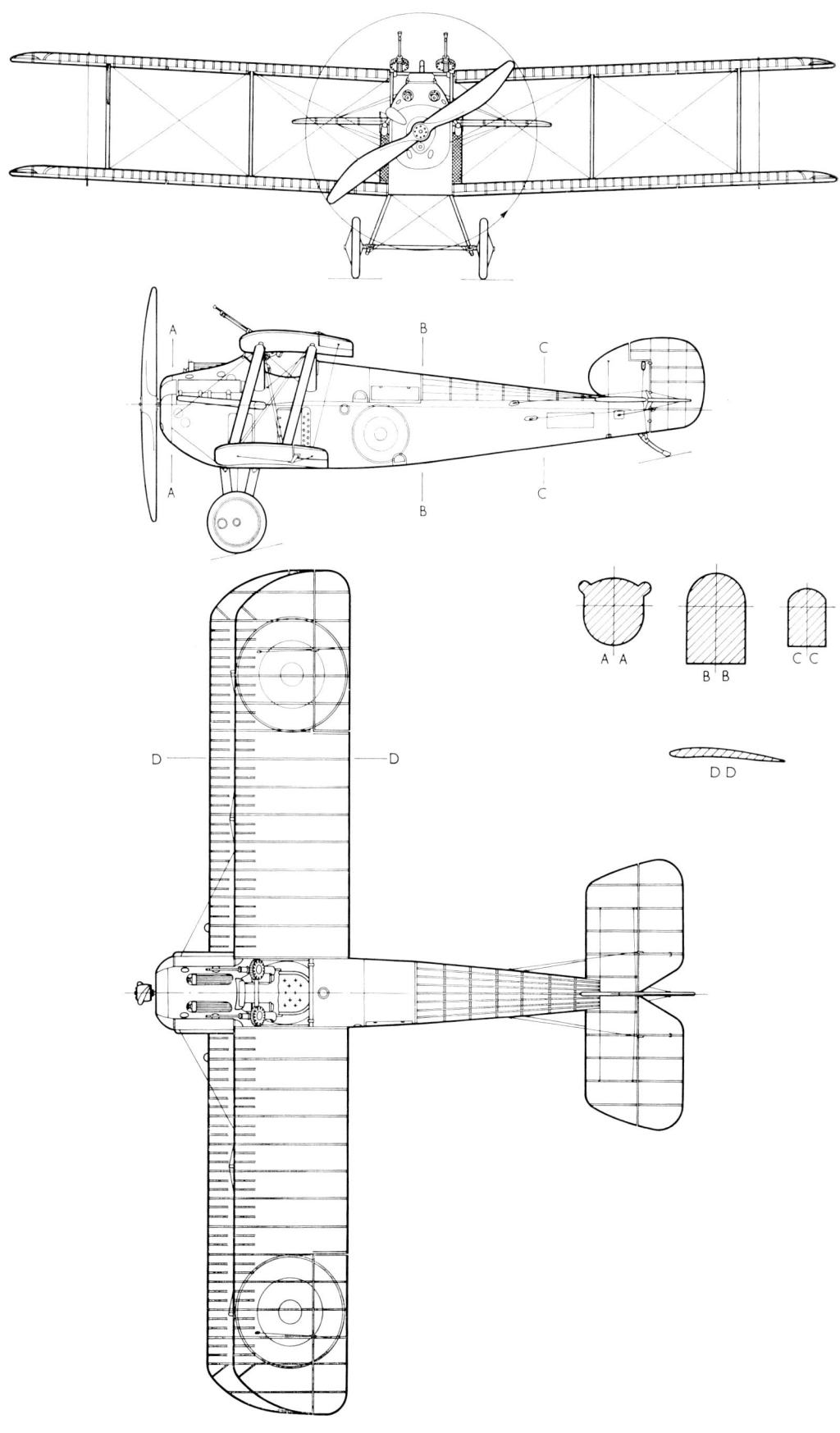

Snipe 7F.1

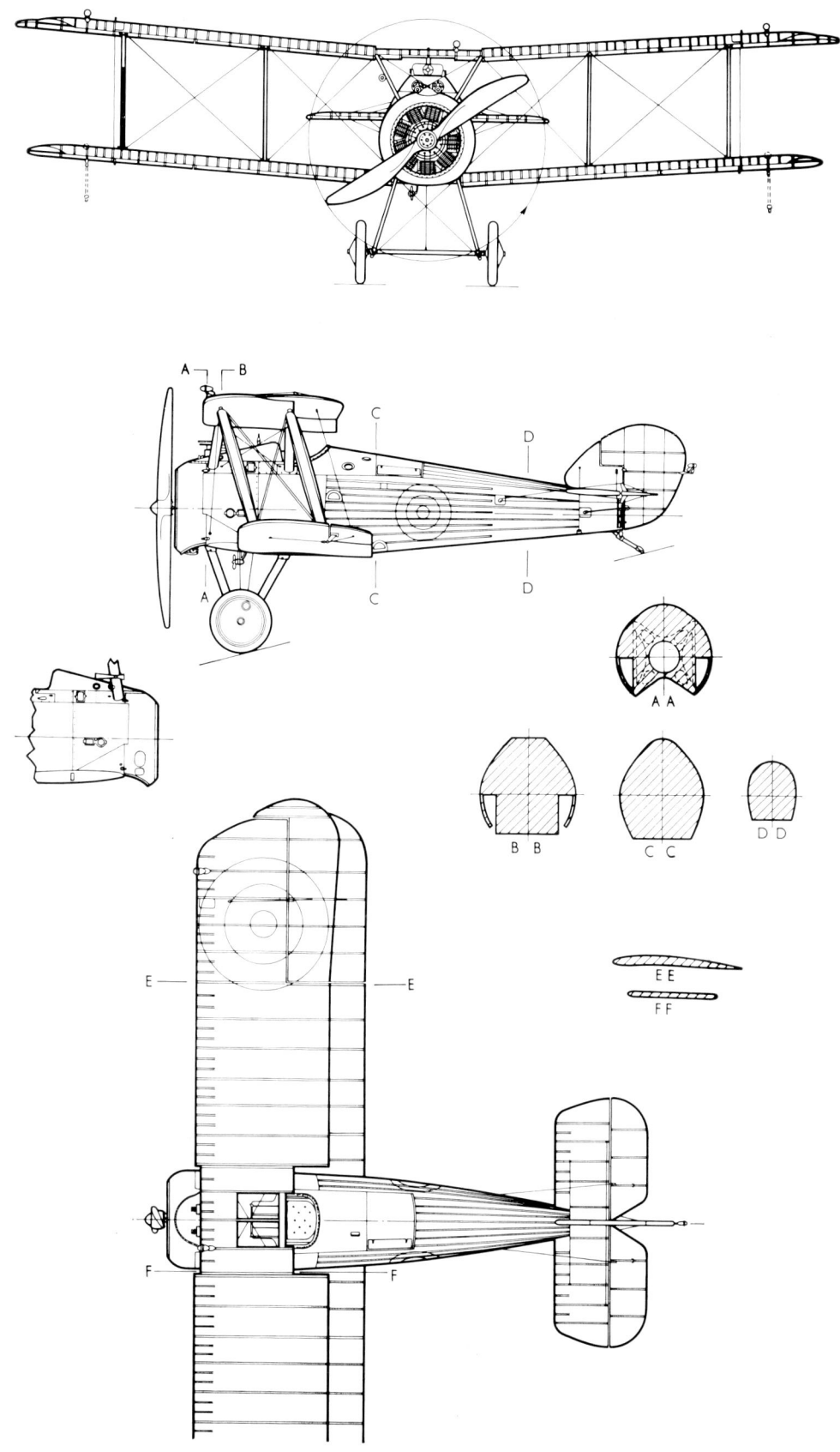

Buffalo

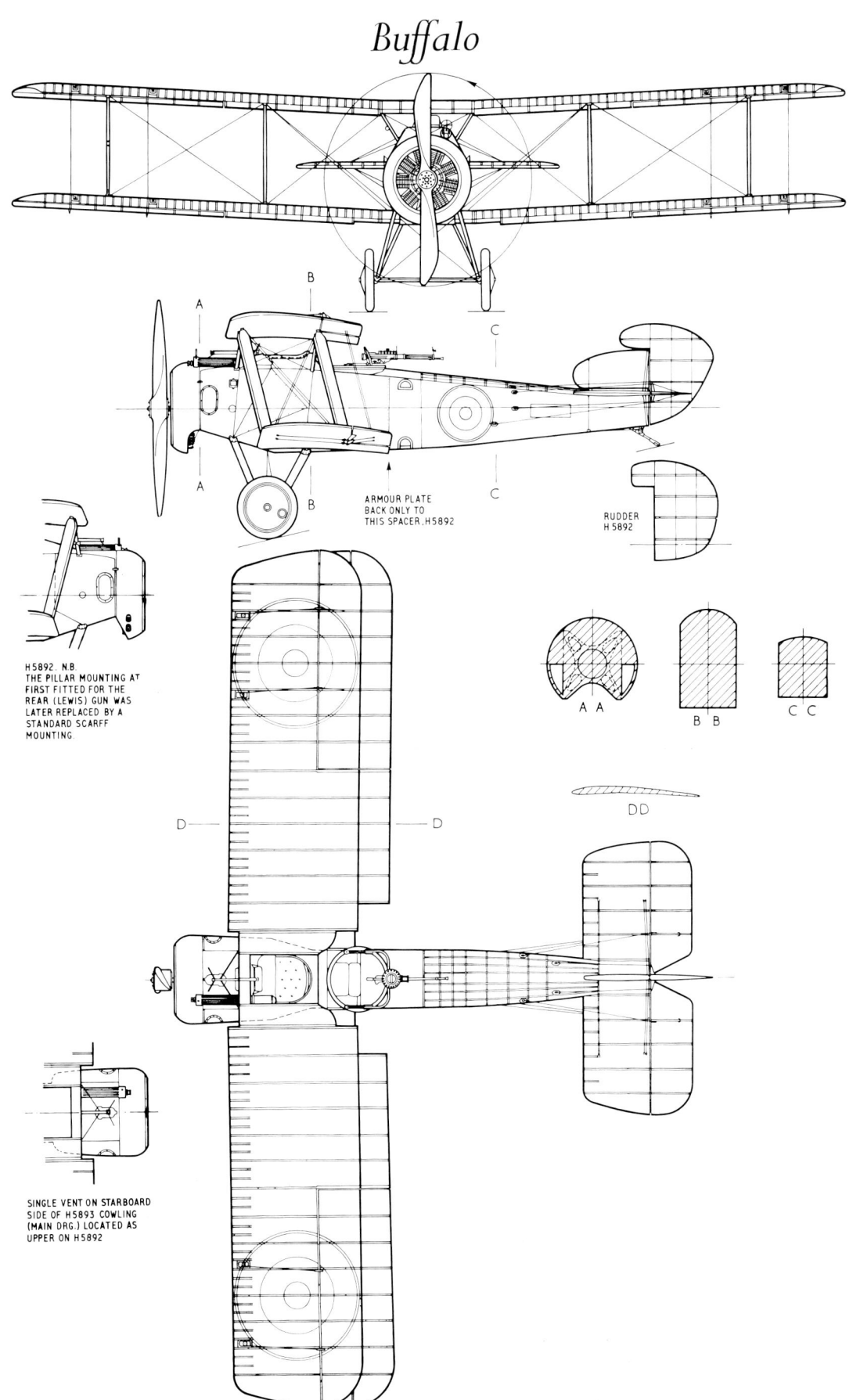

Antelope

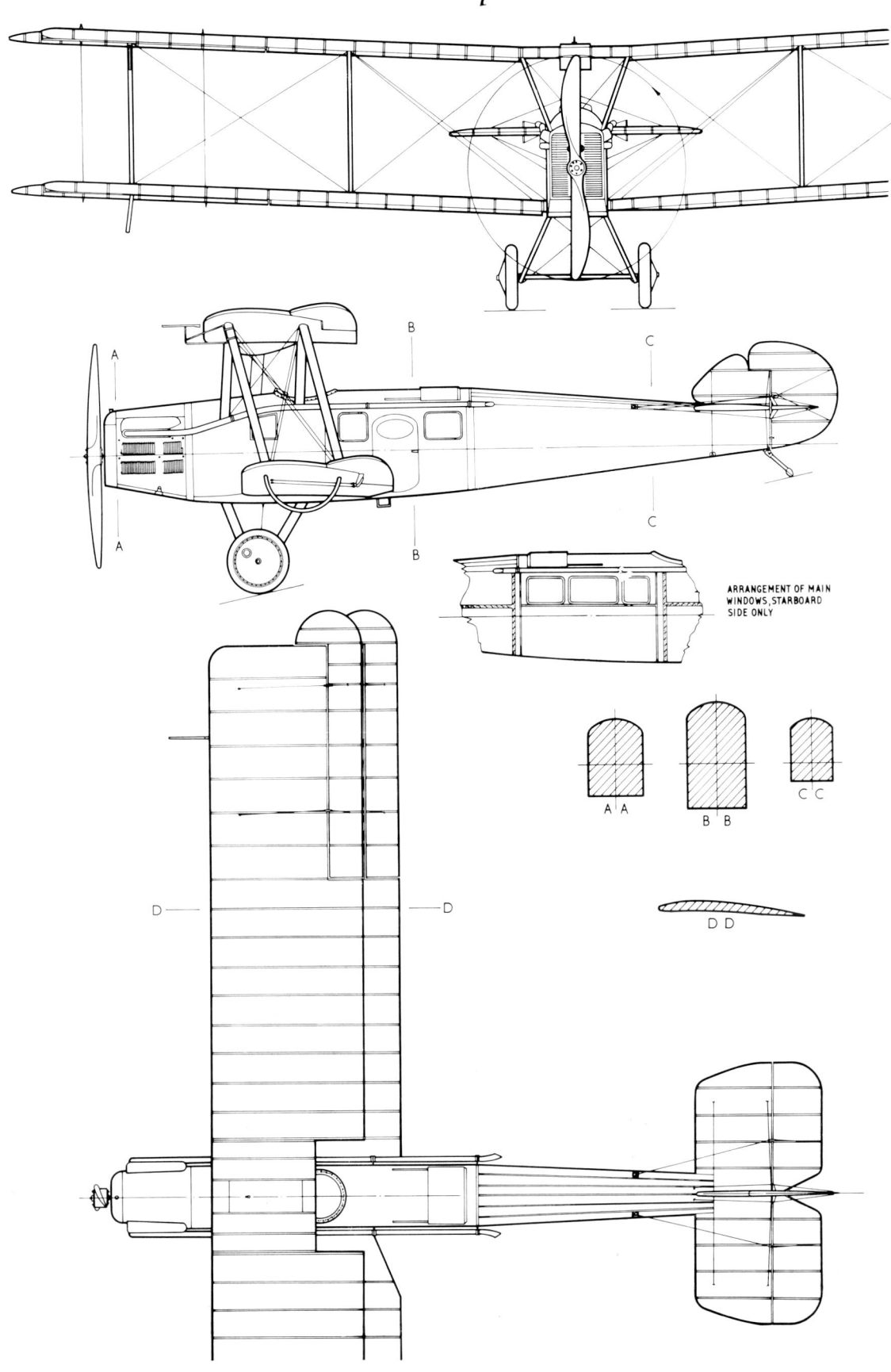

ARRANGEMENT OF MAIN WINDOWS, STARBOARD SIDE ONLY

Swallow

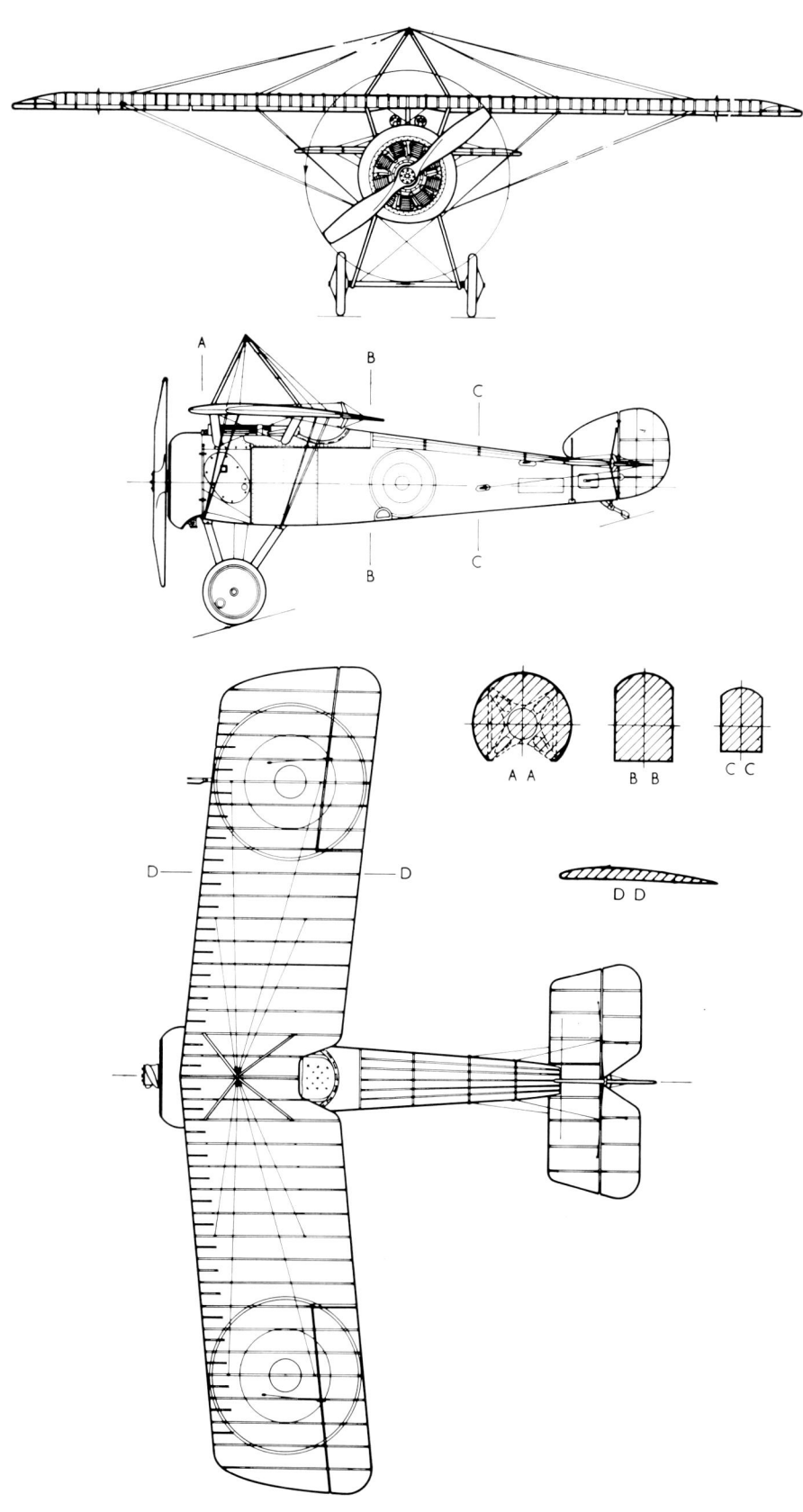

Gnu

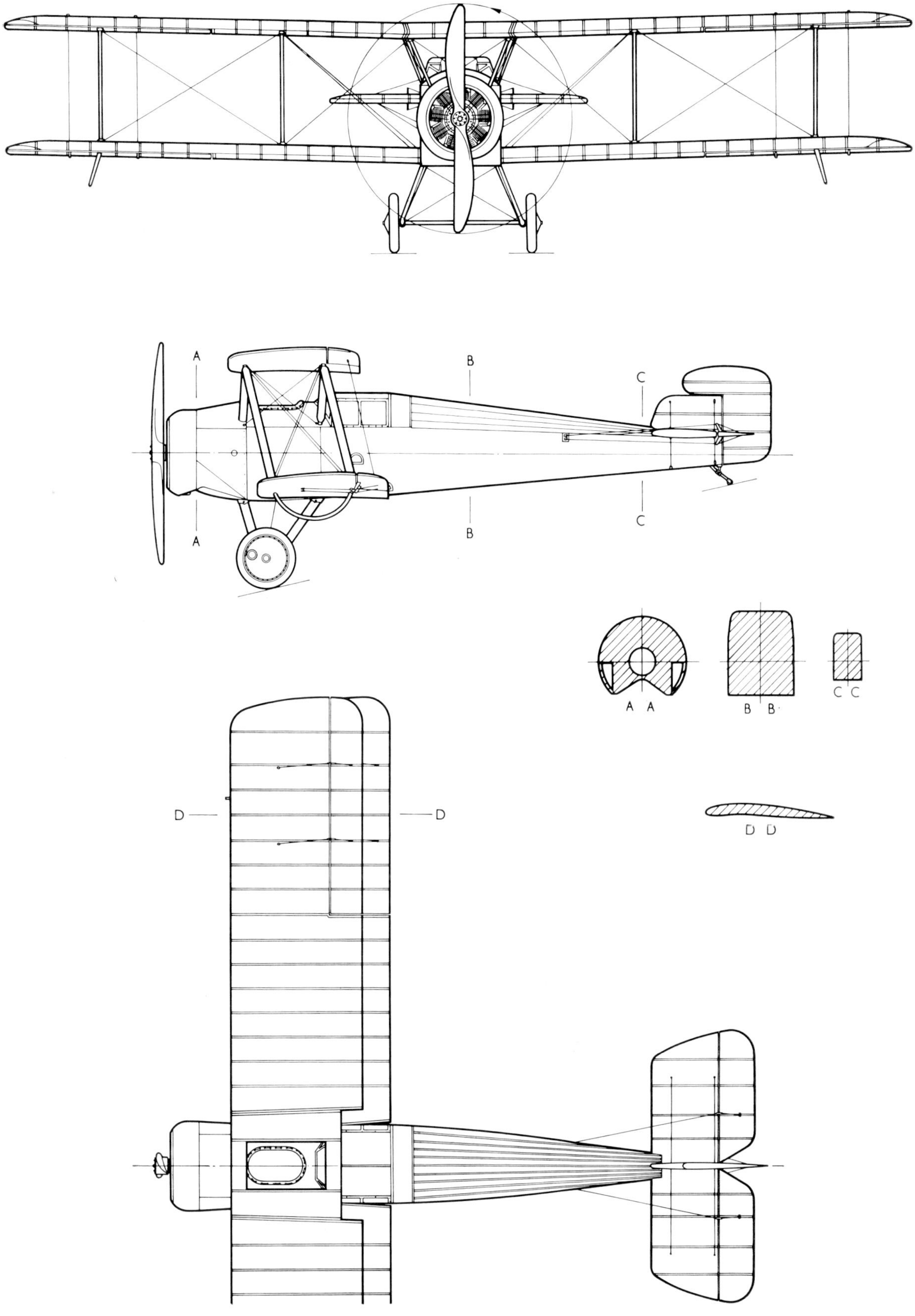

T.F.2 Salamander

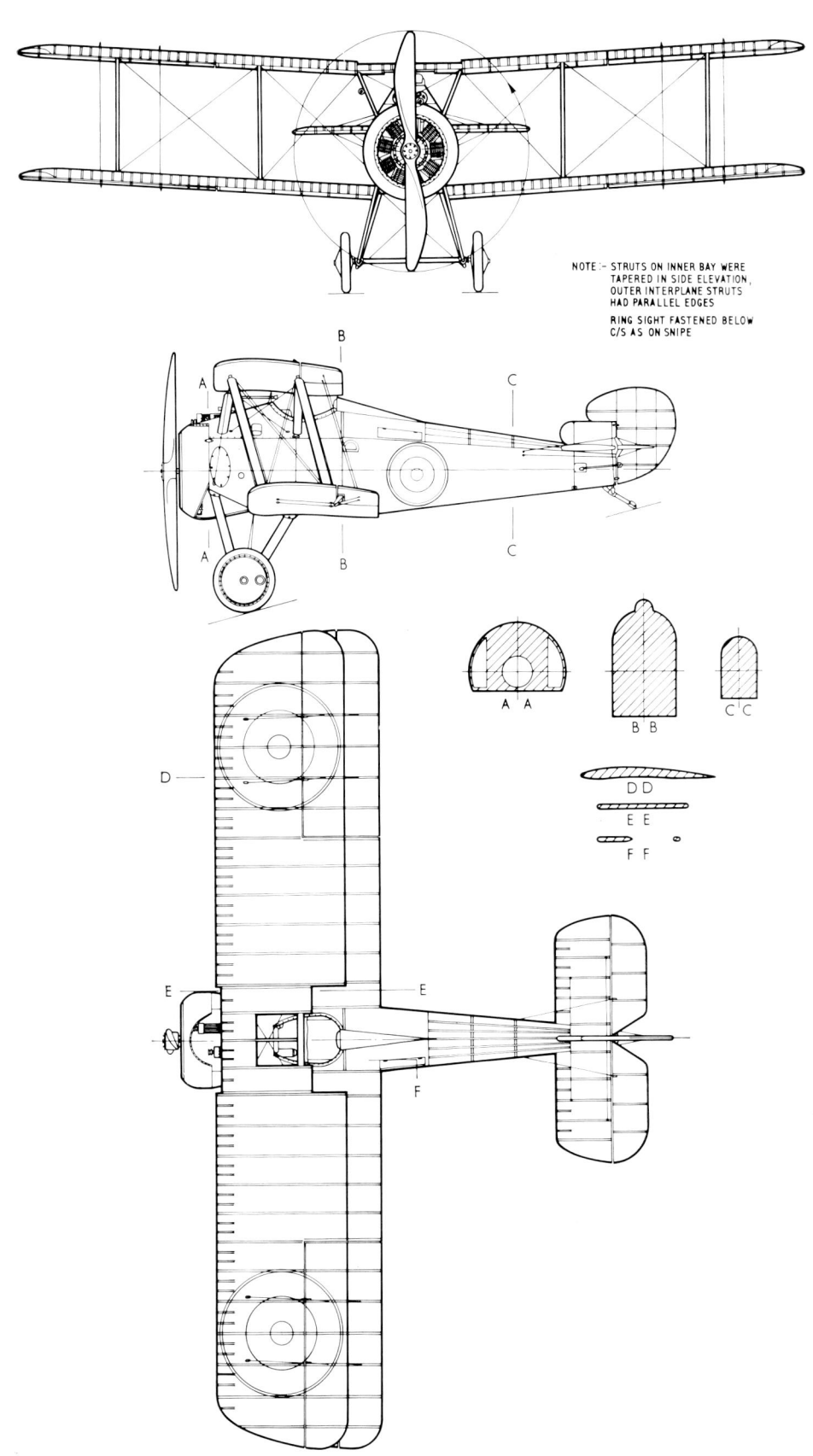

Grasshopper

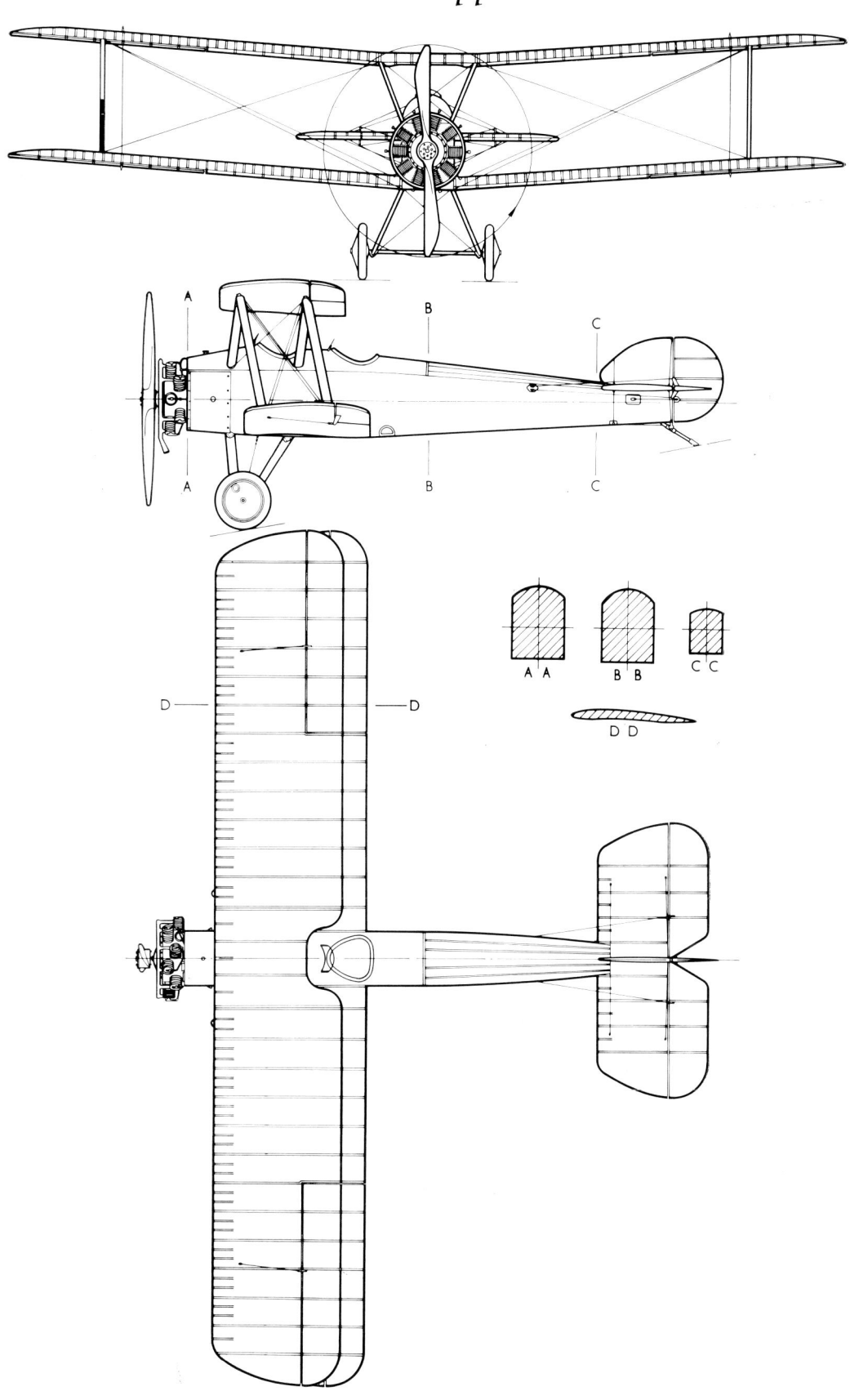

Dove

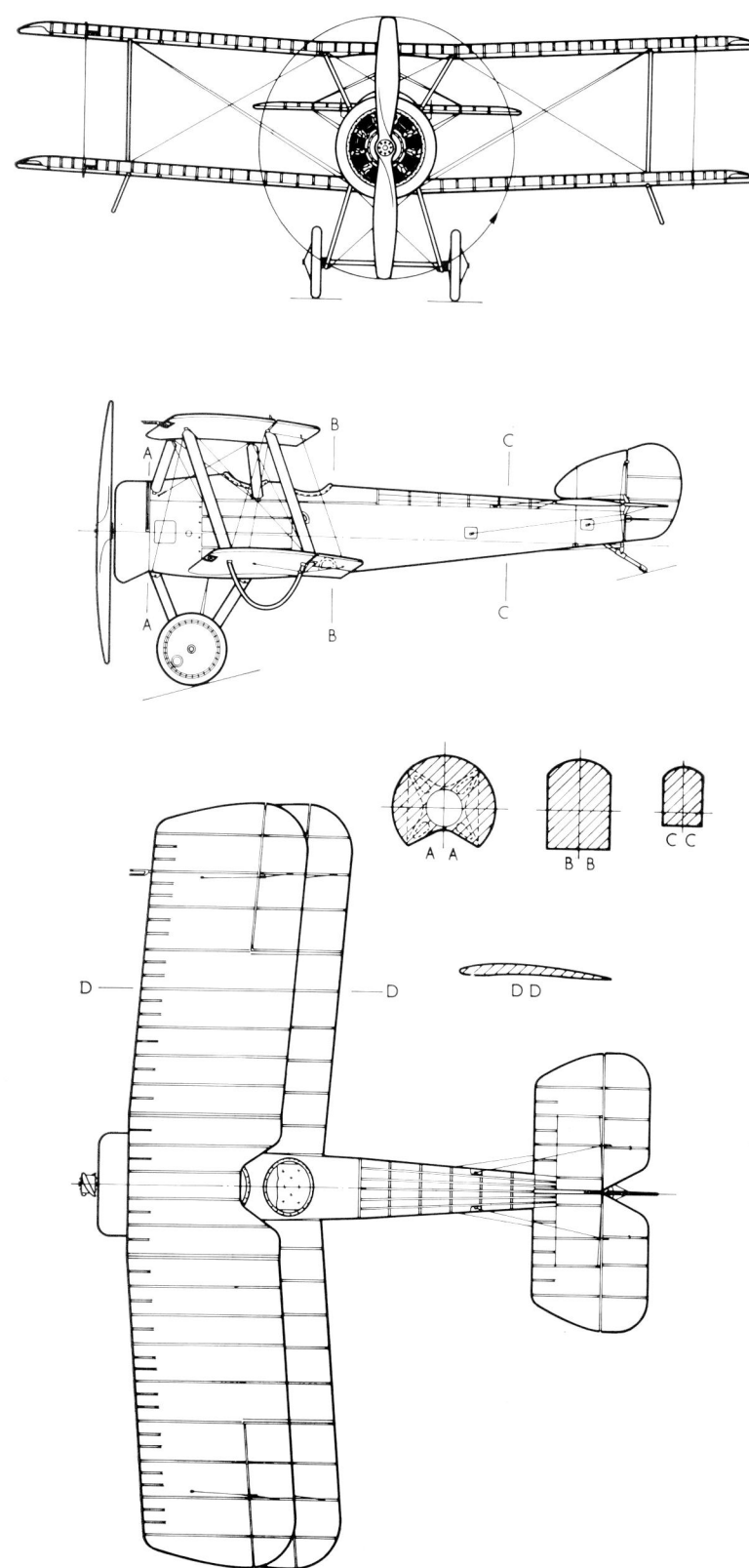

Wallaby

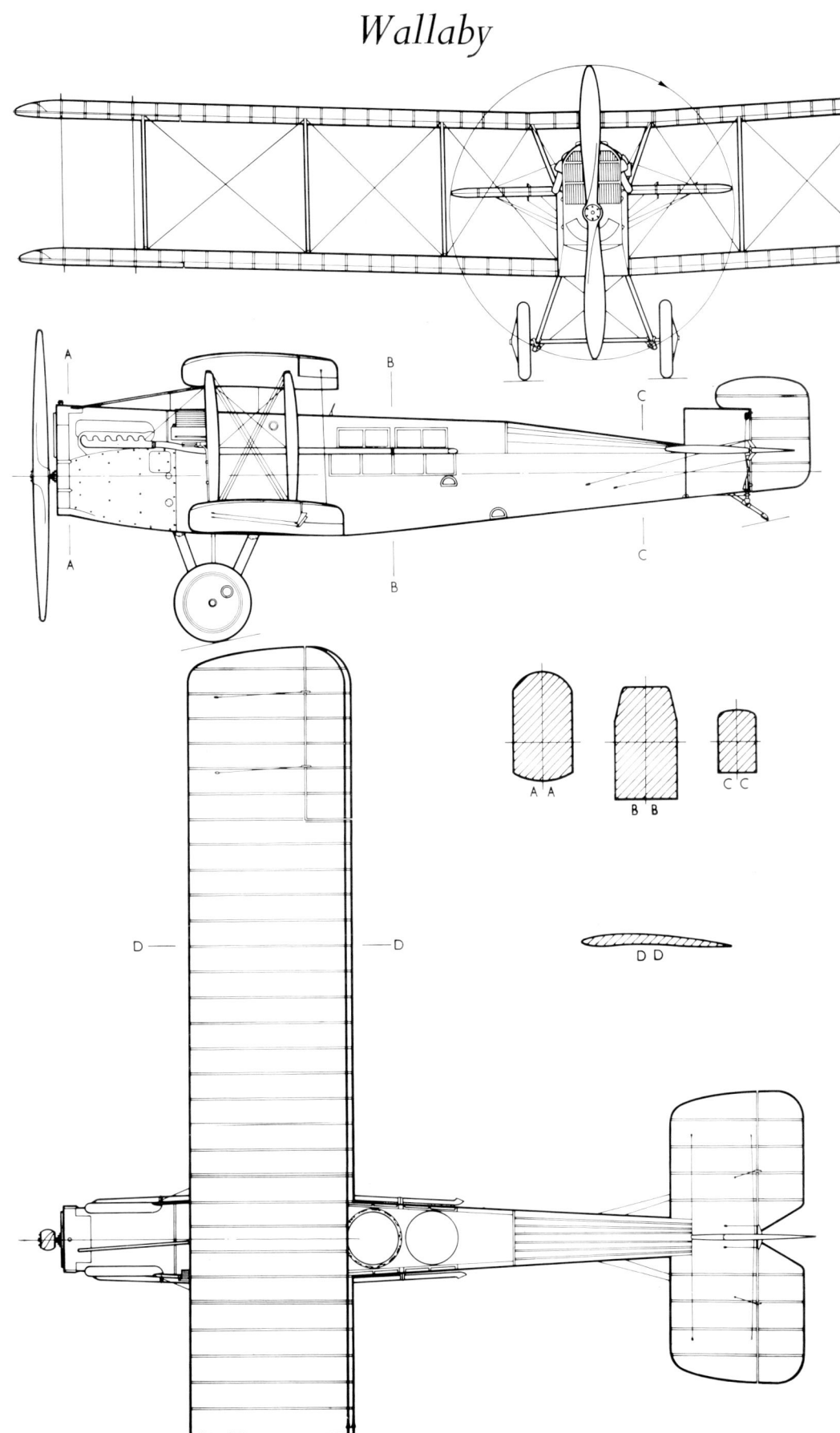

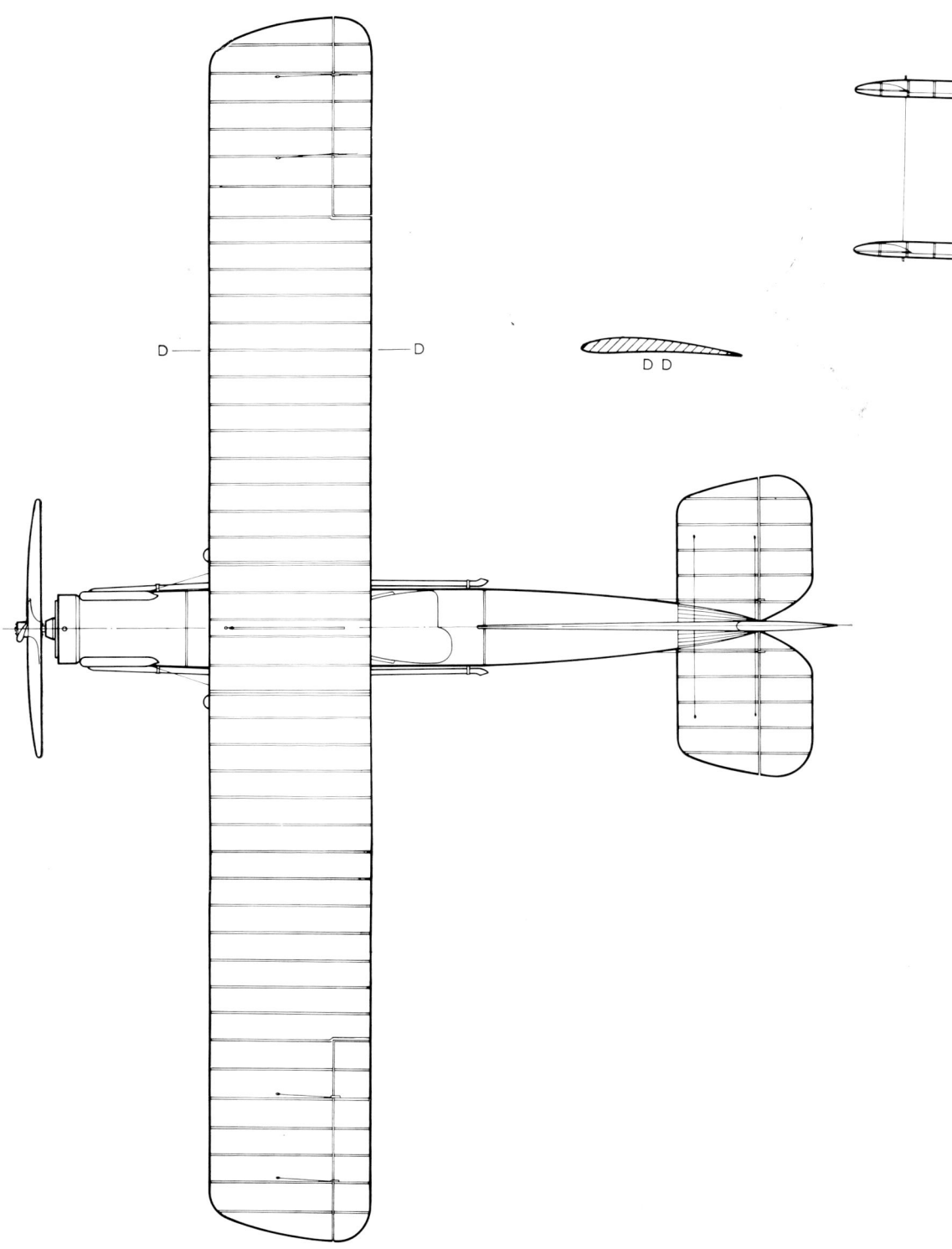

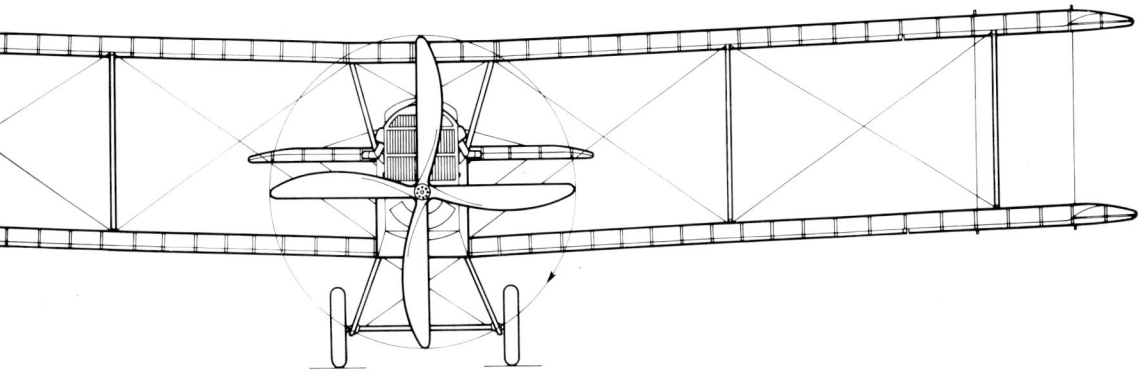
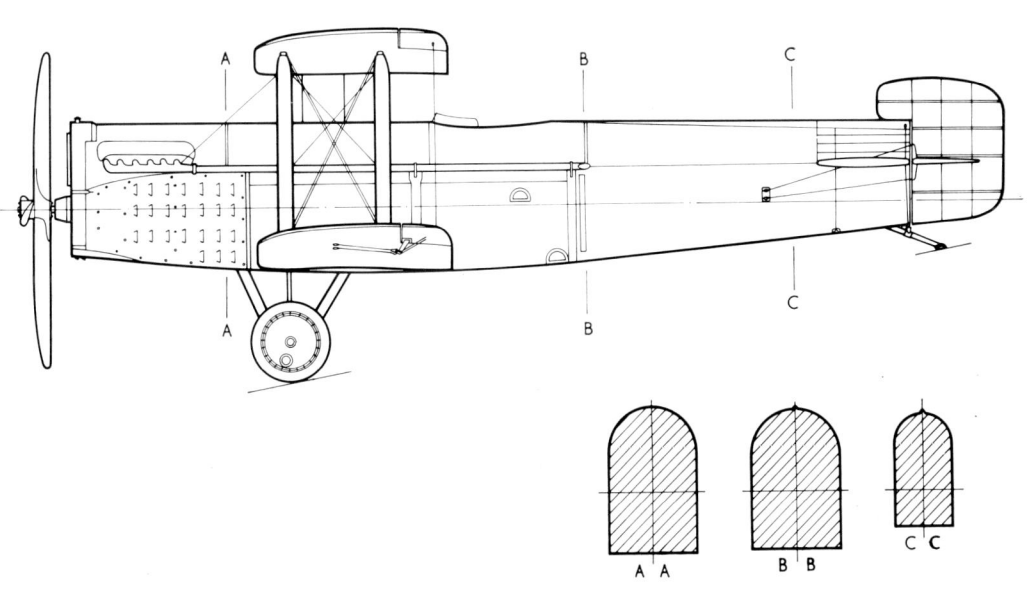

CAMEL CRASHES

Top: Forced landings at Noyelles-sur-l'Escaut by D9638 of No. 70 Squadron, October 8th 1918, and by a No. 208 Squadron machine adjacent a captured German reserve trench, September 4th 1918. Middle: B2321 of No. 210 Squadron forced-landed in a shell-hole, and a No. 73 Squadron Camel with a shattered wing, the result of a mid-air collision; date not known but evidently before March 22nd 1918, when the squadron marking changed to two bars. Bottom: A No. 43 Squadron Camel being examined by Germans on the edge of a cratered British airfield, and an Australian cadet surveying the result of a heavy landing on F1343 at Minchinhampton.

CAMEL CAMEOS

Top: Presentation Camels B3850 'LEPOQO' (left) and 'Manchester, INDIA' (right). Middle: An experiment in creating optical illusions as a form of camouflage, using a Camel fuselage, at Orfordness in 1918 and the standard scheme as represented by Camels of No. 73 Squadron at Humieres, April 6th 1918. (The censor has scratched a serial number from the negative; Bristol Fighter landing in distance). Bottom: King George V with Lord Stanfordham and Lt. Col. Clive Wigram examining a Camel under repair at Valheureux (possibly No. 3 Squadron), March 29th 1918, and mechanics removing the engine from a wrecked Camel at Aircraft Repair Depot.

Camel

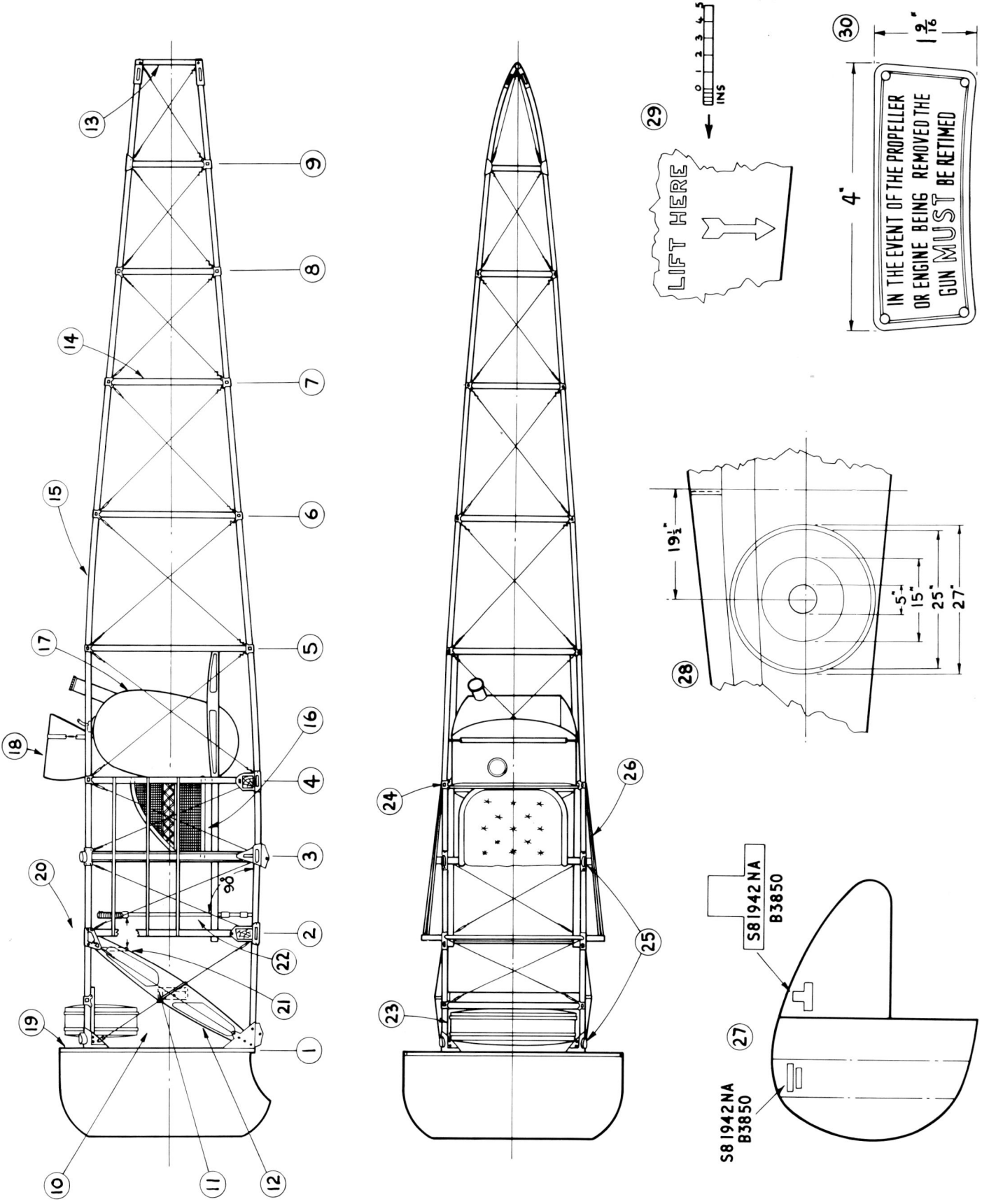

Camel

1-9 Fuselage spacers, wire braced
10 Aluminium sheet
11 Rear engine mounting
12 Wooden diagonal member
13 Steel rudder post
14 Spruce spacers
15 Ash longerons
16 Wicker seat; lap strap or Sutton Harness
17 20 gallon main fuel tank
18 5 gallon gravity fuel tank
19 Steel tube for engine mounting
20 Steel support for gun mounting
21 Instrument panel
22 Control column
23 5 gallon oil tank
24 Steel fittings
25 Location for centre-section struts
26 Stringers covered with plywood
27 Part Number positions, black on white
28 Fuselage roundel dimensions
29 Handling instruction dimensions
30 Brass instruction plate—engine cowling
31 Eyelets, 1½" pitch, starboard only
32 Serial Number, 4¼" high, black and white
33 Patches over control wire fairleads
34 Fabric around sternpost, stitched starboard
35 Fabric joined here
36 Fabric tacked, covered with 1" tape
37 Part Number position, black on white
38 Brass brads, covered by half-round cane
39 Cockpit rim of laced leather
40 Aluminium panels
41 Lug
42 4" × 3" patch
43 9" × 6" patch
44 Rim secured by two rows of rivets
45 Step, port side only
46 Eyelets, 1½" pitch, starboard only
47 Ply over fabric
48 Brass brads, 2" spacing
49 Fabric stitching detail
50 Steel tube
51 Aluminium eyelets
52 Fabric over tubing
53 Fabric patches
54 Fabric doped to final spacer
55 Undercarriage bracing wire slots
56 8G wires
57 Steel tube/wire, between top of u/c legs
58 Hole
59 Hole for wire to spreader bar
60 Fabric tucked under

Camel

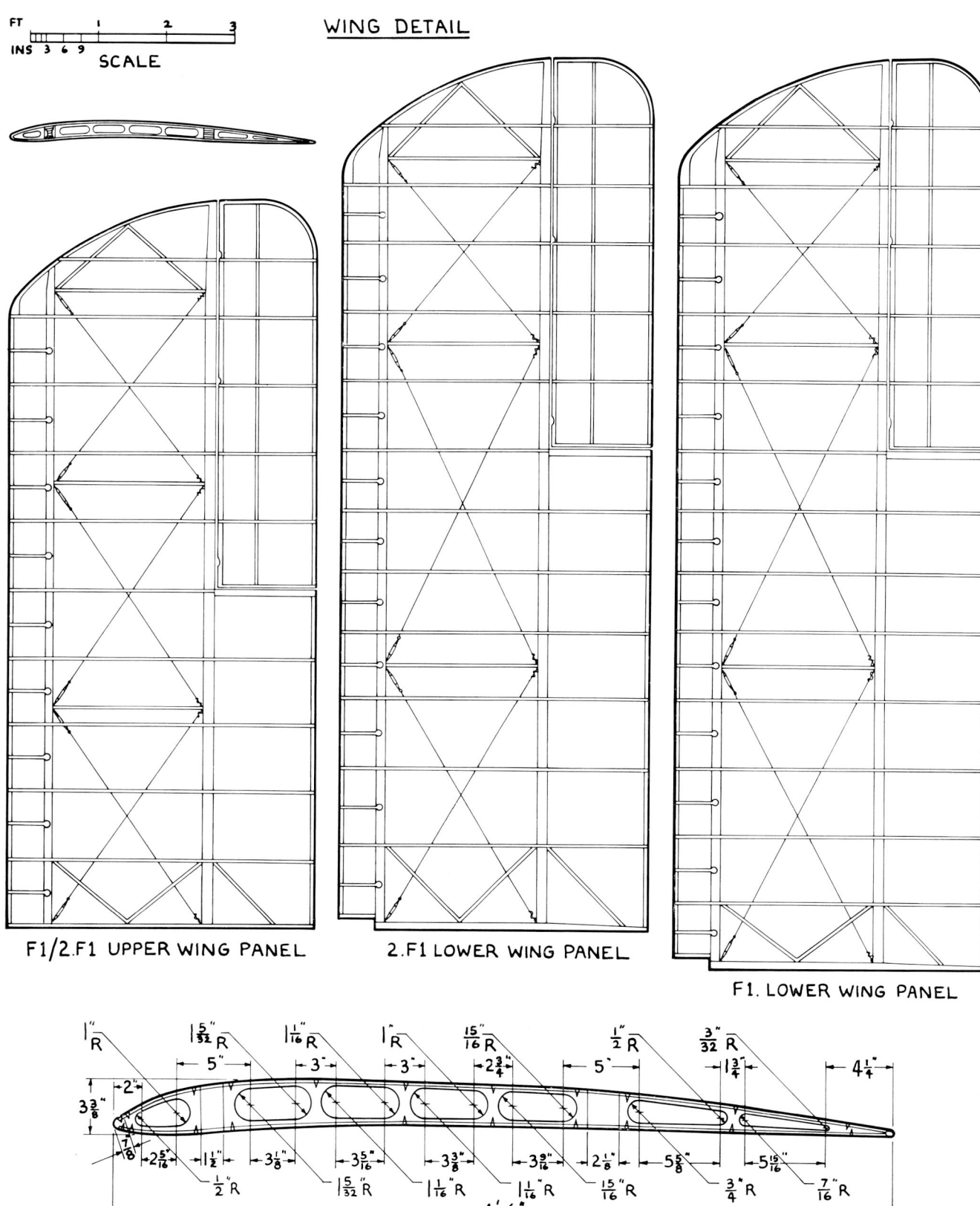

Camel

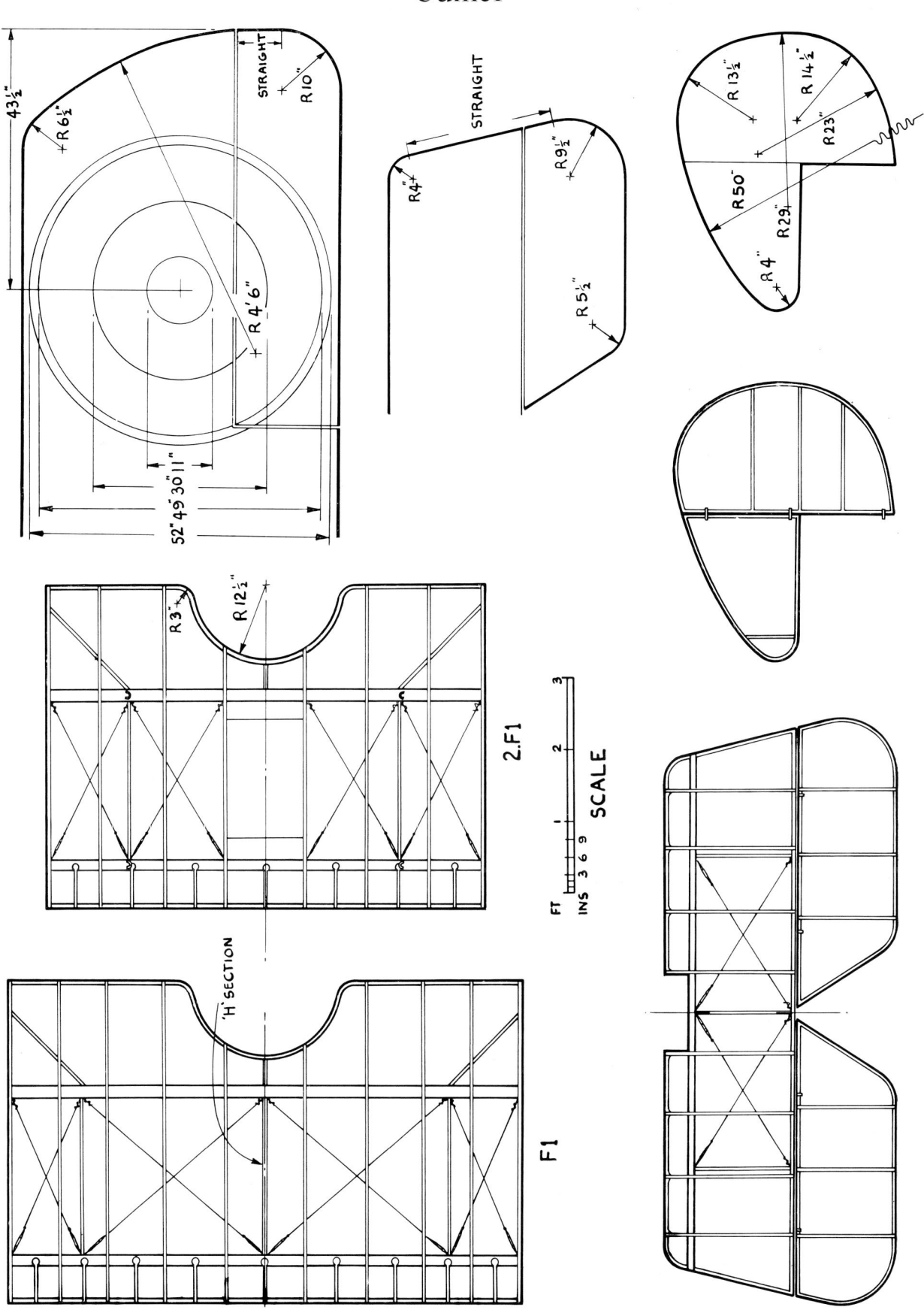

Camel

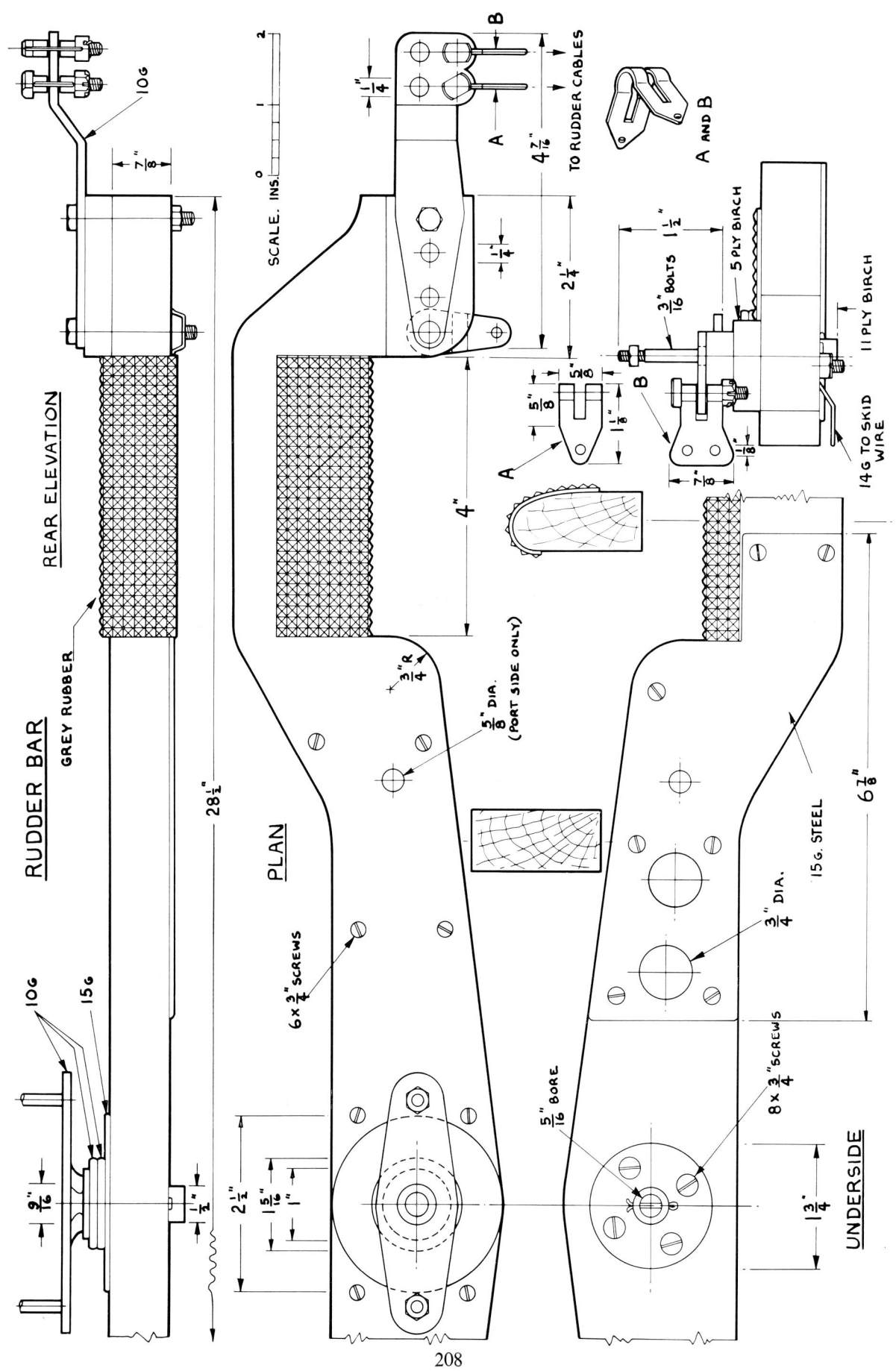

Camel

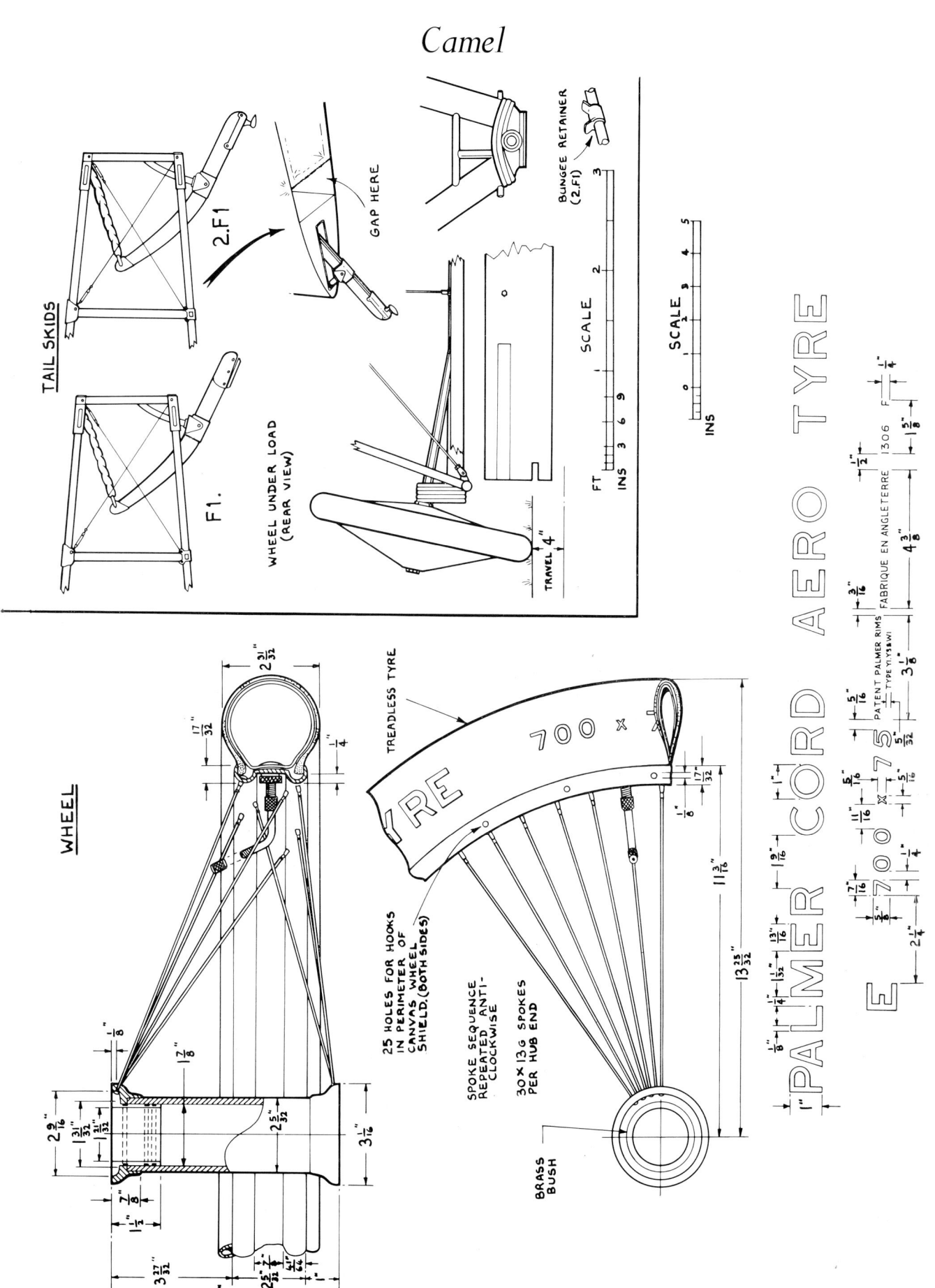

Type-by-Type Review

This review of Sopwith aircraft is designed to give guide to the reader where, in the book, reference information can be found on Sopwith aircraft types, and to include additional detail information not given in the narrative.

Constructional data and performance figures are tabulated on pages 230-237 to enable a comparison between types as well as avoiding repetitive qualifying words. In this way maximum use is made of the space available.

All serial numbers quoted are service numbers as the Sopwith Works did not allocate works numbers. For drawing page references, see Contents List, pages 4 and 5.

SOPWITH-MODIFIED BURGESS-WRIGHT

This Wright Biplane, built by Burgess was purchased by Sopwith in America (see Chapter 4) and formed part of the original Sopwith School of Flying equipment as related in Chapter 5 where it is further illustrated. The original 50 h.p. Gnome engine, was replaced by a 35 h.p. Green and then a 40 h.p. A.B.C. engine. Known dimensions are: Span $38\frac{3}{4}$ ft., Length $29\frac{1}{2}$ ft., Gap $5\frac{1}{4}$ ft.

Sopwith Modified Burgess Wright Biplane at Brooklands.

SOPWITH-WRIGHT BIPLANE

As related at the beginning of Chapter 8, where this aircraft is also illustrated, a tractor biplane was produced, using wings of Wright camber and form, powered by a 70 h.p. Gnome engine from a Bleriot.

Two built as detailed on page 32. The first was rebuilt in the Sopwith Works November 1913 and both, by that time, had 80 h.p. Gnomes.

Sopwith-Wright with covered fuselage, Brooklands, 1912.

TRACTOR BIPLANE

The first aeroplane completely designed by the Sopwith team this tractor biplane evolved from the Sopwith-Wright. Successive improvements included a change from wing-warping to aileron control. Originally called the Three-Seat Tractor Biplane, the seating reference was omitted later since requirements were for two-seaters.

Pre-war construction and flying is dealt with in Chapters 6 and 7 and wartime service in Chapter 11.

Known deliveries are: Nos. 103 and 104, No. 906 (ex-demonstration aircraft) to the Navy and Nos. 248, 315, 319, 324 and 325 to the Army.

The drawings, produced after the prototype was built, were by R. J. Ashfield in 1913.

Tractor Biplane in its revised form with enlarged rudder and tail skid in place of tail guard.

TRACTOR FLOATPLANES

The first of the Sopwith floatplanes was on 100 h.p. Anzani-powered adaptation of the Tractor Biplane to Admiralty requests, the first being delivered in June 1913 as described and illustrated in Chapter 7. Of the 3 built, Nos. 58-60, the first is known to have had war service.

Tractor Seaplane (100 h.p. Anzani) in service.

BATBOAT 1/1A (90/100 h.p.)

The original Batboat, shown being built on page 32 had a 90 h.p. Austro-Daimler and was wrecked as related in Chapter 6. However, the Admiralty were impressed to order two examples, Nos. 38 and 118. On June 28th No. 38 was tested by Lt. A. W. Bigsworth fitted with a 50 candle-power 4-volt car bulb on one wingtip and a landing light in the bow shining downwards and forwards. With

Bat Boat No. 38 of R.N.A.S. in original form.

lights shining, it was flown by Flt. Cdr. Travers over the fleet at midnight July 17th during the Royal Naval Review at Spithead. The aircraft remained based on Calshot until written-off March 13th 1915.

A new amphibious version of the Batboat was planned with a 100 h.p. Green engine as the Type 1A, to Drawing D0001, but evolved with twin fins.

BAT BOAT (200 h.p.)

Sometimes called the Bat Boat No. 2, the 200 h.p. Canton Unné (Salmson) engined flying boat was a complete re-design, and was the third main type of Bat Boat built. Details and illustrations are on page 38.

Bat Boat in its amphibian twin-fin form.

TABLOID

The development and war service of the Tabloid is covered in Chapters 7 to 13. The design and construction owed much to Harry Hawker and Fred Sigrist, but the final version, passed by R. J. Ashfield on June 16th 1914, was known as the Sopwith Scout. This had a more streamlined form, revised fin and rudder and decreased gap.

In February 1915 Commander Samson had a special Lewis gun mounting fitted to the top wing of a Tabloid, and at least one had a Lewis gun fitted to fire through the propeller arc—the airscrew being protected by deflectors.

No documentation of Tabloid production can be traced. Service numbers traced are 167-168, 169 ex-prototype. 1201-1213 developed version for the R.N.A.S. known as Sopwith Scout, and 326 and 394 of the R.F.C.

Improved Tabloid also known as Sopwith Scout.

SOCIABLE (TWEENIE OR CHURCHILL)

So-called by its side-by-side seating, this enlarged Tabloid, R.N.A.S. No. 149, was sometimes known as Tweenie. This aircraft, introduced in Chapter 7, and drawn on page 174, was also known as the Churchill having been sponsored by Winston Churchill when First Lord of the Admiralty. He had minuted the Director of the Air Department on December 2nd 1913 asking if one of the Sopwith biplanes at Eastchurch could be fitted with dual controls. In a follow-up minute on December 21st, Churchill virtually decreed the specification for the Sociable.

The Sociable photographed in March 1914.

GORDON BENNETT RACER

Sopwith had apparently prepared two racers for the Gordon Bennett race. One with a special slim-line fuselage as related on page 46 and a more conventional one based on the Tabloid. The Admiralty raised contract C.P. 60619/14 for these to be completed as scouts as mentioned in Chapter 11. No. 1214 emerged as a modified Tabloid with 100 h.p. Gnôme, armed with a Lewis gun mounted outboard on the starboard fuselage side, and with a propeller fitted with bullet deflectors at Dunkirk. No. 1215 appeared with the thin fuselage as illustrated.

Gordon Bennett racer with R.N.A.S., 1194.

CIRCUIT SEAPLANE 1913

Built hurriedly for the "Circuit of Britain" race in 1913 and wrecked in the race, few details of the construction of this aircraft have survived; its flying life is detailed in Chapter 7. A rebuilt aircraft was ordered by the Admiralty under contract C.P. 31415/14 as illustrated page 39.

FLOATPLANES (120 h.p., 135 h.p., 200 h.p.)

Two floatplanes were ordered from Sopwith under Admiralty contract C.P. 30775/14 and three to contract C.P. 37885/14 with the service numbers 137-138 and 157-160. A 120 h.p. Austro-Daimler was fitted to the first, a 200 h.p. Canton Unné to the second and the remainder had Canton Unnés of 135 or 200 h.p.

The original 1913 Circuit Seaplane.

SEAPLANE TYPE C

The designation Type C to the large seaplane shown at the top of page 39 suggests that the earlier two float-planes listed were Types A and B respectively. This is believed to be R.N.A.S. No. 127 which had a 200 h.p. Canton Unné engine.

CIRCUIT SEAPLANE 1914 (D3 or DAILY MAIL)

For the 1914 "Circuit of Britain" race preparations an aircraft was built to drawings marked D3 and tested as a landplane. (See pages 44 and 46). The outbreak of war led to a cancellation, but the basic design evolved as the Two-Seat Scout and the Type 807 floatplane.

Circuit of Britain seaplane, on test at Brooklands as a landplane.

PUSHER SEAPLANE

A large pusher seaplane, of a type that became known as "Gunbus", was ordered by the Admiralty as their No. 93 under contract C.P. 01717/13 and received the Sopwith designation Hydro Biplane Type S. Powered with a 120 h.p. Austro-Daimler it had span of 80 feet. Although apparently not successful, it engendered an order from the Greek Navy for a modified version powered by a 100 h.p. Anzani engine. (See Chapter 7).

PUSHER SEAPLANES (GUNBUS)

Pusher seaplanes were ordered by the Greek Government and initial deliveries were made (see Chapters 9 and 10). When war came the aircraft to Greek orders were taken over by the Admiralty as Type 880 and altogether eight of this type were taken on strength with 100 h.p. Gnômes.

PUSHER LANDPLANES (GUNBUS)

The first Sopwith aircraft to be sub-contracted (see Robey page 158), six were built by Sopwith as Type 806 (Nos. 801-806). The drawings for Robey's production were made by R. J. Ashford and approved by Herbert Smith May 6th 1915 (see our drawing page 173). In late 1915 Nos. 801, 804-805 were used on training duty at Chingford and then Nos. 804-806 were allotted to the Clement Talbot works, presumably for mechanic training.

Sopwith references to pusher type seaplanes give the designations SPG (Sopwith Gun machine) when fitted for armament.

Sopwith Gunbus, 150 h.p. Sunbeam.

TWO-SEAT SCOUT (SPINNING JENNY)

The Spinning Jenny evolved from the Sopwith Tandem Two-Seater Biplane entered for the 1914 Circuit of Britain race, which was designated D3. Its association with the aircraft designed originally for the *Daily Mail* competition is such that the Admiralty in ordering 24 (Nos. 1051-1074) referred to them as Sopwith Daily Mails.

Winston Churchill examining Two-seat Scout, Hendon, 1915.

Folder Seaplane (100 h.p. Gnôme).

FOLDER SEAPLANE (ADMIRALTY TYPE 807)

Like the Two-seat Scout, the Type 807 evolved from the same D3 machine. Four (Nos. 807-810) were built by

Admiralty Contract C.P. 02142/14 and eight (Nos. 919-926) to C.P. 02155/14. The service of the second of these floatplanes is dealt with in Chapter 12.

TRACTOR SEAPLANE (ADMIRALTY TYPE 860)

Eighteen Sopwith 860s were built (Nos. 851-860 and 927-938 of which 933-934 and 936-937 were cancelled). These two-seat floatplanes, powered by 225 h.p. Sunbeam engines, did not all have equal span wings as illustrated. Nos. 854, 857 and 860 are referred to in Chapter 12, Nos. 851-852 were written off in March 1917 and Nos. 931-932 were reduced to spares in the Supermarine works.

Type 860 seaplane No. 928 (225 h.p. Sunbeam).

Flying shot of Type 860 Seaplane.

SCHNEIDER CUP SEAPLANE

Virtually a floatplane version of the Tabloid, the success of the prototype at the 1914 Schneider Cup contest is dealt with in Chapter 8, and other references to this machine are made on pages 44-45. War service details and illustrations will be found in Chapters 12-13, 15-16 and 26.

BABY SEAPLANE

Developed from the Schneider the history of the Baby is traced and illustrated in Chapters 13, 15-17, 25 and 34 and is drawn on page 179. Production details of all Baby-type floatplanes are given under Blackburn, Fairey, Parnall and Sopwith, pages 151, 154, 157, 159 and 160.

Apart from British service use the airframes of Nos. 8125, 8197, 8204 and 8209 were sent to Canada. Plans were for the aircraft to be held at Halifax for Royal Canadian Navy vessels. The French, who also used a Hanriot single-seat floatplane, took delivery of three Babies, 8128-8129 with 110 h.p. Clergets and 8185 without an engine. Nos. 8214-8215 were shipped to the Italian Government in October 1916 as pattern aircraft pending a decision to licence-build these floatplanes.

Late in 1918 the U.S. Navy Base at Killingholme acquired a few for training and four were sent to America for evaluation, being taken on U.S. Navy strength as Nos. A869-872. Earlier, in 1918, N2121 had been exhibited in America. Post-war the type was also used by Chile.

A late production Sopwith-built Baby by the Thames.

SL.T.BP

See photo and captions in Chapter 21. Not being a service aircraft, this biplane has escaped documentation. It survived post-war fitted with an 80 h.p. Le Rhône engine.

1½ STRUTTER (TYPES 9400 and 9700)

Built in its thousands, the service history of the 1½ Strutter features in Chapters 14, 16 to 22, 26, 32, 35 and 36. The R.F.C. used the Sopwith Two-Seater and a single-seat bomber version was supplied but had limited use. In the R.N.A.S. it served in two versions, the Type 9400 two-seater fighter and fighter/bomber and the Type 9700 single-seat bomber. Engines were 110 h.p. or 130 h.p. Clergets.

1½ Strutter single-seat bomber, built by Hooper.

A Sopwith built 1½ Strutter two-seater of the Belgian Flying Corps.

Production details are given in Fairey, Hooper, Mann Egerton, Morgan, Ruston Proctor, Sopwith, Vickers, Wells and Westland on pages 154 to 161. The contract airframe price (without engines, instruments) was £892.

Original drawings have not been traced and general arrangement drawing D1179 drawn by Feacey, and approved by W. G. Carter January 22nd 1917, was evidently intended for sub-contractors.

Patents granted as a result of developments incorporated in the 1½ Strutter included wind brakes (No. 110419), a device for attaching cable wires (110795) and an adjustable seat for air gunners (No. 11264) all granted November 1st 1916 in the name of the Sopwith Company and T. O. M. Sopwith.

1½ STRUTTER (140 h.p. Clerget)

Late surviving 1½ Strutters may have had 140 h.p. Clergets fitted as an alternative to the 130 h.p. Clergets, but the first was undoubtedly B2566, a Ruston Proctor-built airframe, fitted with experimental Ruston Proctor long-stroke Clerget R1630 on November 19th 1917.

SOPWITH 1A2, 1B1, 1B2 and CIVIL 1½ STRUTTERS

See Chapters 20 and 36 and all French manufacturers between pages 150-159 and data tables. One French example survives today, a 1A2. No. 88 of the Belgian Flying Corps (see Chapter 36) is the only known surviving aircraft of the 1½ Strutter type.

The sole 1½ Strutters reaching the British and Belgian civil registers, G-EAVB and O-BAJN, are both believed to have been ex-French built and both were converted to 3-seaters. Nearly fifty were taken on the French Civil Registers including one cabin conversion.

SHIP STRUTTER (British-built)

Sopwith 1½ Strutters for fleet use were to several standards. At first R.N.A.S. machines were used and by mid-September 1917 No. 9390 had been fitted with a biplane hydrovane and inflatable air bags.

To meet Admiralty requirements the Experimental Constructive Department, Grain, commenced a modification programme of 25 1½ Strutters for ship use. These had 140 h.p. Clergets installed and were fitted with detachable wings, hydrovanes, air bags and skid chassis. The first was delivered in April 1918 and the remainder were programmed for delivery at the rate of two per week.

Efforts were continually made to improve performance and during April 1918 a lightweight rear gun mounting was devised and tests were also carried out with a continuous wave W/T set. A drawing is given on page 180.

A5741, a Ship Strutter of the U.S. Navy.

SHIP STRUTTER (French-built)

Due to limited numbers of 1½ Strutters available for conversion to Ship Strutters, numbers were acquired from the French. These were initially 20 (F2210-2229) and a further 50 (F7549-7596) allotted in September 1918 for conversion at Grain. The U.S. Navy also acquired French examples for ship use in the U.S. Navy as their A5660, A5725-5728 and A5734-5750.

PUP

The Sopwith single-seat Scout of the R.F.C. and Type 9901 of the R.N.A.S., soon became universally known as the Pup (see page 61). The service history of the type is dealt with in Chapters 17 to 19, 22, 25 and 36. Production details are given under Sopwith, Standard and White-

The prototype Pup at Guston Aerodrome, Dover.

Pup pilots. Lieutenants Bush, Metson and C.G.O. MacAndrew.

Mishap to Standard-built A7313.

head on pages 160-162. The basic airframe price was £710/18/0.

The U.S. Navy had two Pups, A5655 and A5656.

SHIPS PUP

Shipboard use of Pups, the Admiralty Type 9901A, is dealt with in Chapter 26 and production details are on pages 150-151 and page 160.

PUP (CIVIL)

Considering that the Pup had been so popular with pilots, it is surprising that more were not offered on the civil market. The Aircraft Disposal Company registered G-EAVV-Z (ex-C440, C312, B1807, C438 and C540 respectively) in November 1920 but only two appear to have been used. Flt. Lt. T. Gran bought C-EAVW which he housed at R.A.F. Andover and G-EAVX was damaged at Hendon, July 21st 1921 and scrapped.

One civil Pup, believed built up from the parts of two, was G-EAVF which was scrapped in 1921. Another, G-EBFJ (ex-C242) was registered in February 1923 and scrapped the following year. Perhaps the most widely used Pup postwar was G-EBAZ ex-C1524 registered to the Whitehead test pilot Herbert Sykes. It was sold to P. T. Capon in June 1923.

BEE

See photo and captions given in Chapter 21.

Harry Hawker's unofficial runabout—the Bee.

BEARDMORE W.B.III (S.B. 3D/F)

Although normally classified as a Beardmore aircraft the W.B.III (W. B. for William Beardmore & Co.) was an adaptation of the Pup for deck use. The first 13 (N6100-6112) were designated officially S.B.3F (Shipboard Type III with folding undercarriage to facilitate stowage) and the following 87 (N6113-6129 and N6680-6749) as S.B.3D (D for Dropping or jettisonable undercarriage).

W.B.III could be distinguished from Pup by lack of dihedral

TRIPLANE (110/130 h.p. Clerget)

Service details are in Chapters 14, 17 and 18, the drawing on page 152 and production details under Clayton & Shuttleworth (pages 152-153), Oakley (page 157) and Sopwith (page 160). Pilots reported on the Triplane in glowing terms. Sqn. Cdr. C. Courtney after flying the prototype N500 on June 22nd at Dunkirk recommended that it be ordered in large quantities at once. Flt. Lt. R. Baudry, having flown it shortly beforehand found it "very light on all controls". Only Flt. Cdr. R. E. C. Pierce who also flew it that same day implied any fault, but it was the common fault of all fighters at that time—the gun; he wrote, "Flies hands off, time to attend to gun jams".

The only Triplane to serve with the R.F.C., N5430 which was later incorrectly marked A5430.

The prototype led a varied career after tests and service use by Flt. Sub-Lt. Dallas, as related, it went to No. 8 (Naval) Squadron where it was crash landed on its first flight. After repair and further service on the Western Front, it was written off at Dunkirk on December 17th 1917.

N504 the second prototype, flying in late August went to France on November 14th 1916; this, too, served for a while in No. 8 (Naval) Squadron. Sopwith's first production aircraft, N5420 went to Clayton and Shuttleworth's at Lincoln as a pattern aircraft for the latter's production.

Triplane at a training unit.

TRIPLANE (150/200 h.p. Hispano)

Two special triplanes were built, N509 and N510 to contract C. P. 133540/16 and were fitted with 150 and 200 h.p. Hispano engines respectively. Details in tables and drawing on page 183. Both aircraft were written off in 1917 at Manston and Eastchurch respectively.

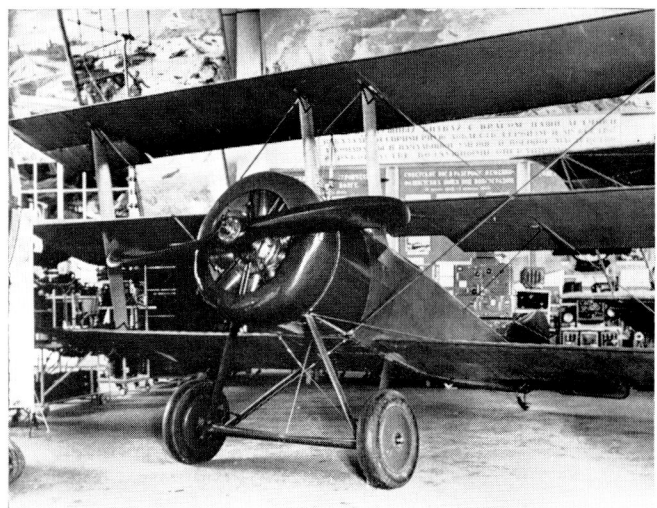
The Triplane illustrated on page 61 seen preserved in Russia.

Triplane 150 h.p. Hispano-Suiza.

Triplane 200 h.p. Hispano-Suiza.

L.R.T. Tr.

This large three-seat Triplane, powered by a 250 h.p. Rolls-Royce Eagle I engine, had a top gun position in which Ernest Newman sat for the first test with Hawker. The structure moved out of true on its first landing and was modified but only the single example was built. It was not accepted for service and the only details known are span of $52\tfrac{3}{4}$ feet and length $35\tfrac{1}{4}$ ft. See photo page 96.

CAMEL F.S.1 or CAMEL FLOATPLANE

Evolved from the Sopwith Baby, the Camel F.S.1 floatplane is the missing link in the chain of Camel development. The drawing, depicted on page 186 was approved

The L.R.T.Tr. triplane home defence project.

by Herbert Smith late in 1916 and such data as is available is given in the tables. A report in April 1917 stated "In order to proceed with the trials of the Sopwith Camel seaplane, the first of which (evidently the F.S.1) was wrecked on trials, it has been decided to convert the land machine of this type, with split fuselage, to a seaplane (presume Camel 2F.1 N5). Handed to Sopwith's for floats to be fitted". Two weeks later it was decided to keep the Camel at Grain as a land machine with plans to convert it to a floatplane for which Sopwith supplied floats. In July arrangements were being made for trials of a Camel floatplane for use on the Belgian Coast, but evidently this project was dropped.

CAMEL F.1 Prototypes

The Camel, as originally devised, was to have dihedral on both wings, but it emerged with a straight top wing and a pronounced dihedral on the bottom planes that became a characteristic of the type. The first F.1 prototype (unnumbered) had a 110 h.p. Clerget and twin Vickers guns. It was passed by the Sopwith Experimental Department on December 22nd 1916 but not until March 1917 was the decision taken to put it in production with twin Vickers. The prototypes are illustrated in Chapter 23.

Another prototype, the F.1/1 (also un-numbered) was known as the Taper-Wing Camel. During testing in May 1917 (figures are in performance tables) a decision was taken not to place it in production.

The F1/1 Taper Wing Camel with I struts.

A further prototype, the F.1/2 is believed to be the first naval prototype N517. This naval aircraft first flew at Brooklands on February 26th 1917 and used for various tests. It was sent to Dunkirk in the summer and was flown on June 6th by Flt. Sub-Lt. Edwards who took a Nieuport up between two short flights on the Camel to gauge its potentialities. It was crashed at Dunkirk August 20th.

A final unnumbered prototype, the F.1/3 powered by Clerget 9B No. 2730, arrived at Martlesham Heath for test March 24th 1917. Later, in May, it was tested at the same location with 110 h.p. Le Rhone 9J No. 100508/WD7917 which in July was changed again to an experimental Long Stroke Clerget engine.

The second naval prototype, N518, was the first Camel with a B.R.1 engine and was tested in May 1917 with LP2850, AB705 and AD644 propellers. From August 25th 1917 it had 150 h.p. Clerget 42953/A35213 and before the following November was fitted with an experimental B.R.1 with a 5:2 compression ratio.

CAMEL F.1 (Production)

The origin and service details of the Sopwith Camel are given in Chapters 23 and 28, with production details in Sopwith's Fortyfold under the various contractors to which the table on page 164 gives a quick guide.

Due to a spate of tail skid breakages in July 1917, the original solid wood skid was replaced in August/September by a skid of laminated plywood. By November 1917, units in the Field had carried out several modifications. Owing to the possibility of the fittings in the centre-section struts slipping outwards, tension wires were fitted between port and starboard struts, and the 3-ply cowling in front of the pilot's seat was cut away to the gun cartridge slides to allow easier access to the guns. Later an official modification gave instructions on how to cut away the decking to allow access to the gun crank handles.

The positioning of the propeller-driven Rotherham air-pressure pumps varied—as photographs show. An instruction as early as April 14th 1917 called for it to be fixed on to the rear starboard centre-section strut, but in late 1917 in France, units were transferring them to an undercarriage strut, as it was considered they created a blind-spot for the pilot. However, some units, especially in the winter, preferred them in the decreed position where they could be seen by the pilot if they stopped functioning—and sometimes could be re-started by hand when they froze up. For almost a year Camels in production had the pumps fitted on the centre-section strut, when the Aeronautical Inspection Department, reported malfunctioning of the pump through vibration and recommended re-positioning on an undercarriage strut, and Camels on production were so fitted from September 1918 onwards.

Unarmed Camel fitted with additional mid-bay rigging wires.

CAMEL F.1 (130 h.p. Clerget)

If any engine could be said to be the standard power unit of the Camel then it was the 130 h.p. Clerget 9B or 9Bf. This engine powered the late prototype Camels and at the end of the war more Camels had Clergets than all other types of power plants, to be precise 1,342 as on October 31st 1918. However, some of these engines had been modified to 140 h.p. Clergets.

To cut down spares problems overseas, only Clerget-engined Camels were supplied to the British Forces in Italy and the Mediterranean areas.

Up until September 1917, a Lang LP2850 propeller had been a standard fitting for 130 h.p. Clerget-engined Camels. But following suggestions that this might not be the most efficient airscrew, a series of propellers were tried on the Ruston-Proctor-built 130 h.p. Clerget R1498/WD11647 fitted in Camel B2312. The AD644 was found to give the best results with the LP2850 running a close second. Both were adopted as a standard fitting for Camels with 130 h.p. Clergets.

Hydrovane fitted to Camel for ditching trials.

The first 130 h.p. Clerget-engined Camel to arrive at Martlesham Heath for testing was the prototype F1/3 on March 24th 1917. Like all early Camels it had a French-built Clerget—No. 2730 in this case. The next, B3751, arrived on June 15th 1917 with French Clerget No. 20521. This aircraft was kept at Martlesham Heath for a series of general tests and it was found that after thirty hours flying, testing a series of propellers, the performance had fallen off. The engine was sent for examination to Gwynne's, the British licensees for the Clerget, and found to be in good condition, and was then fitted in B6218 on

Camel F.1 at a training unit with camera gun fitted.

return. Meanwhile all early production aircraft were being fitted with French Clergets.

First Camel with a British-built Clerget to be tested

was B3851 which arrived at Martlesham Heath August 23rd 1917 with Ruston-Proctor-built engine R1465/WD11614. Following reports of failing power after fifty hours flying a number of modifications were introduced on the Clergets, including uprating to 140 h.p. by using a new crankshaft giving a longer stroke.

CAMEL F.1 (130 h.p. Clerget LS)

The 130 h.p. Clerget Long Stroke engine, was renamed 140 h.p. Clerget and is detailed under that title.

Actual ditching shot of a Camel. See Chapter 26.

CAMEL F.1 (140 h.p. Clerget)

Statistically, it is not possible to differentiate accurately between the quantities of 130 and 140 h.p. Clergets, as most unit returns lump figures for the two engines together. In general it could be said that due to a deterioration of 130 h.p. Clergets during service, an uprating was effected by modification and while in 1917 most Camels had 130 h.p. Clergets, in 1918 most had 140 h.p. Clergets.

A move to increase the power of the Clerget by a longer stroke—necessitating a new crankshaft, and increased compression by modified pistons, was put in hand simultaneously by the French parent firm, Gwynne's the British licensees, and Ruston Proctor a sub-contractor.

The first long-stroke Clerget tested was one of two modified Gwynne-built engines with standard pistons, fitted in the prototype Camel F1/3 on July 1st. On the 25th all the cylinders and pistons had to be replaced after it had seized up with oil starvation. After that it vibrated badly and was changed over to Camel B3835 which had been testing a high compression B.R.1 engine. At first it appeared to run better, but after five more hours flying, by August 6th, it had to be removed and examined. It was found that various parts had become loose and the back plate was moving slightly. Ruston Proctor modified a few of their standard engines with longer stroke and standard pistons giving a 5.3:1 compression ratio; Gwynne's overhauled their original two engines, and a French-built Clerget modified to a similar standard was supplied for test.

Towards the end of 1917, after 200 flying hours with Clerget engines, the go-ahead was given to provide long stroke engines with standard pistons, with a warning to units that the engine should not be opened up to full power for more than a few seconds on the ground and not full out, as a general rule, below 5,000 feet. Development proceeded and with new materials and adjustments the restrictions were withdrawn in February 1918 when the 130 h.p. Clerget Loop Stroke, was named the 140 h.p. Clerget.

Gradually the 140 h.p. Clerget replaced the 130 h.p. Clergets with preference being given to the Western Front where some 375 Clerget-engined Camels were serving at the close of hostilities. In other areas 130 h.p. Clergets still predominated e.g. in October 1918 in the Aegean, there were 67 Clerget-engined Camels serving of which only 20 had the 140 h.p. model and at Taranto, where 14 of the 32 Camels had the 140 h.p. model.

CAMEL (150 h.p. B.R.1) or BENTLEY CAMEL

Such was the distinction of Camels fitted with the B.R.1 (Bentley Rotary No. 1 formerly A.R.1—Admiralty Rotary No. 1) that they were known colloquially as Bentley Camels. The development of this engine has been dealt with in Chapter 24.

The first B.R.1 engine was fitted to a Camel which crashed at Orfordness on May 13th 1917. A B.R.1 was also fitted into an early production Camel, N6336, which arrived for engine reliability trials at Martlesham Heath on May 30th 1917. During its time on the station it was flown 23 hours 48 minutes without a hitch, an amazing feat for an untried new design. On June 12th it was handed over to Hendon Royal Naval Air Station and was flown to France by Wing Captain R. M. Groves, D.S.O., of the Air Board. It returned after six hours flying to Biggin Hill from where it was flown for over 34 hours. Some trouble was experienced with valve timing but this was soon righted and the Camel returned to Hendon on July 3rd. Ten days later the engine was taken out and put into B3888; apart from replacing a spring in the oil pump the engine was untouched. In the new airframe the engine was flown back again to Martlesham Heath in August and was kept flying until the engine had logged 100 hours, when it was stripped down for examination. As a result, a number of modifications were made. but the design of the engine was proven beyond doubt.

A Camel from the rear.

To decide standard propellers, N518 conducted comparative propeller tests with LP2850 (2590D, 2270P), AD644 (2660D, 2650P) and AB705 (2590D, 2730P), but in the standard list promulgated in February 1918, the Bentley Camel Standard propellers were given as LP3510, LP3640 and AB644.

The allocation of Bentley engines in relation to others is touched upon in the Chapters. There were never enough B.R.1 engines to meet the demand. At the end of the war, according to the October 31st 1918 return, only 356 Camels of the total of 2,519 on charge had B.R.1 engines. Of these 210 were serving with the Expeditionary Force or the Dover/Dunkirk area, backed by 84 in store, 35 at home stations (only one of which was at a training unit) and 26 were at Turnhouse for fleet duties.

CAMEL F.1 (110 h.p. Le Rhône) and NIGHT FIGHTER CAMEL

The 110 h.p. Le Rhône engine, built in France by the design firm and under licence in England by W. H. Allen & Sons Ltd of Bedford, was specified as an alternative to the Clerget in some Camel contracts, but the actual allocation depended largely on availability. Up to December 1917, Sopwith's themselves had fitted only six airframes with 110 h.p. Le Rhônes, which included B6322 and B6326 used for tests with Bloctube carburettors. One of the earliest Camels to have a Le Rhône was B2312 which had its initial Clerget changed for one in November 1917.

Initially the Le Rhônes tended to overheat and the cowling was modified with cut-out vents rather in the manner of the earlier D.H.5 which had this engine.

This mishap to D8189 in the Belgian Lines, permits a glimpse of the Camel's wing rib construction.

In March 1918, a 110 h.p. Le Rhône with duralumin induction pipes was tested on a Camel and bestowed an additional 3½ m.p.h. over steel pipes at 15,000 feet and achieved six minutes better time in reaching that height. As a result the lighter pipes became standard.

It is not possible to assess precisely the numbers of 110 h.p. Le Rhône powered Camels in use, but indications are that it was, in the main, restricted to home stations and that at the end of the war, all 181 Camels in Home Defence units were 110 h.p. Le Rhône powered, but hardly any were allotted for training purposes.

CAMEL (180 h.p. Le Rhone)

A French-built 180 h.p. Le Rhône 9R engine, developing 186 h.p. at 1370 revs, was fitted into Camel airframe F6394 which had been delivered to Martlesham Heath on September 25th 1918 to await the delivery of this power unit. The engine was installed during October, but the tests, involving some 17 flying hours, were not completed until early 1919.

The engine had Frigo ventilated sparking plugs of French design and manufacture; it throttled down well and had good slow-running qualities. To install it in a Camel required some modifications such as shaft extension, cowl modifications and detail changes. Had the war continued, this engine may well have become one of the standard alternative power plants.

CAMEL (100 h.p. Gnôme Monosoupape)

The 100 h.p. Gnôme Monosoupape engine, No. 30748/WD1182, was fitted to Camel B3811 in August 1917, possibly with a view to using this engine as an alternative power unit for Camels allotted to training units.

With the Camel so powered, it was found to be tail heavy, due to the lighter engine and it was considered doubtful if this fault could be satisfactorily overcome.

CAMEL F.1 (150 h.p. Gnôme Monosoupape)

A 150 h.p. Gnôme Monosoupape engine, usually known as the 150 h.p. Mono, was first tried in a Camel which force-landed outside Hendon in mid-November 1918, slightly damaging the propeller and lower plane. The engine was then fitted in B6329 for testing at Martlesham Heath the following month; it was found to give the Camel a speed of 115 m.p.h. at 13,000 feet at 1265 revs.

Camel with installation of 150 h.p. Gnôme Monosoupape.

Drawings were put in hand during January 1918 for production Camels with the 150 h.p. Mono and the initial modification work was carried out at Orfordness the following February. To mount the engine, a dished adaptor plate was necessary in place of the flat plate used for the Clerget. There appears to have been little trouble in adapting a standard Clerget cowling, but there was some alteration to instrumentation.

The 150 h.p. Mono had dual ignition with two plugs per cylinder and a selector switch which enabled firing to run full, or restricted to 7, 5 or 3 cylinders. Initially, the revolution counter drive was very close to the rudder bar pillar but this was modified on production models. In this connection, the revolutions of this engine, 1350 r.p.m. normal/1450 r.p.m. maximum, were in excess of all other rotary engines.

Trials of the Selsdon Air Pump were conducted in March 1918 with a Camel (believed B6329) fitted with 150 h.p. Mono No. 52238. Other Camels with a 150 h.p. Mono included B2541 and B2542. The former was tested at Martlesham Heath in April 1918 and sent on to Orfordness the same month. This aircraft, not fitted with armament, had a throttling device in lieu of the cylinder firing selector switch; the object being the saving of fuel. With this device the engine could be throttled down to 600 r.p.m. which was sufficient for taxiing without using the selector switch.

The United States decided to adopt the Camel with the 150 h.p. engine and the first production aircraft for the Americans was expected at Martlesham Heath from the Norwich Acceptance Park in May 1918, but did not arrive until July. The American order was for 300, with delivery at the rate of 20 to 30 a month from May 1918.

The first to arrive was D6567 whose performance had to be partially estimated as it crashed in a forced landing during trials. It was replaced by F1336 whose trials were initially delayed by the 150 h.p. Mono needing overhauling and then by the engine seizing after 11 hours running due to the main oil pipe from pump to engine breaking in flight. With a new Mono engine, F1336 went through full trials including comparison tests with propellers LP3640, AB644, AB705 and AB8553. Full performance tests were carried out with AB644, figures are given in the tables.

CAMEL F.1 (TWO-SEAT)

A training requirement existed for a two-seat dual-controlled Camel and work on a prototype was put in hand by the Service in 1918 as the Sopwith Camel Comic, but drawings initially issued in July 1918 were not fully approved until October 1918. The intention was to convert a limited number of machines at depots in order not to disrupt production lines.

In November 1918 the method of fitting dual control was extensively modified and by this time a two-seat Snipe requirement superseded the Camel requirement.

Two-seat Camel, possibly prototype conversion.

Two-seat Camel built from salvage at No. 43 T.D.S.

CAMEL F.1 (CIVIL)

Two Clerget-engined ex-R.A.F. Camels reached the civil register. The first, H2700 which first flew on December 12th 1919, was registered G-EAWN to Hubert S. Broad in March 1921. It came 6th in the Aerial Derby for 1921 and did aerobatics for the Welsh Aviation Company in August that year. During 1922 it was dismantled at Stag Lane. F6302, which had been sent to Taranto in late 1918 and returned, probably unused, was registered G-EBER to Flying Officer W. J. McDonough in August 1922.

CAMEL T.F.1 or ARMOURED CAMEL
Details and illustrations in Chapter 29.

CAMEL 2F.1 PROTOTYPE and CAMEL 2F.1 (130 h.p. Clerget)

The prototype 2F.1 Camel, N5 fitted with French-built Clerget No. 2671 arrived at Martlesham Heath on March 15th 1917 for tests. Presumably it had been on trials at Grain and was sent to Martlesham for comparison tests with the Camel F1/3. It was sent for further trials in June 1917 with a top Lewis gun and rocket gear, W/T set and airbags in fuselage but on take-off for its second test flight in this condition, the aircraft was wrecked.

The "Improved Baby", the Camel 2F.1 prototype.

While the B.R.1 became the standard power plant for the 2F.1, N6610 and N6613 used for training at Vendome, France, were fitted with 130 h.p. Clergets as were the seven 2F.1s in Canada 1924-1929.

CAMEL 2F.1 (150 h.p. B.R.1) or SHIPS CAMEL

Whereas the F.1 Camel had various engines, the 150 h.p. B.R.1 was standard to the 2F.1—or Ships Camels as it was officially known. It differed dimensionally from the F.1 as the data tables will show, and the wing centre section and struts were built differently. In January 1918 all 2F.1s were modified to have the control wires to the tail carried outside the fuselage to avoid fouling the air bags, and with a ball-end fitting to tail skids for deck work. Four airbags were fitted in the rear of the fuselage.

Ships Camels built by Arrol Johnson and delivered to Renfrew.

Ships Camels were produced by Arrol Johnson, Beardmore, Clayton & Shuttleworth, Hooper and Sopwith. The fourth Sopwith production example, N6603, was tested at the Isle of Grain from November 6th 1917 and afterwards went to general service and served aboard ships.

Blackburn-built Ships Camel lashed at take-off point on a warship.

Wartime service details are dealt with in Chapter 35 and post-war actions in Chapter 38. Following the operations in the Baltic, several Ships Camels were transferred to the Latvian and Estonian forces.

The Ships Camel fuselage was detachable for stowage.

CAMEL 4F.1 or TAPER-WING CAMEL

A taper-winged, streamline strutted, Camel was tested in March 1917 with a French-built 130 h.p. Clerget 9B No. 2730 engine, for which performance figures are tabulated (see Camel F1/1). In December 1917 a proposal to build a "4F.1 Taper Wing Camel" was discussed by the Air Board and four prototypes were ordered under contract AS34594, but their production did not proceed.

SCOOTER (MONOPLANE No. 1)

So far Sopwith had not produced a monoplane and their first venture into aircraft of this configuration was by fitting a standard Camel fuselage with a parasol wing, supported from a cabane by rafwires. Named the Scooter, it provided Hawker with an aerobatic machine untrammelled by development schedules and it also functioned usefully as a transport. First flown in June 1918, it survived the war and was registered to the Sopwith Company provisionally as K-135, and then as G-EACZ in mid-1919.

Hawker in April 1921 bought the Scooter as his personal aircraft but after he was killed three months later it was placed in store. Early in 1925 it was overhauled for C. Clayton of Hendon and a Certificate of Airworthiness was issued August 1st 1925. A year later it was sold to Dudley A. N. Watt but before being scrapped in 1927 it was entered in several sporting events, the last being the Lympne Open Handicap on September 18th 1926.

The Scooter with 130 h.p. Clerget, July 1918.

SWALLOW (MONOPLANE No. 2)

Evidently Hawker's Scooter was noted by the authorities for a Sopwith Monoplane was placed on the Air Board's schedule of "Experimental Machines embodying features of general interest to Aeronautics".

The fuselage of this monoplane prototype was from Camel B9275, probably one of the two originally supplied from Boulton & Paul to the Sopwith works under the T.F.1 armoured Camel project.

A single-seat fighter, its possibilities as a ship-board aeroplane were considered. It was passed for experimental flying early in October and was flown by Hawker at Brooklands before ferrying to Martlesham Heath on October 30th 1918, but performance and simulated deck landing trials were held up by engine trouble during November.

Once the pressure of war was over, testing was carried out far more leisurely. Trouble with the petrol system resulted in the piping being completely replaced, then tests were mooted with a different airscrew and it was May before it had completed tests after which it was discarded.

The Swallow with a 110 h.p. Le Rhone, October 1918.

A.T. (AERIAL TARGET)

Strict secrecy veiled Sopwith activity on the radio-controlled flying bomb, called A.T. for aerial target to disguise its real use. The firm were asked to assist in the radio-controlled experiments of Professor A. M. Low and build an experimental machine to prove the apparatus. A small biplane was erected in sheds at Feltham by a small number of Sopwith employees and Royal Aircraft Factory staff. It had a four-wheeled undercarriage to ensure a straight take-off. The radio apparatus was installed in the fuselage with aerials strung around wings and fuselage. The power unit was a 35 h.p. A.B.C. Gnat horizontally-opposed two-cylinder engine. After being damaged the A.T. was abandoned.

The A.T. believed damaged during erection.

SPARROW

Furthering the A.T. project the Sparrow was larger and evidently based on the earlier SL.T.BP. However, radio control was thought to have a more direct application to motor boats, where failures in experiments did not result in a complete wreck and the stage of attempting tests with more than ballast was not reached with the aircraft. However, four aircraft numbered A8970-8973 officially reported as Sopwith small scouts with warping wings in May 1917 at Brooklands without engines, and 50 h.p. Gnômes reported as their power units, suggests that a small production batch was made.

Sparrow, powered by 35 h.p. A.B.C. Gnat.

CUCKOO T.1

The development and early service of the T.1 is dealt with in Chapter 34 devoted entirely to this torpedo-bomber. There is a mention of deliveries to the Japanese in Chapter 36 and of post-war service in Chapter 38.

Since only the prototype N74 was built by Sopwith, production records are given under the sub-contractors, Blackburn, Fairfield and Pegler; pages 151, 154 and 157.

The official Sopwith general arrangement drawing of the T.1, dated October 23rd 1917, was supplied by C. Mayell and approved by W. G. Carter.

Variations in production models can be seen in photos given in Chapter 34. To avoid any confusion that may arise by Cuckoo power units, these are summarised:

Prototype	Original 200 h.p. Hispano-Suiza replaced by 200 h.p. Sunbeam Arab.
Mk. I	200 h.p. Sunbeam Arab initial production.
Mk. II	200 h.p. Wolseley Viper. Early production aircraft were converted to Mk. II.
Mk. III	Assumed allotted for trial installation of 275 h.p. Rolls-Royce Falcon III engine in N7990.

The splash of a torpedo from a low drop endangered the T.1.

B.1/B.2

Contemporary with the Cuckoo, the B.1 was planned as a 200 h.p. Hispano-Suiza powered bomber for the French.

It was flown at Brooklands in early April 1917 and in mid-May flew to Dunkirk for assessment by both British and French personnel. Delivered later to the Isle of Grain, it was the subject of a discussion for a shipborne aircraft in October 1917 and after modification by the Experimental Construction Depot, Port Victoria, it was considered in March 1918 in competition with Parnall, Handley Page and Fairey aircraft, to Specification AF21 for a two-seat reconnaissance ship aeroplane.

To meet the specification it was changed from a single-seat bomber, to a two-seat observation aircraft, and the power unit, as in the T.1, changed to a 200 h.p. Sunbeam Arab. Seven experimental aircraft were ordered, N100-106, but since these were to the modified design and built at Grain, they were not true Sopwith's—indeed, initially it would not appear that any improvement had been made since the Martlesham test pilot on April 7th 1918 reported on N100 "Terrible on all controls, Rudder useless". The new aircraft with folding wings, skid chassis, hydrovanes and air bags, powered by B.R.2s were initially called PVN50 Type and then the name Grain Griffin was given.

Another basic B.1 bomber still with a 200 h.p. engine, was delivered to Martlesham Heath for test on April 23rd

B.1 bomber, believed the original N50.

1918. Allotted the serial B1496 and generally referred to as the B.1 it appears in contract references as the B.2.

Sopwith Bomber B1496.

DOLPHIN 5F.1

The history of the development and service of the Dolphin as a day fighter is dealt with in the chapter "Discourse on the Dolphin" and its application to night fighting in the chapter "Sopwith Defenders".

Production details are given in Sopwith Fortyfold under Darracq, Hooper and Sopwith. The contractual price for a Dolphin airframe was £1010/13/0 and the purchase price of a French-built 200 h.p. Hispano engine £1004.

Harry Hawker with first prototype Dolphin which had radiators on the top planes and the second with radiators re-positioned.

The first prototype Dolphin (un-numbered) was passed for flying on May 23rd 1917 at Kingston and conveyed to Brooklands for makers trials by Harry Hawker. Fitted with Aries-built 200 h.p. Hispano engine No. 10137 it was forwarded to Martlesham Heath for tests early June, where it was reported as nose heavy and having about 20 lb. of lead in the tail to correct this fault. There was a tendency for it to spin when turning to port and the machine was found tiring to fly owing to strong left rudder having to be applied. A balanced rudder was suggested. The machine flew to France June 13th 1917 for field evaluation.

The second prototype (also un-numbered), initially had radiators fitted on the top plane in July 1917 but on being found unsatisfactory, as the engine overheated and they could not be enlarged without affecting the wing bracing, the radiators were repositioned on the fuselage sides as illustrated. This machine had a balanced rudder and was tested with a four-bladed propeller (T.28079) as fitted to the S.E.5A but as it did not effect any improvement, it reverted to two-blade (T.28063) type. At this time, mid-September 1917, the top decking abaft the cockpit was reduced by the Testing Squadron at Martlesham who sent the prototype back to Brooklands for examination.

The airframe of the third prototype (another un-numbered machine) was ready at the end of August 1917 with Brasier-built Hispano engine No. 16034 supplied, but as no propeller or magneto couplings were supplied testing was delayed.

This was the first machine to have full armament fitted, which Mr. Allman of the Sopwith Works was responsible for arranging. His major problems were with ammunition stowage for the Lewis guns—the belt-fed Vickers were no problem and were fitted much the same as on the Camel.

When tested at the end of November 1917, Hispano No. 16136 was fitted but excessive vibration on this engine further delayed tests. Using this engine, propellers L3610 and L3800 were compared on trials.

The fourth and final un-numbered prototype was produced with the top wing further forward and it was found more tail-heavy on test than earlier prototypes. The cut-down decking as advised by the Testing Squadron was incorporated. It left the Sopwith workshops on October 1917 and was flown from Brooklands to Lympne en route for France for tests by the Expeditionary Force.

C3777, the first production Dolphin was used for development work at Martlesham Heath. It had a 200 h.p. Hispano (Peugeot-built) No. 115131/A14305 engine fitted in November 1917 with which excessive vibration occurred after 5 hours running and a "knock" developed after eleven hours due to teeth in the reduction gear breaking.

The second production aircraft, C3778, also used for development work at Martlesham Heath, had Hispano No. 16034 (ex-third prototype) of which a con-rod broke on November 11th 1917. Replaced by another Brasier-built Hispano, No. 16452/A14335, it had an oil leak develop after 17 hours running and after running 22 hours water was found in the sump and oil tank.

Dolphin showing experimental lower wing cut-outs to improve downward view.

On production aircraft following, several modifications were progressively effected. A steel tail skid was introduced in September 1918 interchangeable with the Snipe and Salamander. Considerable alterations were notified in November 1918 to enable wooden members to be strengthened so that Grade B timber could be used, but with the pressure of war over, quantity gave way to quality and only Grade A timber was used. In October 1918 the U.S. Air Service purchased a number of Dolphins and at least 13 were shipped to America, among them E4642, E4643, E4646, E4647 and E4655 and the Canadian Air Force used E4764, F7076, F7085, J3 and J12 and others in their No. 1 Squadron. A single example was registered as a civil aircraft—D5369 as G-EATC on May 7th 1920. Owned by Handley Page Ltd., it was used for demonstrations.

Dolphin I used at training unit in Britain.

DOLPHIN MK. II

Dolphin Mk. 1 D3615 was flown to France for the installation of the more powerful 300 h.p. Hispano engine under liaison arrangements with the Sopwith Paris office.

The French, impressed with the basic Dolphin, were considering production of the type with the new Hispano engine for both their forces and those of the Americans in France. Production did not reach an advanced state before the armistice, but the few basic Dolphins re-engined with the 300 h.p. Hispano were designated Dolphin Mk. III.

A Dolphin Mk. II photographed in France.

DOLPHIN MK. III

Due to difficulties with the gearing on the 200 h.p. Hispano engines, the Southern Aircraft Repair Depot designed a direct drive modification which was flight tested on an S.E.5A. Tests of the ungeared engine were made on Dolphin C8194 (illustrated page 154) which completed tests at Martlesham Heath and left for Farnborough on November 4th 1918.

Before the war ended, contractors were instructed to fit only ungeared engines with a re-designed cowling and others in service were sent to Aircraft Repair Depots for modification. Dolphins with the ungeared engine were designated Mk. III.

Front and rear views of a Dolphin Mk. III showing the opposite rotation of the airscrew to the Mk. I (compare with "prop-swinging" picture page 106) and the altered thrust line.

Cockpit of Dolphin Mk. III.

DOLPHIN TWO-SEAT

A two-seat conversion of the Dolphin C8022 was reported to have been fitted with a 150 h.p. Hispano engine by the Wing Technical Officer of the 23rd Training Wing at the Wing's Aircraft Repair Depot, South Carlton, and it is possible several other unit conversions were made.

SNIPE 7F.1 PROTOTYPES (B.R.1/B.R.2 engines)

The development of the Snipe is dealt with in Chapters 30 and 31. The early prototypes were un-numbered and it has not been possible to trace individual histories from documents. Of the numbered prototypes, B9962-B9967, there is no record of B9962 flying and this may be because it crashed at an early stage; B9963 was reported as being flown by Commander Allan at Farnborough 23rd November 1917 and B9964 is believed tested to destruction on a sand rig to assess the strength of its structure.

The numbers B9962-9964 may well have been applied, or allotted retrospectively, to the first three un-numbered prototypes. However, the remaining three prototypes are well documented. B9965, reported as repaired after a crash, was flown to Martlesham on January 1st 1918. At this time it had 2-bay wings. It was tested throughout February,

Two views of the first prototype with B.R.1 engine, slab-sided fuselage and single-bay wings.

flew to Sutton's Farm for comparative tests on March 10th and three days later to No. 1 Aircraft Supply Depot, St. Omer, for evaluation. On return to England it went to the Sopwith works for the fitting of propeller Lang L4040 G363N28 which had a special spinner. To assess its effect it

Snipe B9966 fitted with special tail unit for trials September 1918.

was flown alternatively with standard propeller Lang 4040 G363N88. The spinner made no appreciable difference to performance, but since it caused the engine to overheat, extra holes were cut on the bottom of the cowling.

Intended as an engine test-bed, B9966 after being damaged was brought up to the latest standard while B9965 was in France. In June it was fitted with B.R.2 No. 5 driving Lang propeller LP4040 6492N92. Tests were also carried out with Lang propellers LP5250 and 5260 and Air Board propellers AB8651 and AB8661. In the late summer it was tested as described in Chapter 31 and fitted with a new type of tailplane as illustrated.

SNIPE 7F.1 DEVELOPMENT MODELS (B.R.2)

The development of the Snipe continued while production was underway, causing a series of modifications to be issued. E7987, the first Sopwith-built Snipe, was retained for tests. Fitted with B.R.2 Experimental No. 2 engine tuned up at Farnborough, it did performance trials at Martlesham Heath in August 1918. After climb and speed trials with Lang propellers types LP5300, LP5350 and LP 4040, the Ministry of Munitions Supply Department were instructed to provision for LP5300 propellers. This Snipe then returned to Farnborough on September 11th. Back again at Martlesham in November 4th, it commenced carburettor and modified control surface trials.

Fleet use was considered and E8068 fitted with flotation gear and hydrovanes at Grain went to Martlesham for

Snipe, showing fin and rudder of early production models.

Snipe, showing fin and rudder of late production models.

trials on December 1918. At the same time E8085 was fitted with slings to facilitate hoisting aboard vessels.

E8137 was fitted to take a Calthrop A1 Guardian Angel parachute.

SNIPE 7F.1 PRODUCTION (B.R.2)

Service details are dealt with in Chapters 31 and 38 and production details under the manufacturers in Sopwiths Fortyfold to which the table on page 164 provides a quick check. The name Snipe was officially adopted in February 1918.

Production deliveries commenced next Summer 1918 and the first major modification was the introduction of a knife-edge skid introduced by service request to give better

Production Snipe with balanced ailerons as the curve of the top wingtip reveals.

control when taxiing. It was planned to be interchangeable with the Salamander.

The larger rudder and revised fin were introduced in September 1918 together with balanced ailerons as an optional modification.

SNIPE MK. IA

The designation Snipe IA was given to three Snipes ordered as bomber escorts for a Long Distance High Altitude Single-seat Fighter Specification. The serials H9964-9966 were allotted but instead of producing new machines, five airframes on the Sopwith lines, E8089-8091, were earmarked and the first, fitted with a 60-gallon fuel tank and a Dolphin type tail unit, was sent to Martlesham in late 1918 after Sopwith trials at Brooklands. However, during performance and fuel consumption trials, the machine was damaged and had to be rebuilt with a new engine by which time the need for a bomber escort had passed.

SNIPE (200 h.p. Clerget)

Early in 1918 it was decided to utilise the 200 h.p. Clerget as an alternative power unit to the B.R.2 and Snipes had rear engine support plates and cowlings modified to allow the two units to be interchangeable. Snipe F2340 was fitted with a 200 h.p. Clerget 11EB in November 1918 and was tested for 20 hours under ground strafing conditions. This aircraft, which was sent to Farnborough on July 1st 1919, later served in No. 43 Squadron as a standard B.R.2-engined aircraft.

SNIPE POSTAL PROJECT (110 h.p. Le Rhône)

The Post Office had a grandiose scheme for an air letter post as soon as hostilities ceased and the Snipe was considered as a light liaison postal aircraft. Estimates issued in 1919 for a Snipe fitted with a 110 h.p. Le Rhône gave 83 m.p.h. at 5,000 feet, with maximum range of 1850 miles, decreasing to 440 miles with maximum freight load of 1068 lbs. The project was not pursued.

SNIPE TWO-SEAT

At least 40 Snipes were converted to 2-seaters.

SNIPE (Dragonfly)

The A.B.C. Dragonfly engine No. 1 was installed in the Snipe 6th prototype, B9967, and was tested at Brooklands on April 27th 1918 (see photo page 153). However, trouble with the magnetos delayed trials and the engine was taken out and a new magneto was fitted, and by June 12th the engine had flown for only 4½ hours. Trouble was experienced with the engine cutting out or missing at altitudes above 18,000 feet.

A Clyno Engineering Company-built Dragonfly engine WD48204 was then used until September 1918. This engine too, was only used for a short period as the spindle of the port magneto drive broke and fell into the crankcase causing considerable damage. As a Mk. II Dragonfly engine was promised at an early date from Farnborough, the engine was scrapped, but it was November before the engine actually materialised.

This engine was Sheffield Simplex-built No. 4 and to test the exhaust-heated air intake pipes of this engine two flights were made in March 1919 to 21,000 feet. Due to ignition troubles the engine was removed for examination and the fitting of a new magneto. The airframe was fitted with a new set of wings but further development was to the new Dragon standard detailed below.

DRAGON

The Sopwith Dragon was basically a Snipe with a 320 h.p. Dragonfly engine. As related earlier, a Dragonfly engine was fitted into the Snipe prototype B9967 in April 1918. At that time it was confidently expected that the teething troubles of the engine would be overcome and large orders for the engine were placed. Plans were made to equip initially one squadron with this special version of the Snipe and a new contract for an extra thirty (F7001-7030) Snipes was placed with Sopwith Aviation in June. The firm were also asked to prepare one for the U.S.

Dragon prototype E7990 at Brooklands.

The first true Dragon was Snipe airframe E7990 taken off the Sopwith line in July 1918 for the fitting of a Dragonfly, but the airframes for the remainder had to wait. The Wolseley-built Dragonfly Mk. I engine for the American model, broke down on its initial test on July 28th. Due to an oil circulation failure, some bearings were badly damaged and Walton Motors, who had an engine development contract, had their workmen give up their holiday week to make good the damage. Meanwhile the Clyno Engineering Company worked on completing their first thirty engines for the proposed "speed Snipe" squadron.

It was February 14th 1919 before the Dragon E7990 arrived at Martlesham Heath for performance tests and it was returned to the firm by rail for a new machine, F7001, to be allotted, which arrived in July. Trials con-

Production Dragon at Martlesham, believed March 1920, with balanced ailerons.

tinued into 1920 to assess the value of utilising the numbers built (see page 160), and a two-seat version was planned, but the Dragon was declared obsolete for service use in April 1923 and did not enter service.

A Dragon (J3628) was eventually shipped to the U.S.A., being allotted the U.S. Air Corps No. 94106 and the McCook Field project number P-149.

SALAMANDER T.F.2

The history of this basic trench fighter, evolving from the armoured Camel (T.F.1), is given in Chapter 29 and production contracts are detailed on pages 150, 154, 156, 157, 160 and 162 of Sopwiths Fortyfold.

Based on the Snipe, the Salamander carried much less fuel to compensate for the weight of armour. The main tank, under the pilot's seat and protected by armour, held only 16 gallons and fuel was fed from this to a gravity tank close behind the engine.

Salamander cockpit, showing staggered gun breeches.

A prototype Salamander at Brooklands, 1918.

RHINO 2B.2

T. O. M. Sopwith personally presented the details of the 2B.2 triplane bomber to Air Board officials in August 1917 and intimated that, in view of the congested state of his experimental steps, a reduction in the normal order for six prototypes would be welcome.

The Rhino was designed to take the 230 h.p. B.H.P. engine and was thought of as a possible rival to the D.H.9. At the end of 1917 it was also considered, with the D.H.10, as a contender for Specification A.2a for a light bomber. Possibly it was the first of the bombers to carry an armament pack, for to speed up re-arming after a raid a crate-like pack, containing the bombs with release gear, was drawn up into the fuselage.

In late October X7 the first Rhino was on test at Brooklands. It was found to be nose heavy and the engine overheated necessitating a larger radiator which was fitted

Rhino prototype with balanced ailerons.

before it left for Martlesham Heath on January 4th 1918. It cannot be said that its trials were satisfactory, but it was a case of an aircraft designed for one specification, being pushed to a revised specification. As a result, in addition to its fitted tankage of 54 gal. of fuel and 5 gal. of oil it carried ballast of 155 lb. representing 6 gal. of fuel, $2\frac{1}{2}$ gal. of oil, extra ammunition and fittings, in addi-

Rhino prototype with unbalanced ailerons, December 1917.

tion to 400 lb. deadweight representing bomb-load. The highest it reached on test was 13,300 feet taking 55 minutes and its service ceiling was as low as 12,000 feet.

The second Rhino X8, which had a Scarff gun-ring fitted, reached Brooklands in February 1918 and from that time the two machines were used for general development work including propeller comparative tests.

HIPPO 3F.2

The Hippo was designed early 1917 as a 1½ Strutter replacement for French industry, and the drawings were approved on April 30th that year. However, not until September was a 200 h.p. Clerget engine available to the Sopwith works. The first prototype made a short flight in bad weather on September 13th and next day, with improved conditions, tests started in earnest. The machine was very heavy on lateral control and after a recommendation for balanced ailerons and minor adjustments it was flown to Martlesham Heath.

For some reason the engine was taken out and returned to France and another specially boosted to give 260 h.p. was awaited. Meanwhile, to match the machine for the new Specification A.2a, new wings were made. For a time, it was debated whether a B.R.2 or Clerget would be fitted, but by late 1917 a firm decision was taken on the Clerget. Early in January 1918 it was flown at Brooklands and went on to Martlesham where it was in competition with the Bulldog. A decision was taken to abandon the design, and its Clerget was then fitted to the Bulldog.

Engine test of the Hippo first prototype at Brooklands by T. O. M. Sopwith, seen in the front cockpit.

Hippo X11 with Scarff gun ring.

The Hippos were licence-built with X10 allotted to the original prototype; X11 was a revised version with a Scarff gun ring and a third, X18, was flying in June 1918.

BULLDOG 2FR.2

While the Hippo had been an existing design adapted to meet an official specification, the Bulldog was drawn up in September 1917 to meet Specification A.2a for a two-seat fighter; its early testing is dealt with in Chapter 30.

With the issue of a revised A.F. Specification, there was no call for a new two-seat fighter and official interest suddenly waned in 1918. Work was well ahead on the second machine in February when it had been decided to fit the new Dragonfly in the aircraft for test, but work was

Bulldog I X3 (later allotted H4422) with single-bay wings.

stopped at Sopwith's on the third machine in April and the fourth was probably not started. By mid-April, Sopwith engineers had ironed out the snags with the first machine and it was sent, via Hendon, to reach Martlesham on April 22nd.

The second Bulldog (X4) with the first Clyno-built Dragonfly engine (WD48204), was tested at Farnborough mid-1918. After 20 flying hours, early in 1917, the engine was replaced by a Mk. II Dragonfly (Clyno 50008).

In spite of abandoning the design early in 1918, the authorities in Mid-1918 referred to the Bulldog as a possible machine for contact patrol work and for attacking ground targets. A number of mock fights were arranged between X3 and a Bristol Fighter. It was found that their manoeuvrability was about the same and a test pilot reported "the Bulldog is remarkably light on controls being similar to the Avro two-seater in this respect; also it is fairly stable, and might attempt to land itself alone—an important point on machines used for contact work. The Bulldog is a good machine for attacking ground targets with its front guns, but it is considered that the view, and position of the pilot and observer, are better in a machine like the Bristol Fighter". There was not sufficient improvement, however, to consider replacing the Bristol Fighter in production by the new Sopwith, and Sopwith's made their final bid for a contact patrol aeroplane by the Buffalo.

Bulldog II X4 (later H4423) with 2-bay wings.

BUFFALO 3F.2

In July 1918 the Air Board, drawing on Sopwith's experience with the armoured Salamander, requested two armoured two-seat contact patrol aircraft, powered by the B.R.2, embodying Bulldog parts as far as possible. Work was put in hand during August when the name Buffalo was officially bestowed. This project was not to an Expeditionary Force specification, but was classed as "a machine possessing features likely to be of use to the R.A.F.".

The drawings were signed "Eden" and were approved by W. G. Carter. The first of the two, H5892, was ready for trials at Brooklands September 18th and nine days later it was flown to France for evaluation. Reports from the Field recommended a Scarff gun-ring for the Lewis gun instead of the pillar mounting in the rear cockpit, but accepted a single Vickers gun firing forward. Other suggestions were for a larger windscreen. The second machine, H5893, had been held from completion, pending the recommendations which were incorporated. H5892 returned to the Sopwith works for these changes, while H5893 with them effected was flown to Martlesham Heath for trials on November 18th.

In February 1919 while the first machine was on test at Brooklands, H5893 was selected to try out an experimental Humber carburettor. As it could not climb, due to

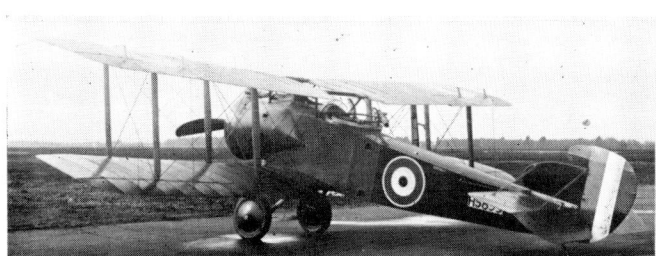

Buffalos H5892 and H5893 with Scarff gun ring in the latter.

the weight of armour, above 10,000 feet, it was not an ideal aircraft for the test, but with other B.R.2-engined aircraft such as the Snipe it would have meant structural alterations.

SNAIL 8F.1

The Snail is introduced in Chapter 30. Six were ordered C4284-4287 of conventional construction and C4288-4289 with monocoque fuselage. Interest focused on C4284 and C4285 being the first of their respective types. C4284 went for trials at Brooklands in March but carburettor troubles delayed flying. C4285 joined it at Brooklands in April but due to further troubles with the Wasp engine further testing was suspended.

Conventional and Monocoque Snails.

SNARK

To meet Specification RAF Type 1 for a high altitude single-seat fighter the Sopwith design staff again tried a triplane form, but utilising this time the A.B.C. Dragonfly engine. A contract was placed in April 1918 for three prototypes F4068-4070.

Due to troubles with the Dragonfly engine, it was decided in October 1918 not to put either the Snark or (see below) Snapper on production but to complete one of each for evaluation. An engine for each was made available in December, but magneto troubles caused a grounding instruction to be issued. The engine was eventually fitted in April 1919. The following September it first flew at Brooklands and went to Martlesham on November 11th 1919. It remained on trials until March 1920. The other two, in spite of earlier decisions, were completed and F4070 flew in early 1921 on trials.

The first Snark.

SNAPPER

Lest the triplane contender for Specification RAF Type I did not have the necessary speed, Sopwith designed a biplane high altitude fighter to utilise the A.B.C. Dragonfly. Three prototypes were also ordered and in late May 1918 the experimental shop at Kingston were completing the first Snark fuselage and covering the first Snapper fuselage, but work on the third Snapper did not commence until September. The type was conditioned by the condi-

The first of the three Snappers.

tions governing the Snark (see above) and of the three that emerged from the works, their histories are briefly as follows: F7031 finished July 1919 and sent to Martlesham for test the following September, in June 1920 it was sent to Farnborough. F7032 was abandoned temporarily in July 1919, but it was finished and flown in December that year. In June 1920 it was at Farnborough. F7033 was still under construction at the end of 1919 and it went to Farnborough for aero research work in 1920.

COBHAM

In June 1918, Sopwith tried the bomber field again, to meet a rather general specification for a twin-engined, fighter reconnaissance or short or long distance bomber using Dragonfly engines. The mock-up was inspected at the Sopwith works in August, when the name Cobham was officially bestowed. Some modifications were necessary to allow the wings to be easily detachable and due to troubles with the Dragonfly the first of the three ordered (H671-673) was fitted with Pumas and became the Mk. II.

Once the war had ended, progress proceeded very leisurely. At the end of March 1919 the first machine, Mk. II H671 with boosted Pumas was still not fully completed and H672 the Mk. I, awaiting Dragonfly engines, still had its framework uncovered. H671 eventually went to Martlesham Heath for tests in mid-1919 and was last recorded at Brooklands Aircraft Acceptance Park for repairs that December. H672 did not do its trials until the Spring of 1920 and on H673, the third machine ordered, work was stopped in July 1919.

ATLANTIC (or TRANSATLANTIC)

The beginning and end of this famous aircraft is related and illustrated in Chapters 37 and 39.

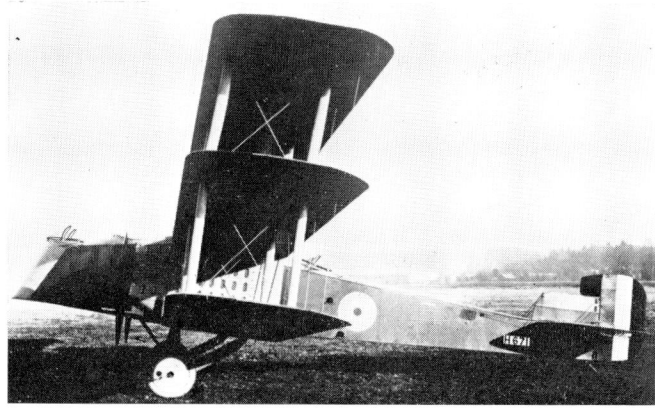

Cobham with Puma (above) and Dragonfly engines.

DOVE

The Dove was a two-seat sports version of the Pup powered with an 80 h.p. Le Rhône. In all ten were built registered as follows: G-EACM, EACU, EAFI, EAGA and EAHP, ex-K-122, 133, 148, 157 and 168 respectively all sold abroad; G-EAJI, EAJJ, EAKH and EAKT were sold to Australia as G-AUDN, AUDJ, AUKH and AUDP respectively and G-EBKY converted to a Pup and is now in the Shuttleworth collection numbered N5180.

The first Dove, Brooklands, March 26th 1919.

GNU

This civil Sopwith is introduced in Chapter 39 where its prototype is illustrated. Thirteen were registered as follows: G-EAAH, EADB, EAEP, EAFR, EAGP, EAGO, and EAHQ ex-K-101, 136, 140, 156, 163, 164 and 169 respectively of which the last went to Australia as G-AUBX followed by G-EAIL as G-AUBY. The four remaining Gnus were G-EAME-H.

The prototype with B.R.2 engine and first production with 110 h.p. Le Rhône are illustrated in Chapter 39. Most subsequent models had the Le Rhône, but G-EAMG had a B.R.2 and G-AUBY in Australia was re-engined with a Wright Whirlwind. The second production Gnu had an open cockpit as did most subsequent.

GRASSHOPPER

The Grasshopper shortly after leaving the works.

A two-seat tourer of new design, powered by a 100 h.p. Anzani, the single Grasshopper was flown from 1920 to 1929 by four different owners.

SCHNEIDER CUP

For the 1919 Schneider Trophy Contest, as related in Chapter 39, Sopwith produced a post-war racing seaplane.

Schneider ready for trials with 450 h.p. Cosmos Jupiter.

RAINBOW

The Rainbow was a racing landplane version of the Schneider which is mentioned in Chapters 39 and 40.

The Rainbow at Brooklands with Dragonfly engine.

WALLABY

The history of this type is dealt with in Chapter 39.

The one only Wallaby which was rebuilt in Australia as an eight-seater and re-registered G-AUDU, shown October 3rd 1919.

ANTELOPE

Evolved from the Wallaby, the Antelope, powered by a 180 h.p. Wolseley Viper, had a cabin for two. Modified with ailerons of reduced chord and flown by Harry Hawker, it won £3,000 in the Air Ministry Competition of 1920 for small commercial aircraft. Another "one-off", it was also shipped to Australia to join the Wallaby on mail routes. Fitted first with a Hispano and then a Puma engine, it survived until 1935.

The Antelope at Brooklands, June 26th 1920. The undercarriage was modified as drawing shows.

A SOPWITH PUP CONTRACTOR

Left to right, top to bottom: The original works at the old Richmond Drill Hall and the Hanworth erecting sheds. Airframes built in the old works were transported to Hanworth in skeleton for covering and erection. A vast engineering scheme was launched to put the Cardinal River, feeding the Hampton Court lakes, underground to provide an airfield for testing the Pups. Military transport waiting to collect spares from the factory and the firm's own fleet of vehicles.

WHITEHEAD AIRCRAFT LIMITED

Left to right, top to bottom: Setting the Constantinesco CC gun synchronising gear and the installation of a 100 h.p. Gnôme Monosoupape engine; engine cowling and off-set Vickers gun with hydraulic lead pipe to engine; views in the erecting shops and the first production Pup by Whiteheads and a standard later production aircraft. All photographs on this page taken at Hanworth. The firm chose to call their contract-built Pups Whitehead Fighting Scouts.

CONSTRUCTIONAL

Notes: Full data is not available for early types produced before a drawing office was formally established.

Aircraft name and/or Designation	h.p.	Engine Type	Span Top ft. in.	Span Bottom ft. in.	Length ft. in.	Height ft. in.	Chord ft. in.	ft. in.	Gap ft. in.	Sweep-back in.	Stagger ft. in.
Tractor Biplane	80	Gnome	40 0	40 0	29 6		5 1½	5 1½	5 3	Nil	1 0
Tractor Biplane	100	Green	40 0	40 0	29 0		5 1½	5 1½	5 3	Nil	1 0
Tabloid	80	Gnome	25 6	25 6	19 6	8 5	5 0	5 1	4 6	Nil	— 11
Tabloid	80	Gnome	25 6	25 6	20 4	8 5	5 1½	5 1½	4 6	Nil	— 11
Sociable	100	Gnome	36 0	36 0	24 3	9 0	5 1½	5 1½	4 6	Nil	1 0
Bat Boat	90	Austro-Daimler	41 0	41 0	32 0	11 6	5 6	5 6	5 6	Nil	Nil
Bat Boat 1A	100	Green	41 0	41 0	30 4	11 6	5 6	5 6	5 6	Nil	Nil
Bat Boat II	200	Canton Unné	55 0	45 0			6 9	6 9	7 0	Nil	Nil
Bat Boat II	200	Sunbeam	55 0	45 0			6 9	6 9	7 0	Nil	Nil
Greek Seaplane	100	Anzani	50 0	50 0	31 0		5 1½	5 1½	5 6	Nil	1 0
Gunbus SPGN	100	Anzani	50 0	50 0	31 0		5 1½	5 1½	5 6	Nil	1 0
Gunbus	150	Sunbeam	50 0	50 0	32 6	11 4	5 1½	5 1½	5 6	Nil	1 0
Circuit 1914	100	Mono Gnome	36 0	36 0			5 1½	5 1½	5 7	Nil	1 0
Type 807	100	Mono Gnome	36 0	36 0					5 6	Nil	Nil
Two-Seat Scout	100	Mono Gnome	36 0	36 0			5 1½	5 1½	5 6	Nil	Nil
Schneider	100	Mono Gnome	25 8	25 8	22 10	10 0	5 2	5 2	4 6	Nil	0 8
Schneider	100	Mono Gnome	25 8	25 8	22 10	10 0	5 2	5 2	4 6	Nil	0 8
Baby	110	Clerget	25 8	25 8	22 10	10 0	5 2	5 2	4 6	Nil	0 8
Baby	130	Clerget	25 8	25 8	23 0	10 0	5 2	5 2	4 6	Nil	0 8
1½ Strutter	100	Clerget	33 6	33 6	25 3¾	10 3	5 6	5 6	5 4¾	Nil	2 0
1½ Strutter	130	Clerget	33 6	33 6	25 3¾	10 3	5 6	5 6	5 4¾	Nil	2 0
1½ Strutter	110	Clerget	33 6	33 6	25 3¾	10 3	5 6	5 6	5 4¾	Nil	2 0
1½ Strutter	110	Le Rhone	33 6	33 6	25 3¾	10 3	5 6	5 6	5 4¾	Nil	2 0
1½ Strutter	130	Clerget	33 6	33 6	25 3¾	10 3	5 6	5 6	5 4¾	Nil	2 0
1½ Strutter	110	Clerget	33 6	33 6	25 3¾	10 3	5 6	5 6	5 4¾	Nil	2 0
1A2	145	Clerget 9Bc	33 6	33 6	25 3	10 3	5 6	5 6	5 4¾	Nil	2 0
1A2	135	Le Rhone 9Jby	33 6	33 6	25 3	10 3	5 6	5 6	5 4¾	Nil	2 0
1B1	135	Clerget 9Bb	33 6	33 6	25 3	10 3	5 6	5 6	5 4¾	Nil	2 0
Pup	80	Le Rhone	26 6	26 6	19 3¾	9 5	5 1½	5 1½	4 5	Nil	1 6
Pup	80	Le Rhone	26 6	26 6	19 3¾	9 5	5 1½	5 1½	4 5	Nil	1 6
Pup	80	Le Rhone	26 6	26 6	19 3¾	9 5	5 1½	6 1½	4 5	Nil	1 6
Pup	100	Mono Gnome	26 6	26 6	19 3¾	9 5	5 1½	5 1½	4 5	Nil	1 6
Pup	80	Clerget	26 6	26 6	19 3¾	9 5	5 1½	5 1½	4 5	Nil	1 6
SB3D	80	Le Rhone	25 0	25 0	20 2½	8 1¼	5 1½	5 1½	4 5	Nil	1 6
Triplane	110	Clerget	26 6	(all 3)	18 10	10 6	3 3	(all 3)	3 0	Nil	1 6
Triplane	130	Clerget	26 6	(all 3)	19 6	9 9	3 3	(all 3)	on each	Nil	on each
Triplane	150	Hispano-Suiza	28 6	(all 3)	23 2	10 6	4 3	(all 3)	3 6	Nil	1 6
Triplane	200	Hispano-Suiza	28 6	(all 3)	23 2	10 6	5 1	(all 3)		Nil	on each
Camel FS1	130	Clerget	26 10	26 10	21 8	11 4	4 6	5 0	5 0	Nil	1 6
Camel F.1/1	150	Clerget 9Z	28 0	28 0	18 10	8 3	4 6	4 6	4 9	Nil	1 6
Camel F.1/3	130	Clerget 9Z	28 0	28 0	18 10	8 6	4 6	4 6	5 0	Nil	1 6
Camel F.1	130	Clerget 9Z	28 0	28 0	18 9		4 6	4 6	5 0	Nil	1 6
Camel F.1	140	Clerget 9Bf	28 0	28 0	18 9		4 6	4 6	5 0	Nil	1 6
Camel F.1	110	Le Rhone	28 0	28 0	18 9	8 6	4 6	4 6	5 0	Nil	
Camel T.F.1	100	Le Rhone	28 0	28 0	18 8	8 6	4 6	4 6	5 0	Nil	1 6
Camel F.1	150	Mono Gnome	28 0	28 0	18 6	8 6	4 6	4 6	5 0	Nil	1 6
Camel F.1	150	Mono Gnome	28 0	28 0	18 6	8 6	4 6	4 6	5 0	Nil	1 6
Camel F.1	180	Le Rhone	28 0	28 0	19 0	8 6	4 6	4 6	5 0	Nil	1 6
Camel F.1	150	Bentley B.R.1	28 0	28 0	18 6	8 6	4 6	4 6	5 0	Nil	1 6
Camel F.1	150	Bentley B.R.1	28 0	28 0	18 6	8 6	4 6	4 6	5 0	Nil	1 6

DATA TABLE

For values annotated, see Remarks column.

Dihedral	Incidence	Total Area	Tailplane Span		Tailplane Chord		Track		Built	Constructors	Remarks
Degrees	Degrees	sq. ft.	ft.	in.	ft.	in.	ft.	in.	Years		
2½	4	397	10	0	7	2	5	8	1912–14	Sopwith Works only	Known as 3-seat Biplane but
2½	4	397	10	0	7	2	5	8	1914	Service modification as above	used as two-seater
1½	1	240	9	6	4	1	4	8	1913	First Sopwith-built scout	Known as Tabloid Baby
1½	1	241	8	3	4	1	4	8	1914	Sopwith works only	Sopwith Drawing D458 16.6.14
1	1		10	6	5	4½	5	1	1914	Sopwith Works only	1914 R. J. Ashfield drawing
2	4	400					4	6	1913	Hulls made by S.E.	One only built
3	4	428	9	0	4	6	4	6	1913	Saunders, Isle of Wight	1913 Hull length 21 ft.
2		600+					Hull		1914	Superstructure by Sopwith	One sold to Germany
2		600+					Hull		1914		One only
3½	4	505					5	10	1914	Sopwith Works only	Greek Government
3½	4	505					5	10	1914	Sopwith Works only	Or 100 h.p. Gnome Monosoupape
3½	4	505	13	7	5	1	5	10	1914–15	Sopwith & Robey built	Track diameter relates to skis
2		330							1914	Sopwith Works only	Built for Round Britain race
2									1914–15	Sopwith Works only	Fins modified in service
2		330+					4	4	1914–15	Sopwith Works only	24 built
3	3	200+	8	0	4	6	4	6	1914	Sopwith Works only	1914 Schneider Cup entry
2	3	200+	8	0	4	6	4	6	1914–15	Sopwith Works only	137 built
2½	3	240	8	0	4	6	4	6	1915–17	Sopwith and Blackburn	Or 100 h.p. Gnome Monosoupape
2	3	240	9	5	4	6	3	6	1915–17		117 of all types built
2⅓	2⅓	349	13	6	4	6	5	6	1915–16	Sopwith, Fairey,	Basic design as per Sopwith
2⅓	2⅓	350	13	6	4	6	5	6	1916–17	Hooper, Morgan,	drawing D1179 dated 22.1.17
2⅓	2⅓	347	13	6	4	6	5	6	1915–17	Mann, Egerton, Ruston	Type 9400 and 9700 in
2⅓	2⅓	347	13	6	4	6	5	6	1916–17	Proctor, Wells and	R.N.A.S. for fighter and
2⅓	2⅓	346	13	6	4	6	5	6	1916–17	Westland	bomber versions
2⅓	2⅓	347	13	6	4	6	5	6	1916–17		
2⅓	2⅓	347	13	6	4	6	5	6	1917–18	Amiot, Bessoneau, Darracq,	80 h.p. Le Rhone 9C alternative
2⅓	2⅓	347	13	6	4	6	5	6	1917–18	Loire et Olivier,	when used as trainer in
2⅓	2⅓	347	13	6	4	6	5	6	1917–18	Hanriot, Sarazin & R.E.P.	French or American service
3	1½	254	10	1			4	7	1916–17		Tailpiece incidence 1½ degrees
3	1½	254	10	1			4	4	1916–17	Sopwith, Beardmore,	Sopwith Drawing D1326 2.12.16
3	1½	254	10	1			4	7	1916–17	Standard and	Director of Research figures
3	1½	252	10	1			4	7	1916–18	Whitehead	Tailplane incidence 2½ degrees
3	1½	254	10	1			4	7	1916–18		Operational 1917–17, Trainer 1918
3	1½	243	10	1			4	7	1916–17	Wm. Beardmore	Beardmore modification
2½	2	234	7	6	3	9	5	5¾	1916–17	Sopwith, Clayton &	Sopwith Drg. D.1720 by B. W. Tyler
2½	2	234	10	1†	3	9	5	3¾	1916–17	Shuttleworth and Oakley	†Later 8 ft.
	2	340+	13	6			5	3	1916	Sopwith Works only	Sopwith Drg. D1618 21.7.16
	2	340+	13	6			5	3	1916	Sopwith Works only	Experimental
5*	2	230	8	2	3	4	5	0	1917	Sopwith Works only	Camel Floatplane
5*	2		8	2½			4	6	1917		*N.B. No dihedral on Camel top wings.
5*	2		8	2½		2½	4	6	1917		100 h.p. Mono Gnome alternative
5*	2	231	8	2½			4	6	1917–18	Sopwith, Boulton	Director of Research figures
5*	2	231	8	2½			4	6	1917–18	Paul, British	French-built Clerget LS No. 2676
		231							1917–18	Caudron, Clayton &	Home Defence version
5*	2	231	8	2½			4	6	1917–18	Shuttleworth, Hooper;	Experimental only
5*	2	231	8	2½			4	6	1917–18	March, Jones & Cribb;	Test Report M216
5*	2	231	8	2½			4	6	1917–18	Nieuport & General,	Also Props AB644, AB705, AB8553
5*	2	231	8	2½			4	6	1917–18	Portholme and Ruston	Report M250
5*	2	231	8	2½			4	6	1916–18	Proctor.	Engine known as A.R.1 at test time
5*	2	231	8	2½			4	0	1917–18		Production

CONSTRUCTIONAL

For values annotated, see Remarks column.

Aircraft Name and/or Designation	Engine h.p.	Engine Type	Span Top ft. in.	Span Bottom ft. in.	Length ft. in.	Height ft. in.	Chord ft. in.	Gap ft. in.	Sweep-back in.	Stagger ft. in.
Camel 2F.1	130	Clerget	26 11	26 11	18 6	9 1	4 6	4 6	4 11	Nil	1 6
Camel 2F.1	150	B.R.1	26 11	26 11	18 8	9 1	4 6	4 6	4 11	Nil	1 6
B.1	200	Hispano-Suiza	40 6	40 6	27 9	10 9	6 3	6 3	6 0	Nil	Nil
B.1	200	Hispano-Suiza	38 6	38 6	27 0	9 6	6 3	6 3	6 0	Nil	Nil
B.1	200	Hispano-Suiza	38 6	38 6	27 6	9 8	6 3	6 3	6 0	Nil	Nil
Cuckoo I	200	Sunbeam Arab	46 8	45 9	28 6	11 6	6 3	6 3	6 0	Nil	Nil
Cuckoo I	200	Sunbeam Arab	46 8	45 9	28 10	11 0	6 3	6 3	6 0	Nil	Nil
Cuckoo II	200	Wolseley Viper	46 8	45 9	28 10	11 0	6 3	6 3	6 0	Nil	Nil
Cuckoo III	255	R-R Falcon III	46 8	45 9	29 0	11 0	6 3	6 3	6 0	Nil	Nil
Dolphin 5F.1	200	Hispano-Suiza	32 6	32 6	22 3	8 6	4 6	4 6	4 3	Nil	1 6*
Dolphin 5F.1	200	Hispano-Suiza	32 6	32 6	22 3	8 6	4 6	4 6	4 3	Nil	1 6*
Dolphin 5F.1	200	Hispano-Suiza	32 6	32 6	22 3	8 6	4 6	4 6	4 3	Nil	1 6*
Dolphin 5F.1	200	Hispano-Suiza	32 6	32 6	22 3	8 6	4 6	4 6	4 3	Nil	1 6*
Dolphin 5F.1	200	Hispano-Suiza	32 6	32 6	22 3	8 6	4 6	4 6	4 3	Nil	1 6*
Dolphin 5F.1	200	Hispano-Suiza	32 6	32 6	22 3	8 6	4 6	4 6	4 3	Nil	1 6*
Dolphin II	300	Hispano-Suiza	32 6	32 6	22 3	8 6	4 6	4 6	4 3	Nil	
Dolphin III	200	Hispano-Suiza	32 6	32 6	22 3	8 6	4 6	4 6	4 3	Nil	1 6*
Snipe 7F.1	230	Bentley B.R.2	31 1	30 0	19 10	9 6	5 0	5 0	4 3	Nil	1 4
Snipe 7F.1	230	Bentley B.R.2	25 9	25 9	18 9	8 3	5 0	5 0	4 3	Nil	1 3
Snipe 7F.1	230	Bentley B.R.2	30 0	30 0	19 2	9 6	5 0	5 0	4 3	Nil	1 4
Snipe 7F.1	230	Bentley B.R.2	31 1	30 0	19 10	9 6	5 0	5 0	4 3	Nil	1 4
Snipe LD	230	Bentley B.R.2	31 1	30 0	19 5	8 11	5 0	5 0	4 3	Nil	1 4
Snipe LD	230	Bentley B.R.2	30 0	30 0	19 5	8 11	5 0	5 0	4 3	Nil	1 4
Bulldog I	200	Clerget 11EB	34 6	34 6	23 0	8 5½	5 6	5 6	4 6	Nil	1 3
Bulldog II	360	A.B.C. Dragonfly	34 6	34 6	23 2	8 9	5 6	5 6	4 6	Nil	1 3
Rhino 2B.2	230	B.H.P.	33 0	(all 3)	27 8	10 11	6 0	6 0	4 0	Nil	0 5
Hippo 3F.2	200	Clerget 11EB	38 9	38 9	24 6	9 4	5 0	5 0	4 6	Nil	2 3
Hippo 3F.2	200	Clerget 11EB	38 9	38 9	24 6	9 4	5 0	5 0	4 6	Nil	1 9½
Buffalo	230	Bentley B.R.2	34 6	34 6	33 3½	9 6	5 6	5 6	4 6	Nil	1 3
Swallow	110	Le Rhone 9J	28 10	— —	18 9	10 2	5 9	— —	— —	0 6	— —
Snail I	170	A.B.C. Wasp	25 9		18 9	8 3				Nil	— 5
Snail II	170	A.B.C. Wasp	25 9		19 0	7 10				Nil	
Snark	360	A.B.C. Dragonfly 1A	26 6	(all 3)	20 9	10 1	4 6	4 6	2 10½*	Nil	1 0½*
Salamander T.F.1	230	Bentley B.R.2	30 1½*	30 1½	19 6	9 6	5 0	5 0	4 3	Nil	1 5
Salamander T.F.1	230	Bentley B.R.2	31 2½*	30 1½	19 9	9 4	5 0	5 0	4 3	Nil	1 5
Snapper	360	A.B.C. Dragonfly	28 0	28 0	20 7	10 0					
Dragon	360	A.B.C. Dragonfly	31 1	30 0	21 9	9 6	5 0	5 0	4 3	Nil	1 4
Cobham I	360	A.B.C. Dragonfly 1A	54 0	(all 3)	38 0	13 0			4 6	Nil	
Cobham II	290	Siddeley Puma	54 0	(all 3)	38 0	13 0			4 6	Nil	
Gnu	230	Bentley B.R.2	38 1	38 1	25 10	9 10	5 6	5 6	5 0	Nil	1 2
Dove	80	Le Rhone	24 9½	24 9½	19 4	9 0	5 1½	5 1½	4 4	0 5	1 6
Grasshopper	100	Anzani	33 1	33 1	23 1	9 0	5 0	5 0	4 9	Nil	1 5
Atlantic	375	R-R Eagle 8	46 6	46 6	32 0	11 0	6 3	6 3	6 0	Nil	0 3
Schneider	450	Cosmos Jupiter	24 0	24 0	21 6	10 0	5 1½	5 1½	4 6	Nil	— 2½
Rainbow	320	A.B.C. Dragonfly	24 0	24 0	18 0					Nil	
Wallaby	360	R-R Eagle 8	46 6	46 6	31 6	10 8	6 3	6 3	6 0	Nil	0 3
Antelope	180	Wolseley Viper	46 6	46 6	31 0	11 3	6 3	6 3			Nil	1 6

DATA TABLE

Dihedral	Incidence	Total Area	Tailplane Span		Chord		Track		Built	Constructors	Remarks
Degrees	Degrees	sq. ft.	ft.	in.	ft.	in.	ft.	in.	Years		
5½*	2		8	2½			4	6	1917–18	⎱ Sopwith, Arrol Johnson,	*No dihedral on Camel top wings
5½*	2		8	2½			4	6	1917–18	⎰ Beardmore and Hooper	
2½	3	488	11	6			5	3	1917	Sopwith Works only	⎱ B1496 only
2½	2	469	11	6			5	3	1918	Modification	⎱ Documented figures
2½	2	475	11	6			5	3	1917	Modification	⎰ vary
2½	3	582+	11	10	5	6	7	7	1917	Sopwith Works only	Re-engined prototype
2½	3	566	11	9	5	6	7	6	1918	Blackburn, Pegler & Fairfield	First production
2½	3	566	11	9	5	6	7	6	1918–19	Service modification	Became standard version
2½	3	566	11	9	5	6	7	6	1920	Service modification	One only
2½	1¾	258	10	6	3	10½	5	0	1917–19	⎱ Over 1559 built by	Sopwith drawing D2267/1 14.11.17
2	1¾	263+	10	10¾	3	10½	5	0	1917–18	Sopwith, Darracq	Special endurance test C3777
2	1¾	263+	10	10¾	3	10½	5	0	1917–18	and Hooper	Test of C3777
2	1¾	263+	10	10¾	3	10½	5	0	1918		Test of C3778
2	1¾	263+	10	10¾	3	10½	5	0	1918		
2	1¾	263+	10	10¾	3	10½	5	0	1918	⎰	
—	—	—	—		—		—		1918		
2	1¾	263+	10	10¾	3	10½	5	0	1918		Ungeared engine
4	2	271	9	2	3	6	5	0	1918		Sopwith Drawing 30 Oct 18
4	2	230+	8	0	3	6	5	0	1918	⎱ Modification as Dolphin 5F.1	First Prototype
4	2	256+	9	2	3	6	5	0	1918–19	Various constructors as given on page 164	Unbalanced ailerons
4	2	256+	9	2	3	6	5	0	1918–19	⎰	Balanced ailerons
4	2	269	9	2	3	6	5	0	1918	Sopwith Works only	⎱ Three only. Tests on E8089
4	2	269	9	2	3	6	5	0	1918	Sopwith Works only	⎰ Snipe (Long Distance) or Mk.IA
2½	2	335+	11	1½	4	6½	5	0	1918	Sopwith Works only	Span 26 ft. 6 in. single bay version
2½	2	335+	11	1½	4	6½	5	0	1918	Sopwith Works only	B.R.2 engine also fitted
2½	3	345	12	0					1918	Sopwith Works only	Two built × 7 & × 8
3	2	340+	11	3			5	0	1917	Sopwith Works only	Clerget 30003/WD/4180 *Negative
5	2	340+	11	3			5	0	1917	Sopwith Works only	*Negative
2½	1¾	375	11	1½	4	0½	5	0	1918	Sopwith Works only	Two built
Nil	2	181	8	2½	3	400			1918–19	Sopwith Works only	Monoplane
		250+	9	0	3	6			1917–18	Sopwith Works only	Conventional fuselage C4284–7
		250	9	0	3	6			1918	Sopwith Works only	Monocoque fuselage C4288–9
		322							1918	Sopwith works only	*top 3 6½ lower—to 1 9
4	2	272+	9	2					1918–19	⎱ Sopwith and	*Unbalanced ailerons
4	2	272+	9	2					1918–19	⎰ Glendower	*Balanced ailerons
		292+							1918–19	Sopwith Works only	High altitude fighter F7031–3 only
4	2	271+	9	2	3	6	5	0	1918–19	Sopwith Works only	Limited production
									1918–19	Sopwith Works only	First prototype H672.
									1918–19	Sopwith Works only	High compression Puma H671.
2½	2	407	11	0½	4	5⅜	5	6	1919–20	Sopwith Works only	4-inch shorter with Le Rhone.
3	2	224					4	8	1919–20	Ten by Sopwith Works	
4	2	312+	10	7	4	5⅜	5	0	1919	Sopwith Works only	Drawing D3208. G-EAIN only
2½	2½	575	11	4	5	2	4	6	1919	Sopwith Works only	Sopwith drawing D0004. 1 only
		222+							1919	Sopwith Works only	G-EAKI only
									1920/22	H.G. Hawker rebuild	G-EAKI modified
2½	1½	583	11	6	5	5	5	9	1919	Sopwith Works only	G-EAKS for England-Australia flight
	1½	100					5	3	1919	Sopwith Works only	G-EASS to Drawing D3250

PERFORMANCE

Notes: Propeller type prefixes as follows: AB = Air Board (later Air Ministry), AD = Air Department (of Admiralty), IPC = Integral Propeller Company, L or LP = Lang Propeller Company, P = British & Colonial Aeroplane Company (Bristol).

Aircraft Name and/or Designation	Engine h.p.	Type	Type or Function	Propeller Type D = Diameter P = Pitch Figures in millimetres	Armament
Tractor Biplane	80	Gnome	Three-seat General Purpose	2-blade 2600D Chauviere	Nil planned
Tractor Biplane	100	Green	Two/Three seat	2-blade Integral	Nil planned
Tabloid	80	Gnome	Single-seat prototype	2-blade	Nil planned
Tabloid	80	Gnome	Single-seat Scout	2-blade 8 ft. 3 in. diameter	Nil standardised one Lewis
Sociable	100	Gnome	Dual control 2-seater	2-blade 8 ft. 6 in. diameter	Nil planned or carried
Bat Boat	90	Austro-Daimler	Two-seat amphibian	2-blade	Nil planned
Bat Boat 1A	100	Green	Two-seat amphibian	2-blade	Nil planned
Bat Boat II	200	Canton-Unne	Two-seat flying boat	2-blade	Nil planned
Bat Boat II	200	Sunbeam	Two-seat flying boat	2-blade	Nil planned
Greek Seaplane	100	Anzani	Two-seat pusher biplane	2-blade	Provision for machine gun
Gunbus	100	Anzani	Two-seat pusher biplane	2-blade	Not standardised
Gunbus	150	Sunbeam	Two-seat pusher biplane	4-blade 2900D 2000P	Single Lewis gun in nose
Circuit 1914	100	Mono Gnome	Contest floatplane	2-blade	Not applicable
Type 807	100	Mono Gnome	Two-seat floatplane	2-blade Chauviere	Not standardised
Two-seat Scout	80	Mono Gnome	Two-seat observation	2-blade Chauviere 2600D	Not standardised
Schneider	100	Gnome Mono	Racing floatplane	2-blade 2650D 2100P	Not applicable
Schneider	100	Gnome Mono	Single-seat floatplane	2-blade	Various light bombs mgs.
Baby	110	Clerget	Single-seat floatplane	2-blade LP690 or AD500	Lewis gun, Rankin dart,
Baby	130	Clerget	Single-seat floatplane	2-blade LP690 or AB555B	Le Prier rockets
1½ Strutter	110	Clerget 92	Single-seat bomber	2-blade LP710C	Fixed Vickers and bombs
1½ Strutter	130	Clerget 9B	Single-seat bomber	2-blade LP710	Fixed Vickers and bombs
1½ Strutter	110	Clerget 9J	Two-seat fighter	2-blade 2740D 2120P	Fixed Vickers and
1½ Strutter	110	Le Rhone 9J	Two-seat fighter	2-blade LP710C	Lewis in gun ring
1½ Strutter	130	Clerget 9B	Two-seat fighter	2-blade LP710C	with provision for
1½ Strutter	110	Clerget 9J	Two-seat fighter/bomber	2-blade LP710C	4 × 65 lb. bombs
1A2	145	Clerget 9Bc	Two-seat *Corps d'Armee*	2-blade	Fixed Vickers and Tour
1A2	135	Le Rhone 9Jby	Two-seat *Corps d'Armee*	2-blade	machine gun in gun ring
1B1	135	Clerget 8Bb	One-seat Bomber	2-blade	Vickers & 150 kgs. bombs
Pup	80	Le Rhone	Single-seat scout	2-blade LP1020	Fixed Vickers
Pup	80	Le Rhone	Single-seat scout	2-blade 2600D, 2200P	Fixed Vickers
Pup	80	Le Rhone	Single-seat scout	2-blade LP1020	Fixed Vickers
Pup	100	Mono Gnome	Single-seat scout	2-blade P3012 or Vickers 57	Fixed Vickers
Pup	80	Clerget	Single-seat scout	2-blade P43	Fixed Vickers
SB3D	80	Le Rhone	Ship's aeroplane	2-blade P43	Lewis gun only
Triplane	110	Clerget 9Z	Single-seat scout	2-blade 2740D 2480P	Fixed Vickers
Triplane	130	Clerget 9B	Single-seat scout	2-blade LP2100 or AD553	Fixed Vickers
Triplane	150	Hispano-Suiza	Experimental triplane	2-blade	Fixed Vickers gun
Triplane	200	Hispano-Suiza	Experimental triplane	2-blade	Fixed Vickers gun
Camel FS1	130	Clerget	Single-seat fighter	2-blade 2600D, 2270P	One Vickers, one Lewis
Camel F1/1	130	Clerget	Single-seat fighter	2-blade	Twin fixed Vickers
Camel F.1/3	130	Clerget	Single-seat fighter	2-blade	Twin fixed Vickers
Camel F.1	130	Clerget	Single-seat fighter	2-blade	Twin fixed Vickers
Camel F.1	140	Clerget 9Bf	Single-seat fighter	2-blade L11927 or AD644	Twin fixed Vickers
Camel F.1	110	Le Rhone	Single-seat fighter		Twin fixed Vickers
Camel T.F.1	110	Le Rhone	Trench fighter	2-blade	Vickers/Lewis combinations
Camel F.1	150	Mono Gnome	Single-seat fighter	2-blade AD644 2660D	Twin Vickers guns
Camel F.1	150	Mono Gnome	Single-seat fighter	2-blade LP3640	Twin Vickers guns
Camel F.1	180	Le Rhone	Single-seat fighter	2-blade L5440 2640D	Twin Vickers guns

DATA TABLE

Notes: Metric figures, as indicated for French-built aircraft, kg = kilograms, kph = kilometres per hour, m = metres.
Full documentation does not exist for early types.

Weights Empty lbs.	Loaded lbs.	Petrol Gallons	Oil Gallons	Endurance Hours	Speeds at height m.p.h./feet	Rate of Climb Mins. to height in feet	Ceiling ft.	Remarks
1060	1810	25	6	$2\frac{1}{2}$	73·6 maximum	$2\frac{1}{3}/1000$		First complete Sopwith design
1100	1550	27	6	$2\frac{1}{2}$	70 maximum	15/12,900		Service modification
680	1060	25	6	$2\frac{1}{2}$	93 maximum	1/1200		Stalling speed 37 m.p.h.
730	1120	25	$5\frac{1}{2}$	3	93 maximum	1/1200		Known as Sopwith Scout 1914-15
960	1640	25	4	3	90 maximum			Built for Winston Churchill
1200	1700				65			
					65			Front elevator 8 × 2 ft.
2300	3180	70	7	$4\frac{1}{2}$	70 maximum			
2300	3180	70	7	5	75 maximum			
				$2\frac{1}{2}$	80 maximum		4000	Alternatively 100 h.p.
				$2\frac{1}{2}$	80 maximum		4000	Mono Gnome engine
				$2\frac{1}{2}$	80 maximum	3500/15	4000	Landplane version
1550	2190	35		$3\frac{1}{2}$	80 maximum			One built for R.N.A.S.
				$3\frac{1}{2}$	80 maximum	30/6500	3000	Known as Sopwith Folders
				$3\frac{1}{2}$	69 Ground Level	20/3000	3000	Known as Spinning Jenny
1220	1700	30	$5\frac{1}{2}$	2	87 maximum	15/6500, 30/100000	7000	Landplane and floatplane
1220	1700	30	$5\frac{1}{2}$	3	87 maximum	15/6500, 30/10000	7000	All floatplanes
1226	1580	34	$6\frac{1}{2}$	2	100 sea level	$1\frac{1}{2}/1000$, 24/10000	8000	Armament varied, 65lb. bombs
1286	1715	25	6	$2\frac{1}{4}$	100 sea level	35/10000	8000	Armament varied
1354	2362	40	$9\frac{1}{2}$	4	101/6000, 94/10000	14/6500, 27/10000	14500	Performance varied
1316	2352	40	$9\frac{1}{2}$	$3\frac{3}{4}$	102/6500, 98·5/15000	24/10000, 42/15000	15500	greatly according
1160	1910	40	$9\frac{1}{2}$	$4\frac{1}{4}$	97/10000, 86/14000	19/10000, 41·2/15000	18000	to equipment for
1281	2205	40	$9\frac{1}{2}$	4	103/10000	$10\frac{1}{2}/6500$, 19/10000	16000	tasks allotted
1305	2150	40	$9\frac{1}{2}$	$3\frac{3}{4}$	97/10000, 87/15000	17·8/10000, 42/15000	19000	
1195	2205	43	$9\frac{1}{2}$	4	100/600, 94/10000	14/65000, 30/10000	15600	Director of Research figures
626 kg.	926 kg	40	$9\frac{1}{2}$	$4\frac{1}{2}$	161kph/3000m, 150kph/4000m	24/3000m 41/4000m	1500	French production model
626 kg	926 kg	40	$9\frac{1}{2}$	$4\frac{1}{2}$	156 kph/3000m	11/2000m, 18/3000m	1500	French production model
600 kg	1000 kg	40	$9\frac{1}{2}$	5	102/6500, 99/15000	24/10000, 40/15000	16000	French production model
787	1225	$19\frac{1}{4}$	4	3	103/7000, 101/1100	$6\frac{1}{3}/6000$, 14/10000	17500	Test 21.10.16
787	1225	$19\frac{1}{4}$	4	3	$104\frac{1}{2}/10000$, 94/15000	14·5/1000, 30·1/15000	20000	Official figures released to French
780	1234	$19\frac{1}{4}$	4	$2\frac{1}{2}$				or 110 Le Rhone with LP710C
856	1297	$19\frac{1}{4}$	4	$1\frac{3}{4}$	104/10000, 100/15000	12·5/10000, $23\frac{1}{2}/15000$	21000	Official figures released to French
850	1290	$19\frac{1}{4}$	4	$2\frac{1}{2}$				LP1020A or IPC2360 props
890	1289	17	4	$2\frac{1}{2}$	89 kts/20000, 79 kts/10000	31/6/2000, 42/6500	12400	Grain test 12.1.18 of N6708
993	1415	20	4	$2\frac{1}{2}$	$112\frac{1}{2}/6500$ 95/1500	$6\frac{1}{2}/6500$, 105/15000	20500	110 Le Rhone alternative
1103	1543	20	4	2	116/6000, 114/10000	$6\frac{1}{2}/6500$, 12/10000	22000	Few had twin fixed Vickers
					120 maximum	9/10000		One only N509
					120 maximum	9/1000		One only N510
								Floatplane
950	1482	37		$2\frac{3}{4}$	$112\frac{1}{2}/10000$, 106/15000	6/6500, $10\frac{1}{2}/10000$	19000	Non standard
929	1453			$2\frac{1}{2}$	113/1000, $106\frac{1}{2}/15000$	6/6500, $10\frac{1}{2}/10000$	19000	Clerget No. 2730
956	1455			$2\frac{1}{2}$	105/10000, 1·3/15000	8·5/10000	24000	Director of Research figures
928	1452			$2\frac{1}{2}$	101/10000, 98/15000	$10\frac{1}{2}/10000$, 20/15000		N518 tested Jan. 1918
889	1422			$2\frac{1}{2}$	$118\frac{1}{2}/1000$, $111\frac{1}{2}/15000$	5/6500, 91/10000	24000	
		37	$2\frac{1}{2}$		118 ground level	N/A		400 lbs. armour plate
	1505				114·5/10000, $106\frac{1}{2}/15000$	9·3/1000, $17\frac{1}{2}/15000$	20500	Test of D657 July 18
993	1523	$27\frac{1}{2}$	$5\frac{1}{4}$	$2\frac{1}{4}$	112/1000, 107/15000	11/10000, 21/15000	19700	Test of F1336 Sep. 18
1048	1567	$26\frac{1}{4}$	$5\frac{1}{4}$	$2\frac{1}{4}$	113/10000, 108.5/15000	9·5/10000, $17\frac{1}{2}/15000$	21780	Test of F6394 Nov. 18/Mar. 19

PERFORMANCE

Aircraft Name and/or Designation	Engine h.p.	Engine Type	Type or Function	Propeller Type D = Diameter P = Pitch Figures in millimetres	Armament
Camel F.1	150	B.R.1	Single-seat fighter	2-blade LP3640 2630 2500P	Twin fixed Vickers
Camel F.1	150	B.R.1 HC	Single-seat fighter	2-blade	Twin Vickers mgs.
Camel 2F.1	130	Clerget	One-seat shipborne fighter	2-blade	Fixed Vickers plus Lewis
Camel 2F.1	150	B.R.2	One-seat shipborne fighter	2-blade	Fixed Vickers plus Lewis
B.1	200	Hispano-Suiza	Bomber	2-blade 3050D, 2590P	Fixed Lewis mg., and up to 560 lb. bombs
B.1	200	Hispano-Suiza	Bomber	2-blade L.5150	
B.1	200	Hispano-Suiza	Bomber	2-blade L.P. 3280	
Cuckoo I	200	Sunbeam Arab	Torpedo ship aeroplane	2-blade AB8210	18 in torpedo of 1099 lbs.
Cuckoo I	200	Sunbeam Arab	Torpedo ship aeroplane	2-blade 2590D 2980P	18 in. torpedo of 1099 lbs.
Cuckoo II	200	Wolseley Viper	Torpedo ship aeroplane	2-blade AB6623 2550D	18 in. torpedo of 1099 lbs.
Cuckoo III	275	R-R Falcon III	Torpedo ship aeroplane	2-blade	18 in. torpedo of 1099 lbs.
Dolphin 5F.1	200	Hispano Suiza	Single-seat fighter	2-blade LP3500A	Twin fixed Vickers plus 1 or 2 Lewis (Performance given for various combinations)
Dolphin 5F.1	200	Hispano Suiza	Single-seat fighter	2-blade Lang 3800	
Dolphin 5F.1	200	Hispano Suiza	Single-seat fighter	2-blade Lang L3800	
Dolphin 5F.1	200	Hispano-Suiza	Single-seat fighter	2-blade Lang L.3610	
Dolphin 5F.1	200	Hispano Suiza	Single-seat fighter	2-blade Lang L361Q	
Dolphin 5F.1	200	Hispano Suiza	Single-seat fighter	2-blade AB7673 D2414	
Dolphin II	300	Hispano Suiza	Single-seat fighter	2-blade	Twin Vickers and Lewis
Dolphin III	200	Hispano Suiza	Single-seat fighter	2-blade AB662c 2400D 1750P	Twin Vickers and Lewis
Snipe 7F.1	230	Bentley B.R.2	Single-seat fighter	2-blade L400	Twin Vickers plus Lewis
Snipe 7F.1	230	Bentley B.R.2	Single-seat fighter	2-blade L4040, 2780D	Twin Vickers plus Lewis
Snipe 7F.1	230	Bentley B.R.2	Single-seat fighter	2-blade L4040	Twin fixed Vickers
Snipe 7F.1	230	Bentley B.R.2	Single-seat fighter	2-blade LP5300 2780D 2920P	Twin fixed Vickers
Snipe LD	230	Bentley B.R.2	Long distance fighter	2-blade Lang LP4040	Twin Vickers 90 lb. ammunition
Snipe LD	230	Bentley B.R.2	Long distance fighter	2-blade Lang LP5300	Twin Vickers 90 lb. ammunition
Bulldog I	200	Clerget 11EB	Two-seat armed reconnaissance	2-blade LP4030 8 ft. 6 in. dia.	Twin Vickers and two Lewis guns
Bulldog II	360	A.B.C. Dragonfly		2-blade LP	
Rhino 2B.2	230	B.H.P.	Two-seat bomber	2-blade BR62627 9ft. 6in. dia.	Vickers & Lewis plus bombs
Hippo 3F.2	200	Clerget 11EB	Two-seat fighter	2-blade Lang L4030 G.122	Twin Vickers and Twin Lewis
Hippo 3F.2	200	Clerget 11EB	Two-seat fighter	2-blade	Twin Vickers and a Lewis
Buffalo	230	Bentley B.R.2	Two-seat contest aircraft	2-blade 8 ft. 6 in. L.5390	Vickers and Lewis
Swallow	110	Le Rhone	Single-seat fighter	2-blade 7 ft. 5 in. diameter	Twin fixed Vickers
Snail I	170	A.B.C. Wasp	Single-seat fighter	2-blade	Fixed twin Vickers + 1 Lewis
Snail II	170	A.B.C. Wasp	Single-seat fighter	2-blade	Planned as above
Snark	360	A.B.C.Dragonfly1A	Single-seat fighter	2-blade	Twin Vickers, 4 Lewis
Salamander T.F.2	230	Bentley B.R.2	Single-seat armoured trench fighter	2-blade LP5300 2780D 2920P	Two-fixed Vickers, 4 × 20 lb. bombs. Lewis additionally
Salamander T.F.2	230	Bentley B.R.2		2-blade	
Snapper	360	A.B.C.Dragonfly1A	Single-seat fighter	2-blade L5140	Twin fixed Vickers
Dragon	360	A.B.C.Dragonfly1A	Single-seat fighter	2-blade LP5140A 2670D	Twin fixed Vickers
Cobham I	360	A.B.C.Dragonfly1A	Three-seat bomber	2 × 2 blade	Two Lewis and bombs
Cobham II	290	Siddeley Puma	Three-seat bomber	2 × 2 blade	Two Lewis and bombs
Gnu	230	Bentley B.R.2	Three-seat Transport	2-blade 8 ft. 6 in. diameter	Not applicable
Dove	80	Le Rhone	Two-seat tourer	2-blade 8 ft. 7 in. diameter	Not applicable
Grasshopper	100	Anzani	Dual-control Trainer	2-blade 8 ft. 3 in. diameter	Not applicable
Atlantic	375	R-R Eagle 8	Transatlantic Biplane	2-blade 9 ft. 6 in. diameter	Not applicable
Schneider	450	Cosmos Jupiter	Racing floatplane	2-blade	Not applicable
Rainbow	320	A.B.C. Dragonfly	Racing aeroplane	2-blade	Not applicable
Wallaby	360	R-R Eagle 8	Two-seat transport	2-blade 12 ft. diameter	Not applicable
Antelope	180	Wolseley Viper	Three-seat Commercial	2-blade	Not applicable

DATA TABLE

Weights Empty lbs.	Loaded lbs.	Petrol Gallons	Oil Gallons	Endurance Hours	Speeds at Height	Rate of Climb Mins. to Height in feet	Ceiling ft.	Remarks
977	1508	37	5¼	2½	111/10000, 103/15000	9·8/1000, 20/15000	20000	Also AB644, LP3510 propellers
946	1470	37	5¼	2½	121/10000, 114½/1500	8·3/10000, 16/15000	23500	Official figures released to French
956	1523	37	5½	2½	114/10000, 104/15000	6¼/5500, 11½/10000	12000	Detachable fuselage
036	1530	37	5½	2½	122/11000, 117/15000	6/6500, 11½/10000	17300	Detachable fuselage
700	3000	55	9	3½	118/10000, 95/15000			Drawing by Comford
710	3035	56	9		110/10000, 98·5/15000	16½/10000, 34/15000	21000	Engine H.S.17530/A36273
700	2945	56	9	3¾	118½/10000, 97½/15000	15½/10000, 29½/15000	22000	Engine H.S.7245.130
013	3711	57	6	4	81 kts/5000, 79 kts/10000	8½/5000, 22/15000	13950	Test of N74
199	3883	56	6	4–6	90 kts/5000, 85 kts/10000	11·1/5000, 31/10000	12100	Test of N6954
233	3974	56	6	4	90 kts/6500	30/10000	12000	Or prop AB8224 2940D
585	4350	56	6		101/6500, 95/10000	12½/6500, 23½/10000	13400	Tested October 1919
406	1881	27	4	1¾	128½/10000, 124/15000	8½/10000, 15/15000	25000	Official figures released to French
406	1911*	27	4	2	117–119	18½/14000		January 1917. *Up to 2068
400+	2003	27	4	2	120/10000, 123/15000	11/10000, 20/15000		Engine 115131/WD14305
400+	1990	26½	4¾	2	120½/13000, 116½/15000	16½/10000, 120/15000	21000	Engine 16452/WD14335
400+	1925	27	4	2		10·0/10000, 18/15000		Engine (Brazier) 16136
400	2000	27	4	2	117/10000, 110/15000	11/10000, 22/15000	19000	Ungeared engine
566	2358	27	7	2	140/10000, 133/16400	5/6500 8½/10000	24600	Limited numbers
466	2000	27	7	2	117/10000, 110/15000	6½/6500, 11½/10000	19000	Late production
200+	1674	36	7	2	125/12000, 119/15000	7½/10000, 14 5/6/15000		Dec. 17/Jan. 18 tests
250+	1964	36	7¾	2¼	115/13000, 110/15000	9/10000, 18/15000	19400	Test of B9965 June 1918
300+	1950	43	7½	2	118/10000, 112½/15000	8/10000, 18/15000		Test of B9966 June 1918
262	2028	40	7	2½			19000	Director of Research figures
329	2271	60	10½	4½	114/10000, 103/15000	5/5000, 12½/10000		BR2 32990A/WD48700
329	2271	60	10½	4½	117/1000, 103/15000	5¾/5000, 14½/10000		BR2 80A/WD48272
441	2495			2	109/10000, 101½/1300	8½/6500, 15½/10000	15000	Sopwith 2 F.R.2 Also B.R.2 engine
	3100			2		5½/1600, 9½/10000		Dragonfly Mk. 1A
590	3590	60	7½	3¾	103/1000, 91/13000	24¾/10000, 49½/13000	12000	As Report M167B on X7
867	2590				115/10000, 93/15000	13·4/10000, 28·6/15000	18000	Preliminary trials Jan. 1918
481	2590	40½	9		115½/1000, 106·5/13000	2/5000, 28½/15000	17000	General trials
175	3071	25			114/1000, 105½/6500	5/3000, 17/6500	9000	Tested February 1919
889	1420				113½/10000, 105/15000	5½/6500, 9/10000	18500	Monoplane, Camel fuselage
	1478				114½/60000, 112½/10000	6½/60000, 13/10000		Sopwith 8F.1
390	1920				124½/10000, 121/15000	10/10000, 194/15000		Sopwith 8F.1
	2283				130/3000			High Altitude fighter F4068–70
844	2513	29		1½	123½/6500, 117/10000	9½/6500, 17½/10000	13000	Composite figures. Test
852	2613	28		2	124½/5000, 117/10000	6½/5000, 17/10000	13000	of E5451 and D. of R. figures
462	2190				140/3000, 139/6500	4½/6500, 8/10000	23000	Spinner fitted on one
	2132				150 maximum	7½/10000	25000	Based on Snipe
								Triplane configuration
	6300							Triplane configuration
	3350*			3½	93/91 (230 or 110 h.p.)	645 ft./min. initial 7/500		*1350/2160 lbs. Le Rhone
065	1430			2½	100 maximum	7½/5000		W. G. Carter approved 3.6.19
	1670				90 maximum			W. G. Carter
000	6·50	330	24		118 max. 105 Cruising	N.B. rigged for	13000	Transport was Sopwith name
	2200	400			175 maximum	racing		For 1919 Schneider Trophy
					165 to 175 maximum*			*With 500 h.p. Bristol Jupiter
780	5200	200			115 max. 107 cruising			Drg. D0003 for Australia
387	3450			4	110 max. 84 cruising	7·5/5000		G-EASS, became G-AUSS

SUMMARY OF SOPWITH SURVIVORS AND REPLICAS

TABLOID
No surviving example, but a replica was built by personnel at R.A.F. Finningley, Yorkshire.

BABY
No whole surviving example, but from parts of two that had been stored in England, a replica-cum-restoration Blackburn Baby was built 1969-1970 at the Fleetlands Base, Gosport, for exhibition in the Fleet Air Arm Museum.

1½ STRUTTER
A French built example has survived in Belgium, see Chapter 36 and Type-by-Type Review.

PUP
Again no genuine survivor, but the last Dove G-EBKY was rebuilt as a Pup by R. O. Shuttleworth in 1937-1938. This is the aircraft currently owned by the Shuttleworth Trust, marked N5180. Ten replicas have been built by various enthusiasts, chiefly in America. Of these perhaps the most famous is the one on display at Cole Palen's Museum, Rhinebeck, New York.

Pup replica built by Richard King.

TRIPLANE
N5912, retained postwar, participated in several R.A.F. Pageants at Hendon between the wars; it is also illustrated on page 157. After being stored during the Second World War at Kemble, it was withdrawn in recent years, restored at Henlow, and is being exhibited at the R.A.F. Museum, Hendon. The single example supplied to Russia is still preserved, see Type-by-Type Review. A replica of N5492, built by Carl Swanson in 1966, is in the Canadian National Aeronautical Collection.

Triplane N5912 at the R.A.F. Royal Review, Abingdon, 1968.

CAMEL
One F.1 survives in Brussels (see page 152) and F6314 at the R.A.F. Museum Hendon. Another in America was formerly with the Tallmantz collection at Santa Ana, California. There is a replica at the U.S. Air Force Museum and some eight other replicas owned by enthusiasts in the U.S.A.

The most famous of the Sopwith survivors in the U.S.A., the Frank Tallman Camel in 1957.

Two 2F.1s survive. N6812 (see page 129) was stored on the East Coast after the war, exhibited at the Schoolboy's Own Exhibition 1933-1934, and is today in the Imperial War Museum London, and N8156 in the Canadian War Museum.

A Slingsby Camco 2A built recently for filming purposes to portray a Camel flown by 'Biggles' of boys' fiction.

SNIPE
Snipe E6938 went to America in the 'twenties for film work and was later restored by Jack Canary. It is now part of the Canadian National Aeronautical Collection and was recently on exhibition in England. The Collection also have the fuselage of Barker's Snipe E8102 (see page 119). Another Snipe in the Cole Palen Collection is marked E8100.

DOLPHIN
Parts of a Dolphin have been collected and stored at R.A.F. Henlow for reconstruction and eventual exhibition.

BRED IN TRADITION

The same site in a shifting scene. Snipes and Salamanders late 1918 at the Sopwith Ham Works, and Hawker Siddeley Harriers being built for the R.A.F. and U.S. Marine Corps shown in a photograph specially taken from the same viewpoint in October 1970. An electrically-driven 4-ton travelling crane replaces the 10-cwt chain lifting tackle, the roof trusses have been reinforced and the men and their tools have changed, but the excellence of the product is in the Sopwith tradition.

Index of Personalities

AIRD, H. R. — 86
Alcock, Sir John — 88, 136
Alston, Capt. R. C. W. — 30-31
Alston, Reginald — 51
Armstrong, Capt. D. V. — 121
Ashfield, R. J. — 32, 100, 116
Atholl, Duke of (Marquess of Tallibardine) — 19, 41
BALDWIN, T. S. — 24
Bankes-Price, Flt. Lt. — 64-65
Bannerman, Sir Alexander — 19
Bannerman, Capt. R. B. — 106
Barker, Maj. W. G. — 107, 119, 159
Barnwell, R. H. — 40, 41, 45
Bass, Capt. E. C. — 46
Beachey, Lincoln — 25-26
Beatty, G. W. — 25, 27
Beck, Capt. P. W. — 25
Bell, Lt. C. G. — 48
Bell, Maj. V. D. — 92
Bentley, W. O. — 7, 94-95
Betts, Flt. Sub-Lt. — 70-71
Bikanir, Maharaja of — 85
Blain, 2/Lt. C. W. — 73
Bott, 2/Lt. A. J. — 73
Boyd, Maj. O. T. — 76
Brady, Pet. Off. B. J. — 54
Brand, Maj. C. J. Q. — 120
Brandon, Lt. A. F. — 99
Breadner, Flt. Lt. L. S. — 75
Briggs, Sqn. Cdr. E. F. — 57
Bromet, Sir Geoffrey — 55, 73, 76
Brooke, Flt. Lt. — 64, 66
Brown, Lt. A. G. — 53
Brown, Capt. A. R. — 109
Bruce, R. A. — 161
Bryan, A. Loftus — 46
Burgess, F. T. — 94
Burgoine, S. F. — 34
Busteed, Air Cdr. H. R. — 42
Butler, Sqn. Cdr. C. H. — 96-99, 109
Butler, F. Hedges — 12
CAMM, Sydney — 146
Campbell, R. D. M. — 58
Carbery, Lord — 41, 46, 50
Carr, R. H. — 91
Carter, Capt. A. D. — 104
Carter, Capt. A. W. — 85
Carter, W. G. — 116, 145
Cary, R. O. — 33, 37, 39, 59, 81, 84, 145, 167
Charteris, J. — 31
Churchill, Sir Winston — 37, 38, 49, 211
Claire, Lt. Newton — 49-50
Clarke Hall, Cdr. R. H. — 54-55
Clemson, Flt. Lt. A. W. — 66
Cobby, Capt. A. H. — 130
Cody, S. F. — 13, 17, 28, 30, 35
Coffyn, F. T. — 25
Collet, Flt. Lt. C. H. — 48-50
Collet, Flt. Sub-Lt. R. H. — 69
Collishaw, AVM R. — 108, 137
Compston, R. J. O. — 74
Constantine, King — 44
Courtney, Capt. — 49
Cresswell, C. H. — 15
Croft, Flt. Lt. J. C. — 74
Cruikshank, Capt. G. L. — 73
Culley, Lt. S. D. — 129
Cust, Sir Charles — 20
DALLAS, Flt. Lt. R. S. — 67, 70-71
Dashwood, R. J. — 86
Davies, Sqn. Cdr. R. B. — 48-49
d'Esperey, General — 83

de Forest, Baron — 19
de Linge, Prince — 50
Delacombe, Harry — 41
Derrington, Victor — 7-8, 147
Douglas, Flt. Lt. N. S. — 53
Dowding, Lord — 72
Draper, C. D. — 79, 108, 101
Drexel, J. Armstrong — 21
Drogheda, Lady — 84
Dunlap, Miss Margaret — 22
Dunn, Sgt. W. H. — 75
Dunning, Sqn. Cdr. E. H. — 53, 55, 100, 101
EBERHARDT, Dr. J. C. — 23
Edgar, E. Mackay — 31
Elder, Capt. W. L. — 77, 79-80
Ellington, Sir Edward — 30
Ely, Eugene — 25-27
England, Harry — 21-22, 25
Eschwege, Rudolph von — 86
Eyre, V. W. — 11, 145
FALL, J. S. T. — 75
Fenn, A. R. — 34, 81, 82
Fenn, M. H. — 81, 134
Festing, Capt. F. — 57
Finbow, CPO — 55
Foster, Sir Robert — 75, 109
Fowler, F. B. — 46
Fox, Flt. Cdr. F. M. — 63, 101
Fraunfelder, A. — 93
GALBRAITH, D. M. B. — 74
Garnett, Flt. Lt. W. H. S. — 53-54
Gaskill, H. L. — 87
Gathercole, Capt. G. W. — 95
George V, King — 19-20, 25, 85
Gerrard, Sqn. Cdr. C. L. — 48-51
Gerrard, Flt. Lt. T. F. N. — 68
Gill, Howard — 25
Gillett, Capt. F. W. — 106
Gilmore, Grahame — 12, 28
Goble, Flt. Sub-Lt. — 70, 74
Grace, Cecil — 16
Grahame-White, Claude — 15-17, 21, 26-27
Grahame-White, M. — 17
Grange, E. R. — 74
Greswell, C. — 16-17
Grey, C. G. — 28, 46
Grey, Lt. Cdr. Spenser — 33, 35, 37-38, 48, 50
Grubb, Gen. & Mrs. E. B. — 21
Guillaux, Maurice — 30
HAMEL, Gustav — 20, 30, 37
Hammond, Lee — 25, 27
Hardman, Capt. J. D. — 106
Harkness, Flt. Sub.-Lt. D. E. — 69
Harris, Sir Arthur — 92
Harrower, G. S. — 77, 99
Hartney, Maj. H. — 121
Haskins, Sqn. Cdr. F. K. — 76
Hawker, H. G. — 30-38, 42-43, 45-48, 51, 53, 56-60, 67, 85, 102, 115-117, 134-137, 141-145
Hay, Flt. Lt. J. F. — 62
Hendry, PO J. C. S. — 52
Herbert, Lt. P. L. W. — 30
Hervey, C. E. — 68, 74, 99
Hicks, E. P. — 87
Higgins, Gen. R. F. A. — 121
Hinckler, H. J. L. — 78, 80, 167
Holden, Flt. Lt. H. G. — 71
Hollinghurst, Sir Leslie — 105
Holt, Capt. F. V. — 47
Hope, Maj. Linton — 62
Hore-Ruthven, Hon. B. — 52
Horn, Maj. K. K. — 76
Hucks, B. C. — 30
Hudson, F. N. — 75
Huskisson, B. L. — 67-74

INGHAM, J. N. — 86
JAMIESON, C. W. — 69
Jarrard, Lt. Alan — 107
Jordan, W. L. — 132
KAUPER, Harry — 32, 34-37, 42, 67, 117
Keeble, Flt. Sub-Lt. — 70
Kerby, Flt. Lt. H. S. — 99
Kerr, Adm. Mark — 44
Ker-Seymer — 11
Kershaw, Lt. R. H. — 53-55
LAMBE, Wg. Cdr. C. L. — 79
Lawrence, Maj. G. A. K. — 72
Lawson, Flt. Sub-Lt — 75
Lester, Flt. Sub-Lt. H. L. — 101
Lilley, L. Moore — 47
Little, R. A. — 74, 109, 153
Loraine, Robert — 15-17
Lubbock, Capt. E. — 84
MACKENSIE, C. R. — 74
Mackensie-Grieve — 134-136
MacLaren, D. R. — 109
Macmillan, Norman — 92
Magor, G. A. — 86
Mahl, V. — 37, 40, 43-47, 52-53, 55
Malone, J. J. — 79
Malone, Lt. L. — 33
Mann, Flt. Sub-Lt. — 65
Manning, W. O. — 29
Marchant, Petty Off. — 53-54
Marix, Lt. — 49-51
Martin, James V. — 25
Mary, Queen — 85
Maxwell, J. E. — 86
May, Capt. W. R. — 107
McCurdy, J. A. D. — 25
McClean, F. K. — 35
McCudden, Maj. J. B. — 116
McMaking, Lt. — 92
McMullen, 2/Air Mech. — 75
Miller, H. C. — 117
Moisant, John B. — 11
Moore, Flt. Cdr. W. G. — 101
Moorhouse, W. — 30
Morrison, Flt. Cdr. — 85
Mulock, R. H. — 76
Murison, 2/Lt. — 75
Murlis-Green, Maj. G. W. — 120
Musgrave, H. P. — 145, 167
NEILL, Norman Clark — 31
Norman, Ernest — 27
Nightingale, Flt. Sub-Lt. — 65
Northcliffe, Lady — 37
OGILVIE, Alec — 15, 17, 112
Openshaw, Flt. Lt. — 62
Ovington, Earle — 22, 25-26
PADDON, A. — 11
Pannier, M. — 22-24
Parmelee, P. O. — 25
Pattison, Flt. Sub-Lt. — 78, 80
Peal, Lt. E. R. — 79-80
Perrin, Harold — 13, 17, 41
Perry, Copland — 30
Pickles, Sydney — 35, 46, 161
Pinder, Capt. J. W. — 114
Pixton, H. — 37, 40-41, 44-47
Pizey, C. P. — 29, 42
Pollard, Jack — 6-7, 21-22, 25, 116, 147, 167
Potts, Flt. Sub-Lt. F. — 67
Pretyman, Maj. E. R. — 104
QUIGLEY, F. G. — 93
RAIKES, Edward — 20
Raikes, Mrs. E. — 18, 20
Raynham, F. P. — 29, 46, 134, 246
Read, Maj. W. R. — 73, 76
Rodgers, C. P. — 25-26

Rolls, Hon. C. S. — 11-12
Ross, Capt. — 76
Rouse, Flt. Sub-Lt. — 59
Rutland, F. J. — 100-101
SAMSON, C. R. — 48, 51, 59, 64, 66, 129
Sanday, Capt. W. D. S. — 73
Savage, R. — 34
Saxty, A. G. — 75
Scarff, F. W. — 60, 103
Schwann, Cdr. O. — 32
Schneider, Jaques — 41
Scott, R. F. — 29
Sharman, Flt. Sub-Lt. J. — 78
Sholto-Douglas, N. — 54, 55, 76
Sigrist, Fred — 11-12, 16-17, 19-23, 28, 30, 33-35, 37, 39, 42, 46, 56, 79, 85, 144-146, 167
Simpson, G. G. — 74
Singer, A. Mortimer — 33
Smart, B. A. — 101, 129
Smith, Flt. Sub-Lt. — 66
Smith, Herbert — 70, 90, 102-103, 116, 132
Smythe-Pigott, J. R. W. — 86
Soar, R. R. — 58, 67, 74
Sopwith, May — 13-17, 20-22, 24, 26, 39, 84, 148-149
Sopwith, Sir Thomas — 2, 4-7, 10-46, 52-53, 57-64, 81, 84-85, 93, 102-103, 107, 109, 115-117, 125, 134-136, 141-149, 153, 156
Sopwith, Lady — 4-7, 147-149
Sopwith, T. E. — 147-149
Spratt, Lt. N. C. — 48
Spriggs, Sir Frank — 34, 146
Spottiswoode, J. H. — 34
Starbuck, N. H. — 86
St. John, Lt. — 67-68
Stone, Arthur B. — 25-26
Stuart, Lt. — 93
Sturrock, J. D. — 21-22, 26
Sueter, Capt. Murray — 44, 125
Sweetman-Powell, H. — 30
Sykes, Sir Frederick — 28
TAFT, Henry W. — 25
Taylor, Sir Gordon — 75
Tedder, Lord — 76
Tizard, Sir Henry — 102
Todd, A. S. — 74
Trapp, S. V. H. — 70-74
Trenchard, Lord — 30, 72, 73, 138
Trollope, Capt. J. — 107
VALENTINE, J. — 28-30
Vaucour, 2/Lt. A. M. — 73
WADHAM, V. H. N. — 30
Ward, James — 25
Warminger, Air Mech. H. — 73
Watkins, Lt. H. E. — 15
Watts, J. — 87
Webb-Bowen, Gen. — 30
Wedgwood Benn, Lt. — 64-66
Welsh, W. L. — 57-58, 100
Whistler, Capt. H. A. — 92
Whitehead, J. A. — 161
Whitehorn, Jack — 6-7, 167
Wilcox, Phillip W. — 21
Williamson, Flt. Cdr. — 54
Wilson, Lt. C. W. — 39
Woollett, Capt. W. H. — 109
Woods, N. E. — 74
Wright, Howard T. — 11-12, 29-30, 34, 153
Wright, Orville — 25-27
Wright, Wilbur — 25-27
Wyatt, Flt. Sub-Lt. — 71
YOUNG, D. G. — 29
Yuille, Capt. A. B. — 121

244